Aerosol Technology in Hazard Evaluation

AMERICAN INDUSTRIAL HYGIENE ASSOCIATION

and

U. S. ATOMIC ENERGY COMMISSION
(Office of Information Services)

MONOGRAPH SERIES ON

INDUSTRIAL HYGIENE

MONOGRAPH TITLES

Pulmonary Deposition and Retention of Inhaled Aerosols
Beryllium—Its Industrial Hygiene Aspects
Thorium—Its Industrial Hygiene Aspects
Particle Size Analysis in Industrial Hygiene
Aerosol Technology in Hazard Evaluation

(In Preparation)

Radioiodine—Its Industrial Hygiene Aspects
Uranium—Its Industrial Hygiene Aspects

59956

Aerosol Technology in Hazard Evaluation

THOMAS T. MERCER

Department of Radiation Biology and Biophysics
The University of Rochester
Rochester, New York

Prepared under the direction of the American Industrial Hygiene Association for The Office of Information Services, United States Atomic Energy Commission

ACADEMIC PRESS New York and London 1973
A Subsidiary of Harcourt Brace Jovanovich, Publishers

ACADEMIC PRESS, INC.
111 Fifth Avenue, New York, New York 10003

United Kingdom Edition published by
ACADEMIC PRESS, INC. (LONDON) LTD.
24/28 Oval Road, London NW1

Library of Congress Cataloging in Publication Data

Mercer, Thomas T
Aerosol technology in hazard evaluation.

Includes bibliographical references.
1. Air–Pollution–Measurement. 2. Aerosols. 3. Particle size determination. 4. Respiratory organs–Foreign bodies. I. Title. [DNLM: 1. Aerosols–Poisoning. 2. Air pollution. QV310 M554c 1973]
TD890.M47 614.7′1′028 72-12189
ISBN 0–12–491150–1

PRINTED IN THE UNITED STATES OF AMERICA

Contents

4. Measurement of Concentration

5. Sampling for Geometric Size Measurement

6. Measurement of Aerodynamic Diameter

7. Measurement of Other Diameters Related to Particulate Properties

8. Respirable Activity Samplers

9. Special Problems

Foreword

The tremendous technologic achievements in the past few decades have intensified the need for a comprehensive treatment of information on a single specific topic in occupational disease detection and control. In the early 1960's the American Industrial Hygiene Association inaugurated a program of preparation of a series of monographs under the auspices of the Division of Technical Information (now the Office of Information Services) of the Atomic Energy Commission on subjects pertaining to industrial hygiene.

The author of this monograph, the fifth in the series, has consolidated information with references to original reports appearing in the literature supplemented by a considerable amount of unreported work on the subject of aerosol technology. It is hoped that this treatise will not only provide a convenient source of information to those concerned in industrial hygiene but also will stimulate the interest of those involved in all phases of environmental health.

We are indeed grateful to the author for his dedicated efforts in developing the information in this publication, to the competent reviewers, to the members of the AIHA Monograph Committee who assisted in the mechanics of the manuscript preparation, and to the Atomic Energy Commission for its support of the AIHA Monograph Program.

PAUL D. HALLEY, *President*
American Industrial Hygiene Association

Preface

A scientist of some eminence once pointed out to me that my interpretation of something he had said was limited in its validity by the level of competence I had reached in the subject under discussion—a level, his tone implied, that was not very high. He had reason to be annoyed with me at the time and spoke, no doubt, out of pique; nevertheless, I could discern an element of truth in what he said. When I was asked to prepare a monograph on aerosol technology, the memory of his observation prompted me to add the qualifying phrase "in hazard evaluation," thus limiting the scope of my discussion to those areas of aerosol technology with which I am most familiar. With that restriction, I have attempted to present, as quantitatively as I could, the principles underlying the various techniques involved in the production and characterization of aerosols.

A quantitative discussion requires equations and equations require symbols, which are always a source of difficulty. Rather than trying to find enough symbols to give a unique meaning to each, I have permitted certain symbols to have more than one meaning, defining them each time they are used, if there appeared to be any risk of ambiguity. Units, too, pose a problem, which I have tried to avoid by adopting the following convention: if units are not mentioned, the equation is valid for any consistent set; if units are given only for certain quantities in an equation, all other quantities are in cgs units.

The manuscript for this monograph was reviewed in whole or in part by reviewers for the AIHA, who are identified elsewhere; by Jess Thomas, David Sinclair, Myron Robinson, and Lyle Schwendiman for the AEC; and by Marvin Tillery and Randall Mercer for me. I am indebted to all of them for their comments and suggestions. I am also indebted to my wife, who did the lioness's share of the work of preparing the manuscript.

T. T. Mercer

1

Introduction

1-1 The Nature of an Inhalation Hazard

Airborne toxic particles, swept into the respiratory tract during inhalation, may be deposited there and subsequently give rise to undesirable biological effects. Since the complete removal of contamination from the air is sometimes impossible and is always unattractive economically, it is rarely attempted. Instead, the hazards associated with the inhalation of toxic particles are minimized by maintaining the concentration of the contaminant below some level which has been deemed unlikely to cause detectable biological damage in people exposed over a long period. The two quite distinct processes of establishing the criteria that define that level and of monitoring atmospheres of questionable character to see if they conform to the established criteria constitute hazard evaluation.

Many factors that are not directly related to the chemical or radioactive properties that make a substance toxic play important roles in determining the probability that a toxic agent carried by an airborne particle will reach a site appropriate for causing biological damage. An appreciation of what these factors are and how they contribute to the problem of hazard evaluation can be obtained from a consideration of the processes of deposition in, and clearance from, the respiratory tract.

1-1.1 Deposition of Particles in the Respiratory Tract

The conducting airways of the human lung extend to the end of the terminal bronchioles. Beginning with the trachea, they split into increasing numbers of tubes of ever-diminishing cross section. The inside surfaces of these airways are lined with a ciliated epithelium and the rhythmic beating of the hair-like cilia continually moves a thin sheet of mucus toward the entrance to the trachea. Insoluble particles deposited in these tubes are rapidly swept up through the trachea and swallowed. In the nomenclature of the International Commission on Radiological Protection (ICRP) Task Group on Lung Dynamics [1], these airways collectively comprise the tracheobronchial (T-B) compartment. Beyond them, the tubes are not ciliated and they lead through respiratory bronchioles and alveolar ducts to the alveoli. Gas exchange takes place throughout this volume, which is designated the pulmonary (P) compartment. Some of the insoluble particles deposited in this region remain for very long periods.

Figure 1.1 is a stylized portrayal of a portion of the lung structure. It is obviously an oversimplification, which is necessary for modeling purposes because the actual lung arrangement is much too complex for theoretical calculations. Moreover, despite its many shortcomings, it has permitted calculations that yield deposition values in quite respectable agreement with experimental results. The calculations are based on a regular system of branching at a fixed angle into two tubes of equal diameter, as shown in Fig. 1.1. It is

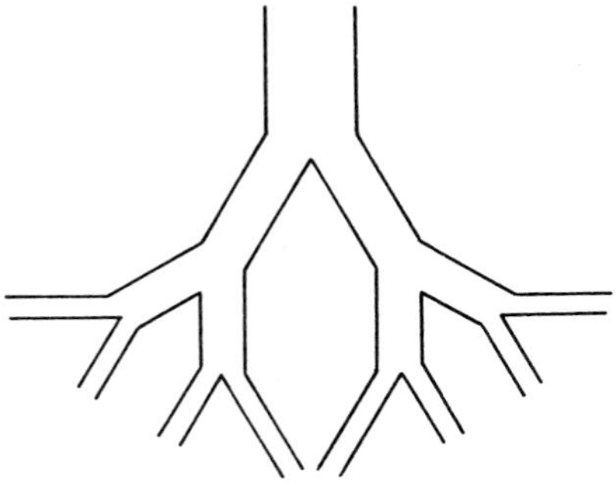

FIG. 1.1. Schematic diagram of the lung conducting airway.

assumed that the tubes have inflexible walls and lead ultimately into spherical alveoli. Considering only particles that are too insoluble to exhibit any hygroscopic action in the humid atmosphere of the lung, four processes may be effective in bringing about deposition:

(1) Particles may be intercepted as they pass from one tube into a smaller one because the streamline along which they move enters the second tube at a distance from its wall that is less than the radius of the particle.

(2) Particles having sufficient inertia may deviate far enough from the air streamlines to impact near the tube bifurcation.

(3) Particles moving through tubes that are not perfectly vertical have a component of their sedimentation velocity that is normal to the tube surface and some of them fall onto the surface.

(4) Very small particles are buffeted about sufficiently by the random thermal motion of the air molecules that they are deposited on the tube walls by the process of diffusion.

The effect of turbulent flow, which occurs in the larger airways, usually is not taken into consideration.

As the air moves deeper into the lung, where the tubes are more numerous and have smaller diameters but have a greater total cross-sectional area, the average velocity through a tube decreases. This favors deposition due to sedimentation and diffusion, but diminishes the effect of impaction. The smaller diameters of the tubes also favor deposition due to interception and sedimentation.

Figure 1.2 brings out the manner in which these processes cooperate to bring about deposition of particles from air moving through the airways. The divisions along the top indicate the level of the lung that is under consideration. They are based on the original Findeisen [2] model of the lung structure. The ordinate is the ratio between the number of particles remaining in a given volume element of air at the indicated segment and the number it contained when it entered the trachea. For an upright man, deposition in the trachea is negligible except for particles of very high diffusivity. At each bifurcation there is a considerable loss of larger particles due to impaction. As the air flows through a tube, there is a continuous loss of particles due to sedimentation and diffusion, an effect that becomes more pronounced as the air moves into the smaller airways. Air that penetrates all the way to the alveoli is cleared of the larger particles and of most of the very small particles. During exhalation, additional deposition occurs. The percentage of deposition, relative to the particles remaining airborne at the start of exhalation, is very nearly the same for the processes of sedimentation and diffusion, but deposition due to impaction is greatly reduced.

It must be emphasized that this refers to deposition of particles after they enter the trachea. During nasal breathing, unit density particles of about 10 μm and larger are completely removed from the incoming air as it passes

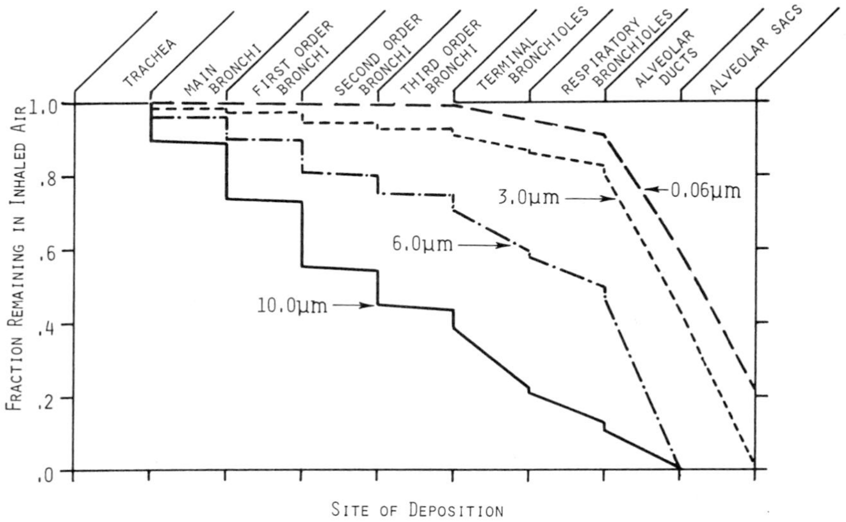

FIG. 1.2. Lung deposition, during inspiration, of particles that enter the trachea. Abrupt changes represent impaction at airway entrance. Other changes represent diffusion and sedimentation losses. Losses in the alveolar sacs represent changes with time rather than distance traversed.

through the nose. Calculations of nasal deposition, which is primarily an impaction phenomenon, have not been entirely satisfactory, but there are quite good empirical relationships available to describe it. Pattle [3] studied the nasal deposition N of monodisperse methylene blue particles and obtained the following relationship:

$$N = -0.62 + 0.475 \log D_A{}^2 F \qquad (1.1)$$

where D_A is the aerodynamic diameter of the particle in micrometers and F is the inspiratory flow rate in liters/minute. Hounam *et al.* [4] felt the correlation could be improved by the use of pressure drop in place of flow rate, but Lippmann's [5] data were in excellent agreement with Pattle's equation. Deposition in the mouth can be correlated with $D_A{}^2F$ in a similar manner [1].

Figure 1.3 shows an example of the overall deposition patterns derived from calculations of the type described above. It is apparent at once that if it is necessary for a particle to get into the pulmonary regions to cause biological damage, then a great many particles are completely irrelevant to hazard evaluation. Of those particles that are relevant, some are more important than others. If we are to have a definition of hazard that takes account of this fact, our criteria must be based on the factors that determine relevance to hazard. Since the regional deposition apparent in Fig. 1.3 is the result of the action of

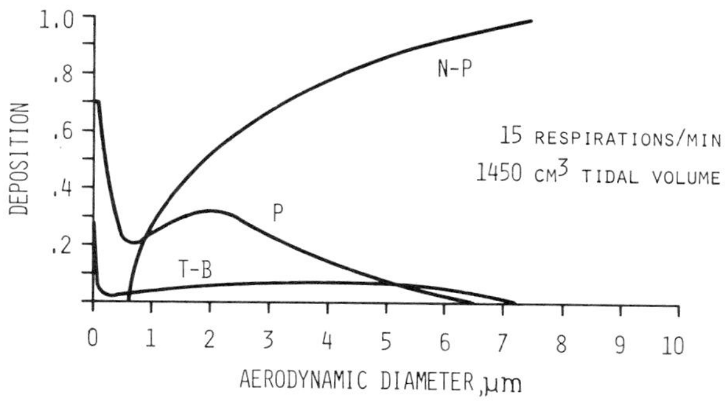

FIG. 1.3. Calculated deposition of particles in the nasopharyngeal (N–P), tracheobronchial (T–B), and pulmonary (P) compartments, relative to number inhaled.

the four processes described above, we must look for our criteria in the physical characteristics of the inhaled particles.

1-1.2 Criteria of Particle Deposition

The probability that a particle will be removed by interception upon entering a tube is related to the ratio of the particle diameter to the tube diameter. Except for particles of extreme shapes, such as long fibers, or of very low density, this effect can be ignored because particles large enough to show significant interception effects are certain to be caught by impaction or sedimentation.

The probability that a particle of diameter D moving with an air stream of average velocity U in a tube of diameter W will impact at a bifurcation is related to a parameter, Stk, called the Stokes number:

$$\mathrm{Stk} = \rho K_s D^2 U / 9\eta W, \tag{1.2}$$

where η is the viscosity of air, ρ is the particle's density, and K_s is its slip factor, a quantity that is very nearly equal to unity for spherical particles if D is greater than 1 μm but increases rapidly when the diameter is comparable to the mean free path of the air molecules. As far as particulate characteristics are concerned, the significant quantities are ρ, K_s, and D^2.

The probability that a particle will settle out in a tube is related to the particle's terminal settling velocity U_G given by

$$U_G = \rho K_s D^2 g / 18\eta, \tag{1.3}$$

where g is the acceleration due to gravity. Again the significant particulate quantities are ρ, K_s, and D^2.

Particle deposition due to diffusion is related to the coefficient of diffusion or diffusivity Δ of the particle, which depends on the particle's diameter:

$$\Delta \propto K_s/D.$$

Deposition due to diffusion does not become very effective until particle size gets below about 0.5 μm.

For most purposes, then, we can say that the probability a particle will deposit depends on its terminal settling velocity. In the size ranges of interest to us, most particles—spherical or otherwise—having the same terminal settling velocity U_G will be distributed throughout the lung in a similar manner. The common deposition parameter for such particles is the aerodynamic diameter, defined as the diameter of a sphere of unit density having the given value of U_G.

In practice, estimates of deposition must take into consideration the fact that aerosol particles have a wide range of aerodynamic diameter. The Task Group on Lung Dynamics (referred to above) applied curves similar to those of Fig. 1.3 to the calculation of the total deposition in each of the three compartments for fixed breathing patterns and for a given aerosol in which the toxic agent was distributed lognormally with respect to the aerodynamic diameter of the particles. The results showed, as Schulte had observed earlier [6], that deposition is closely correlated with the median aerodynamic diameter of the distribution, but is not much affected by its geometric standard deviation. Moreover, as long as the distribution is lognormal, the results are applicable regardless of the relationship between the toxic material and the physical properties of the particle. If we refer to the amount of toxic substance as "activity," the deposition will be related to the activity median aerodynamic diameter (AMAD) of the distribution. If the toxic agent in a given particle is proportional to the particle mass, the AMAD is also the mass median aerodynamic diameter; if the particles are of an inert material and have a toxic substance deposited on the surface in proportion to its area, the AMAD is also the surface median aerodynamic diameter; if the toxic effect of a particle depends only on its presence in the lung, the AMAD is also the count median diameter; and even if the toxic substance is distributed among the particles in an unknown and indeterminate manner, the relationship between deposition and AMAD is still valid. There are available a number of instruments with which AMAD's can be estimated, provided the sample is analyzed for "activity."

In addition to the obvious criterion of gross concentration of toxic material in the air, the consideration of deposition of particles in the respiratory tract has led to the aerodynamic diameter as the significant criterion of hazard, with the need for equivalent diffusion or interception diameters in rare, but not insignificant, cases. Other particulate characteristics that might be expected to affect deposition include the particle's electric charge and its hygroscopicity. So far, theoretical and experimental evidence support the view that electric charge is ineffective after a particle has entered the respiratory tract. Still, it is an interesting fact that in the size range of minimum deposition spherical particles, produced by methods that invariably yield charged particles, consistently show greater overall deposition than do particles produced by condensation methods, which yield particles that are seldom charged. The effect of hygroscopicity varies with the solubility and molecular weight of the compound of which the particle is composed. Its significance depends on the degree to which accretion of water by a particle alters its probability of deposition and the site at which it deposits, relative to a similar insoluble particle.

1-1.3 Factors in Lung Clearance

Additional factors related to the clearance of particles from the lung depend on the relationship between physiological processes and the physical and chemical characteristics of particles. Figure 1.4 shows some of the clearance pathways that might be taken by the toxic agent. Those that indicate transfer to blood represent dissolution of the particle; the others represent the transfer of particles, as such, out of the respiratory tract. The rate of clearance of

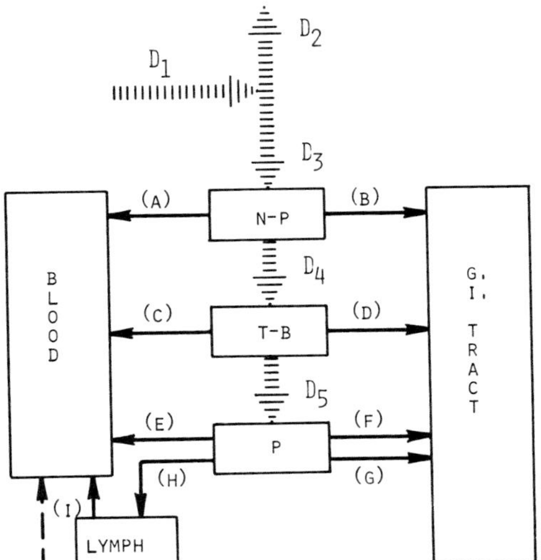

FIG. 1.4. Some possible pathways for clearance from the lung.

particles from the N-P and T-B regions to the gastrointestinal (GI) tract is related to aerodynamic diameter because particles of large aerodynamic diameter deposit higher up in the respiratory tract and, consequently, have less distance to be carried before entering the esophagus. The transfer of particles from the P region to the GI tract and to lymph nodes has not yet been related unequivocally to the physical or chemical properties of particles, and no useful criterion related to this mechanism of clearance is yet available. The transfer of toxic agent to blood by the dissolution of particles, however, is a function of the specific surface of the particulate mass and of the solubiltiy rate of the toxic agent, both quantities that could conceivably be included as criteria in the definition of hazard.

1-1.4 The Definition of Hazards

The factors considered above do not include the all-important biological effects that are responsible for the existence of a hazard. In the field of industrial hygiene, the toxic nature of a particle derives from its chemical composition and must be determined separately for each compound. Sources of information with which to assess the inhalation toxicity of a given compound include: laboratory studies in which animals are exposed to various levels of the toxic agent and any biological effects are recorded; plant surveys in which the concentration of toxic agent is correlated with medical records of exposed personnel; epidemiological studies of humans who have been exposed, more or less routinely, to relatively low levels of the toxic agent; analogy with compounds of similar chemical character, the toxicity of which has already been established; and, one suspects, inspired intuition. In the field of health physics, however, the toxic nature of a particle derives from physical processes associated with its isotopic composition and, presumably, can be related ultimately to the single criterion of radiation dose. For this reason, much of the experimental work on the inhalation of radioactive materials has been directed at providing quantitative data concerning the many factors described above that have important roles in bringing the toxic agent to an appropriate site of biological action. Given these data, and an adequate criterion relating radiation dose to biological damage, it is possible to calculate the air quality criteria defining the inhalation hazard associated with any particular radioisotope.

In many cases, the definition of hazard has been based on the single criterion of gross concentration of toxic material in the air. While this is usually adequate for particles that are readily soluble in body fluids, it has serious shortcomings in the definition of hazards due to insoluble particles, as is apparent from the discussion above. To minimize these shortcomings, instruments have been developed that permit measurement of the concentration of "respirable" toxic material in the air, thus effectively introducing particle size as a second criterion in the definition of hazard. The determination of the distribution of toxic material as a function of the aerodynamic diameter of toxic particles can provide information in addition to the concentration of respirable material in the air. At present, air monitoring practices seldom go beyond the application of these methods. The introduction of other particulate characteristics as criteria in the definition of hazard must await both the experimental demonstration that they will serve a useful purpose and the development of measurement techniques suitable for use in the field.

1-2 Sampling Procedures

Given a definition of hazard, an appropriate system of sampling must be established. A sampling program may be designed to provide data with which to estimate the exposures of personnel or it may be designed merely to establish that safe conditions prevail. The estimation of exposures for correlation with medical records is an important source of information for the definition of hazard. Sterner [7] has pointed out that, in any working environment subject to contamination by some airborne toxic material, man himself is an experimental animal. In fact, man may be the only satisfactory experimental animal with which to study biological damage that appears only after long exposure to levels of a contaminant that may have been deemed safe on the basis of studies of other animals. For this reason, control sampling to demonstrate safe conditions should not be simply a matter of applying a "go, no-go" gauge to the contamination collected in the sampling process.

1-2.1 Sampling to Estimate Exposures

We would like to know the amount and location of toxic material deposited in each exposed individual. Since we cannot generally obtain this information from measurements made on the individual, we must infer it from measurements of the activity to which he was exposed. The following procedures, listed in order of preference, are available to us:

(1) Measure the activity concentration and deposition characteristics of the particles available for inhalation by the individual.

(2) Measure the activity concentration and deposition characteristics of particles in the breathing zone of the individual.

(3) Measure the activity concentration available for inhalation by the individual.

(4) Measure the activity concentration in the breathing zone of the individual.

(5) Measure the activity concentration and deposition characteristics of particles in the general area of the individual.

(6) Measure the activity concentration in the general area of the individual.

The first procedure implies the use either of a personal sampler that collects the respirable mass of toxic material or of one that separates the particles on the basis of aerodynamic diameter so that a size distribution can be estimated. The phrase "available for inhalation" carries an implied assumption, the validity of which has not yet been established, that activity collected by a personal sampler adequately represents that inhaled by the individual wearing it during the full period of exposure. The order of the second and third procedures is debatable; certainly there are circumstances when a gross measure of the activity available for inhalation, made by a nonselective personal sampler, is more useful than a more detailed breathing zone sample that may represent only a fraction of the total exposure.

Since personal samplers are in a hazardous atmosphere only when the individual is, they provide directly a measure of the total activity to which the individual was exposed. Other samplers, however, probably do not sample air of the same activity concentration as that available for inhalation, and may not cover the same time period as the individual's activities. Their data will have to be corrected for the differences in sampler location and weighted according to the time the individual spent in their vicinity.

The use of time-weighted average concentrations as the basis for estimating exposure is an accepted principle of air sampling [8]. Its validity requires that cumulative exposures calculated in this way show a close correlation with the incidence of biological damage. A number of early studies by the U.S. Public Health Service [9–13] demonstrated that such a correlation actually existed in the dusty trades. Roach [14] showed that for workers in certain coal mines, the relationship between the incidence of pneumoconiosis and the cumulative exposure in particle-years followed the classical dose–response curve so often encountered in experimental toxicology; the curve is sigmoid when the data are plotted on rectangular coordinates and linear when they are plotted on logarithmic-probability coordinates. Hatch [15] made similar analyses of data from South African mines and from several of the surveys of dusty trades referred to above, obtaining the curves shown in Fig. 1.5. The incidence of silicosis correlates so well with the accumulated exposure based on overall average concentrations and time at risk that there also must have been a close correlation between exposure and dose to the lung. This indicates that the samples collected must have included a relatively constant fraction in the respirable size range.

Wright [16] pointed out that despite the evidence supporting the use of time-weighted average concentrations, several authorities in this field believe that high concentrations over short periods make significantly greater contributions to the lung burden than they do to the average concentration. This belief is based on the assumption that the deposition of a large number of

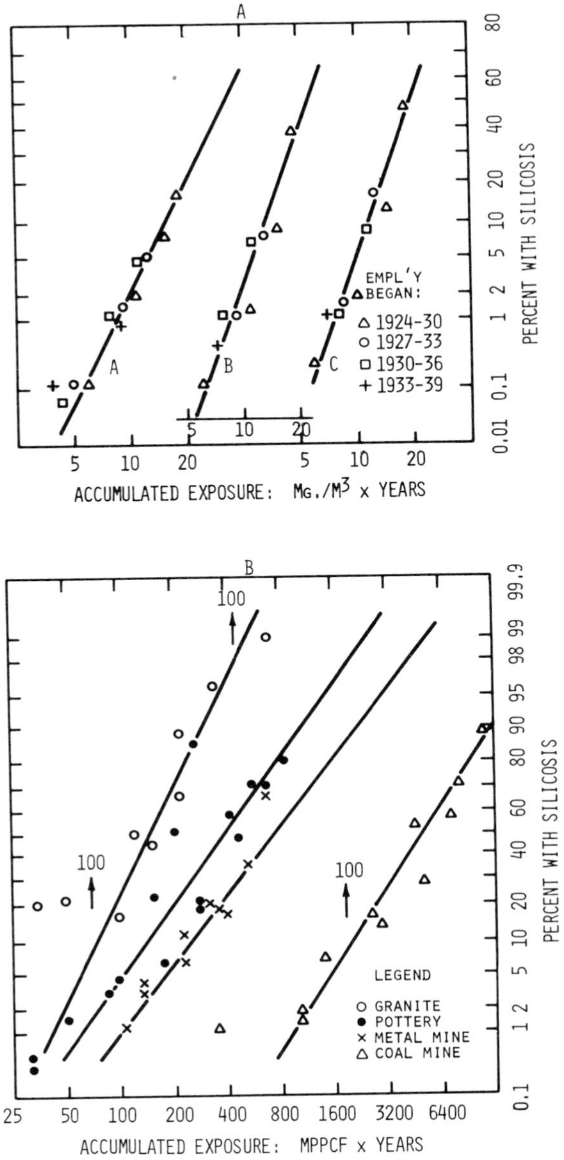

FIG. 1.5 Development of silicosis in relation to accumulated dust exposure: (A) South African gold miners; (B) workers in dusty trades (U.S.) [15]. Courtesy of the American Industrial Hygiene Association.

particles in the lung in a brief period impairs the efficiency of the clearance mechanism; consequently, a larger fraction of the particles is retained. In his comparison of the relative importance of time and concentration in the production of pneumoconiosis, Wright stated that there is no evidence from practical situations to support this assumption. More recently, Reisner [17] has cited evidence from which he concludes that it is "not out of the question" that high concentrations have effects exceeding those expected on the basis of exposure alone. However, it does not seem that any effect of high concentrations, beyond their expected contribution to the time-weighted average, is sufficiently marked to raise questions about the principle of sampling to obtain time-weighted average concentrations. It must be emphasized that the principle does not apply to substances that may produce an acute response, where "acute" means "occurring in a time that is short compared to the duration of sampling."

A constant surveillance of the level of contamination to which each individual is exposed would be a very expensive process. To acquire the desired information by sampling techniques, it is necessary either to have average concentrations for each area in which contamination occurs and the time spent by each individual in each area, or to have an estimate of the average exposures to individuals grouped according to their occupational routine. The data of Fig. 1.5 for the dusty trades were obtained by the former method. Average concentrations associated with a given operation were combined with estimates of the time that each worker spent at each operation to yield estimates of exposures covering periods up to thirty years. It is interesting that for each trade the average concentrations were obtained from single, short-term surveys and were applied under the apparently reasonable assumption that nothing had been done during thirty years to alter the conditions of dustiness. The latter method was used by Oldham and Roach [18], who wished to estimate the exposure to respirable dust experienced by colliers in a group under study by a medical team. Using a random-selection method, they chose a number of colliers to be included in the study, the number depending on the size of the sampling team. Each chosen collier was observed during two work shifts, his exposure being estimated from a number of samples taken in his vicinity at randomly selected times throughout the shift. Oldham [19] has shown that his exposure provided a satisfactory estimate of the exposures of the group as a whole.

The "random colliers" method of Oldham and Roach minimizes the risk of sampling bias, which Tomlinson [20] points out is a particularly serious source of error because it is so difficult to detect and eliminate. To avoid bias, he recommends the following steps in sampling for hazard evaluation:

(1) Enumerate the factors which may affect the answer, e.g., worker, place of work, day of week.

(2) So far as possible, cover these factors systematically and completely.
(3) Where this is not possible, randomize.

1-2.2 Sampling for Hazard Control

Levels of airborne contamination vary with time and location depending on the frequency of occurrence of processes which give rise to contamination, the distance between the sampler and the source of contamination, and the intervening patterns of ventilation. In some cases, however, the nature of the work is sufficiently repetitive and routine that samples taken in a fixed location may provide an adequate assessment of the safety of the environment. The sampling location chosen may be the one at which concentrations are generally highest or it may be one at which the concentration is related in a reasonably consistent manner to the concentrations to which individual workers are exposed. In either case, a program of exposure measurement must be carried out for a sufficiently long time to establish sound baseline conditions from which to derive sampling criteria that will ensure adequate protection for those individuals exposed to the greatest risk.

1-2.3 Sample Variability

Three types of samples were mentioned above: personal samples, collected by devices worn by the individual; breathing zone samples, collected by devices arranged to sample air as close to the individual's face as working conditions permit; and general area samples collected by devices in fixed locations where, it is hoped, the concentrations will be representative of the average concentration in the area under surveillance. General area samples are the most convenient and economical to obtain, but concentrations calculated from them may differ seriously from those derived from breathing zone or personal samples. Where comparisons of the different types of samples have been made, it appears that useful correlations can be found when routine, repetitive work is carried out in a limited area; when workers do a variety of jobs in different locations, however, correlations are found to be very poor.

Breslin and his co-workers [21] compared the three sampling methods during a survey of air contamination in a uranium extrusion plant. Samples of a given type were paired in order to estimate the error of a single random air sample. Three work areas in a single room were monitored and in each area a worker wore a pair of personal air samplers. About 500 samples were collected over a three-day period. Sampling precision errors were 18% for the general area and breathing zone samples and 19% for the personal samples. Exposures based on time-weighted average concentrations for the general area and breathing zone samples agreed well with those calculated from the personal

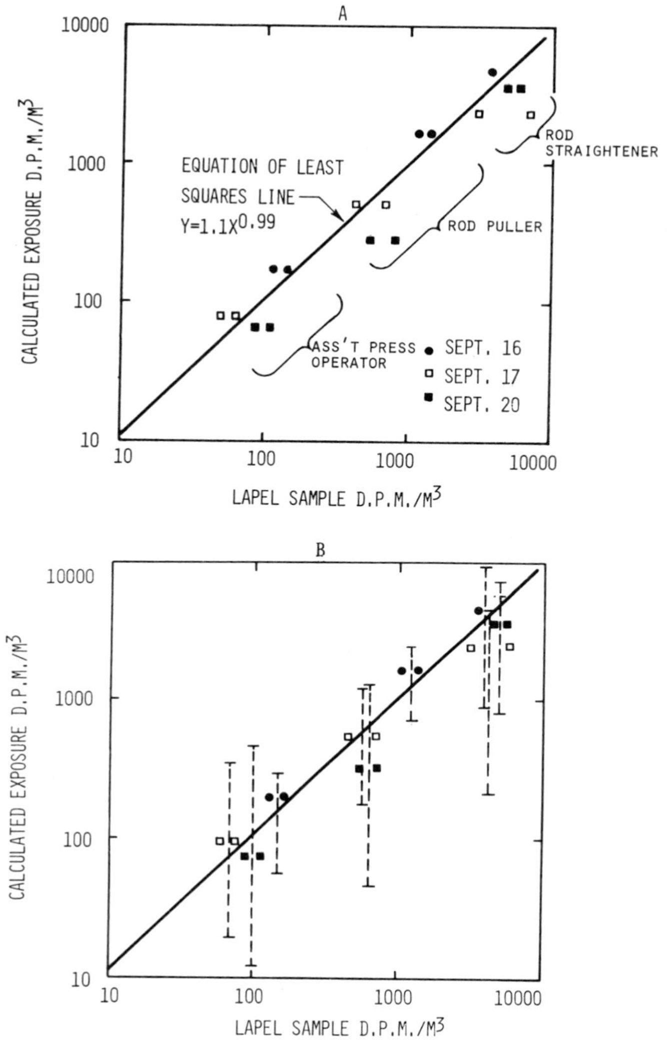

FIG. 1.6. Comparison of exposures based on different sampling methods. (A) Average exposures; (B) extreme values of exposures [21]. Courtesy of the American Industrial Hygiene Association.

samples. Figure 1.6 shows their results for average exposures and for extreme values of the exposures. Similarly, Jacobson [22] found a linear relationship between general area respirable mass samples collected behind the MRE horizontal elutriator and personal respirable mass samples collected behind a

10-mm nylon cyclone. In this case, the former samples indicated concentrations 1.6 times as great as did the latter samples. In a plant handling uranium tetrafluoride pellets, Langmead [23] found that a stable relationship existed between uranium concentrations measured with personal air samplers and with static (general area) samplers. The concentration ratio between personal and static samples was lognormally distributed with a median value of about 2.8 and a geometric standard deviation of about 1.9.

On the other hand, in areas where air concentrations were generally very low but showed occasional high levels due to localized releases of contamination, Langmead found the ratio was again lognormally distributed, but with a median value of 6.6 and with a geometric standard deviation of 4.4. Similarly, Stevens [24] found that the ratio was as high as 40 in decontamination areas where each worker produced an aerosol in his own vicinity.

Oldham [19] was apparently the first to observe that the distribution of concentrations, as measured by air samples of a given type, is often lognormal. He found that the distributions of concentrations from samples taken near one randomly selected collier during one work shift were not only lognormal but also had standard deviations that were essentially the same from shift to shift, despite wide variations in the shift mean values. His results for three surveys in two different mines are shown in Fig. 1.7, in which the deviations of the sample values from their shift means, expressed logarithmically, are shown to fit a normal curve with a standard deviation of 0.22 (geometric standard deviation of 1.66). Oldham pointed out that this made it possible to predict the number of samples needed to achieve a specified confidence interval on the mean.

The concentrations with which Oldham dealt were the numbers of particles between 0.5 and 5 μm in diameter per cubic centimeter of air, as determined from thermal precipitator samples. He was also able to demonstrate that published data on samples of the mass concentration of dust in mines showed distributions of the typical lognormal form with a constant geometric standard deviation of about 1.88. Breslin and others [21] have also found lognormal distributions of sample values in uranium processing plants. They seldom saw geometric standard deviations larger than 2; in many instances the deviations must have been considerably smaller, since statistical tests did not indicate a choice between normal and lognormal distributions.

The lognormal distribution has been found to apply also to the concentrations of airborne radioactivity as determined from samples collected over a period of a year at a reactor site [25] and from samples taken in laboratories and buildings at the Atomic Energy Research Establishment (AERE) [24]. In the former case, sampling was carried out in a single location where the work in progress did not change with time; the geometric standard deviation σ_g of the distribution was $\simeq 2.6$. In the latter case, concentrations based on samples of 7-day duration were collected with a static sampler (cascade centripeter) over

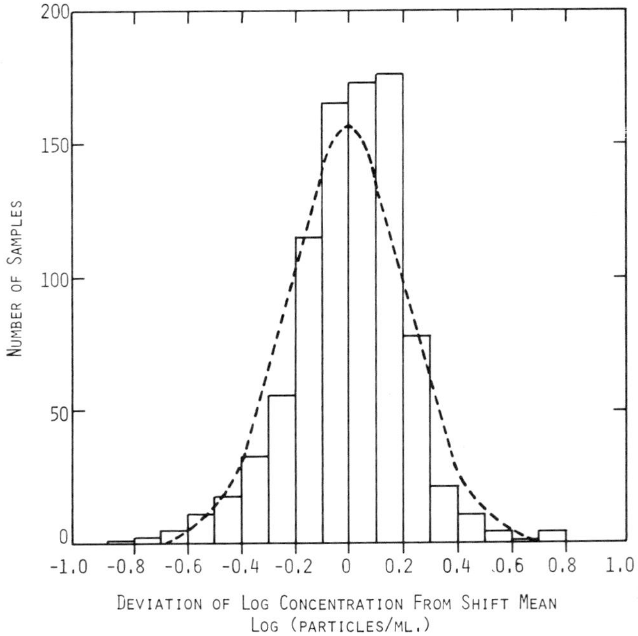

FIG. 1.7. Distribution of relative sample mean concentrations over one work shift in a coal mine [19]. Courtesy of *British Journal of Industrial Medicine*.

a period of a year. These concentrations showed lognormal distributions for which σ_g was between 2.4 and 7.0, although only two had values of $\sigma_g > 3.8$. In the same study, concentrations from personal air samples, also collected on a weekly basis, were lognormally distributed with σ_g's ranging from 2.6 to 6.5 except for one distribution for which $\sigma_g = 17$.

The fact that sampling distributions may be lognormal is important when it becomes necessary to calculate arithmetic mean values, which must be used when making comparisons with standards. For a very large number of samples the arithmetic mean concentration A_a is related to the geometric mean concentration A_g by

$$A_a = A_g \cdot \exp(\ln^2 \sigma_g/2). \tag{1.4}$$

For a small number of samples, however, the arithmetic mean calculated in this way is too large. Finney [26] has shown in that case that the proper estimate of the arithmetic mean of a sample of size n, for which the estimated variance is $s^2 = \ln^2 \sigma_g$, is

$$A_a = A_g \cdot G(t), \tag{1.5}$$

where

$$G(t) = 1 + \frac{n-1}{n} \cdot t + \frac{(n-1)^3}{n^2(n+1)} \cdot \frac{t^2}{2!} + \frac{(n-1)^5}{n^3(n+1)(n+3)} \cdot \frac{t^3}{3!} + \cdots \quad (1.6)$$

and $t = s^2/2$. Oldham [19] provided a modified form of this equation to be used when s^2 could be based on a large number of previous samples. His modified equation was also an infinite series, but it can be expressed more simply as

$$G(t) = e^x, \quad (1.7)$$

where

$$x = \frac{n-1}{n} \cdot \frac{m+1}{m} \cdot \frac{s^2}{2}$$

and s^2 is based on m degrees of freedom, which very nearly equals the total number of previous samples. Similarly, the proper estimate of the arithmetic standard deviation is

$$s(\text{arith}) = A_g \cdot [G(t_1) - G(t_2)]^{1/2}, \quad (1.8)$$

where $t_1 = 2s^2$ and $t_2 = [(n-2)/(n-1)] \cdot s^2$.

On the other hand, if samples from a lognormal distribution are treated as if they were from a normal distribution, the estimates of the arithmetic mean are within a few percent of the proper estimates [Eq. (1.5)]. However, estimates of the arithmetic standard deviation may be seriously in error, even at relatively small values of the population geometric standard deviation.

1-2.4 Sampling Duration

The threshold limit values (TLV's) against which the industrial hygienist must compare concentrations in his own environment are based on time-weighted average concentrations for a normal workday [27]. Roach [28] finds that this introduces difficulties into the interpretation of sampling results for substances that produce an acute response, and for which peak concentrations may be of real significance. He assumed that the body burden x builds up in a simple exponential fashion,

$$x = kC/a + \{x_0 - (kC/a)\} \exp(-at), \quad (1.9)$$

where C is the air concentration of the contaminant, k is the product of the volume of air inhaled per unit time and the fraction of inhaled contaminant that is deposited in the body, x_0 is the body burden at time $t = 0$, and a is an elimination rate related to the half-life of the substance in the body by $a =$

0.693/T. Under these circumstances, if x_0 is negligible,

$$x \propto C \cdot t$$

when T is very long and $t \ll T$, but

$$x \propto C$$

when T is very short and a is very large. In the former case, the time-weighted daily average concentrations can be compared unequivocally with the TLV; in the latter case, however, the presence of peak concentrations during the averaging period confuses the issue.

Taking the reasonable position that the significance of peak concentrations depends on the extent to which they are reflected in changes of the body burden, Roach calculated the variance of the body burden as a function of the variance of the air concentration. He found that for a time interval, t, which is small compared to the elapsed time since the start of exposure, the relationship is

$$\mathrm{Var}(x) = (\bar{x}/\bar{C})^2 \cdot \frac{\exp(at)-1}{\exp(at)+1} \cdot \mathrm{Var}(C). \tag{1.10}$$

Expressing the result in terms of coefficients of variation (standard deviation/mean), he obtained

$$\mathrm{Coef} \cdot \mathrm{Var}(x) = \left[\frac{\exp(at)-1}{\exp(at)+1}\right]^{1/2} \cdot \mathrm{Coef} \cdot \mathrm{Var}(C) \tag{1.11}$$

$$= \left[\frac{(2^{t/T}-1)}{(2^{t/T}+1)}\right]^{1/2} \cdot \mathrm{Coef} \cdot \mathrm{Var}(C). \tag{1.12}$$

Using two sets of experimental data of strikingly different patterns to provide examples of the variation of concentration with time, he demonstrated that for t/T much less than unity, theoretical values of Coef·Var·(x) were in satisfactory agreement with experimental values. He also showed that the maximum value of the coefficient of variation for the body burden during a sampling period of duration t was

$$\mathrm{Coef} \cdot \mathrm{Var}(x) = 0.59(t/T)^{1/2} \cdot \mathrm{Coef} \cdot \mathrm{Var}(C).$$

On the basis of these results, Roach recommended that sampling times should be one-tenth of the biological half-life of the contaminant in the body. For this period, the coefficient of variation of the body burden would not exceed 19% of that of the air concentration, and the average body burden would equal that attained on exposure to a constant concentration at the average level for that period. He then pointed out that by maintaining the average air concentration one standard deviation below the TLV, the average body burden could be kept five standard deviations below the corresponding threshold body burden. He suggested the following criterion be compared with the TLV:

$$\text{average} + \text{range}/(n-1)^{1/2},$$

where the second term is an estimate of the standard deviation for n air samples. Roach's recommendations should be valid even for airborne particles, which have biological half-lives much longer than a normal work shift.

References

1. Task Group on Lung Dynamics: Deposition and Retention Models for Internal Dosimetry of the Human Respiratory Tract, *Health Phys.*, *12:* 173–207 (1966).
2. W. Findeisen, Concerning the Deposition of Small Airborne Particles in the Human Lung During Breathing, *Pfluger's Arch. Ges. Physiol.*, *236:* 367–379 (1935).
3. R. E. Pattle, The Retention of Gases and Particles in the Human Nose, in *Inhaled Particles and Vapors*, C. N. Davies, Ed. Pergamon, Oxford, 1961.
4. R. F. Hounam, A. Black, and M. Walsh, Deposition of Aerosol Particles in the Nasopharyngeal Region of the Human Respiratory Tract, *Nature* (*London*), *221:* 1254 (1969).
5. M. Lippmann, Air Sampling for Hazard Evaluation, *in* T. T. Mercer, P. E. Morrow, and W. Stöber (Eds.), *Assessment of Airborne Particles*, Thomas, Springfield, Illinois, 1972.
6. H. G. Schulte, Particle Size Vs. Inhalation and Sampling, *A.E.C. Health and Safety Personnel Meeting, Berkeley, Nov. 30, 1960.*
7. J. H. Sterner, The Experimental Animal–Man in Industrial Hygiene, *Amer. Ind. Hyg. Ass. Quart.*, *16:* 103–107 (1955).
8. H. F. Schulte, Modern Concepts of Air Sampling and Problems for the Future, *Amer. Ind. Hyg. Ass. J.*, *23:* 20–25 (1962).
9. R. H. Flinn *et al.*, Silicosis and Lead Poisoning Among Pottery Workers, *Public Health Bull. No. 244*, Washington, D.C., 1939.
10. A. E. Russell *et al.*, The Health of Workers in Dusty Trades: II. Exposure to Siliceous Dust (Granite Industry), *Public Health Bull. No. 187*, Washington, D.C., 1929.
11. W. C. Dreesen *et al.*, Health and Working Environment of Non-Ferrous Metal Mine Workers, *Public Health Bull. No. 277*, Washington, D.C., 1942.
12. L. R. Thompson *et al.*, The Health of Workers in the Dusty Trades: I. Health of Workers in a Portland Cement Plant, *Public Health Bull. No. 176*, Washington, D.C., 1928.
13. R. R. Sayers *et al.*, Anthraco-Silicosis Among Hard Coal Miners, *Public Health Bull. No. 221*, Washington, D.C., 1935.
14. S. A. Roach, A Method of Relating the Incidence of Pneumoconiosis to Airborne Dust Exposure, *Brit. J. Ind. Med.*, *10:* 220–226 (1953).
15. T. Hatch, Permissible Dustiness, *Amer. Ind. Hyg. Ass. Quart.*, *16:* 1, 30–35 (1955).
16. B. M. Wright, The Importance of the Time Factor in the Measurement of Dust Exposure, *Brit. J. Ind. Med.*, *10:* 235–240 (1953).
17. M. T. R. Reisner, Results of Epidemiological Studies on Pneumoconiosis in West German Coal Mines, *in* W. H. Walton, Ed., *Inhaled Particles III*, Unwin, Woking, England, 1971.

18. P. D. Oldham and S. A. Roach, A Sampling Procedure for Measuring Industrial Dust Exposure, *Brit. J. Ind. Med., 9:* 112–119 (1952).
19. P. D. Oldham, The Nature of the Variability of Dust Concentration at the Coal Face, *Brit. J. Ind. Med., 10:* 227–234 (1953).
20. R. C. Tomlinson, Sampling Programmes and Sampling Instruments, *Instrument Practice*, June 1957.
21. A. J. Breslin *et al.*, The Accuracy of Dust Exposure Estimates Obtained from Conventional Air Sampling, *Amer. Ind. Hyg. Assoc. J., 28:* 56–61 (1967).
22. M. Jacobson, Respirable Dust in Bituminous Coal Mines in the U.S., *in* W. H. Walton, Ed., *Inhaled Particles III*, Unwin, Woking, England, 1971.
23. W. A. Langmead, Air Sampling as Part of an Integrated Programme of Monitoring of the Worker and His Environment, *in* W. H. Walton, Ed., *Inhaled Particles III*, Unwin, Woking, England, 1971.
24. D. C. Stevens, The Particle Size and Mean Concentration of Radioactive Aerosols Measured by Personal and Static Air Samples, *Ann. Occup. Hyg., 12:* 33–40 (1969).
25. H. J. Gale, Some Examples of the Application of the Lognormal Distribution in Radiation Protection, *Ann. Occup. Hyg., 10:* 39–45 (1967).
26. D. J. Finney, On the Distribution of a Variate Whose Logarithm Is Normally Distributed, *J. Roy. Statist. Soc., Suppl., 7:* 155–161 (1941).
27. American Conference of Governmental Industrial Hygienists: 1947 M.A.C. values, *Ind. Hyg. Newsletter, 7:* 15 (1947).
28. S. A. Roach, A More Rational Basis for Air Sampling Programs. *Amer. Ind. Hyg. Ass. J., 27:* 1–12 (1966).

2

Properties of Aerosols

2-1 Particle Dynamics

To understand the generally unstable nature of aerosols, the methods by which particles may be removed from the air (either for sampling or for cleaning purposes), and the behavior of particles in moving air streams, it is necessary to have a basic knowledge of particle dynamics. This subject has been dealt with in detail by others, notably Fuchs [1]. It will be discussed here primarily to bring out the nature of the fluid–particle interaction, the character of the forces acting on particles, and some of the physical parameters that are useful in data interpretation.

According to Newton's second law of motion, when a body of mass m is acted upon by a system of forces it experiences an acceleration according to the equation

$$m\,d\overline{U}/dt = \sum_i \overline{F}_i, \tag{2.1}$$

where the bars indicate vector quantities. The acceleration will have the same direction as the resultant of the vector sum of the forces acting on the body. It is possible that the resultant will be equal to zero, in which case

$$d\overline{U}/dt = 0,$$

where $\overline{U}$ is the instantaneous velocity of the body. When this occurs, $\overline{U}$ is a constant and may, or may not, equal zero. If the body under consideration is an airborne particle, then the forces acting on it will include a drag force F_R, which air exerts on any particle in motion relative to it, the force of gravity F_G, and a fluctuating, irregular force, F_B, which is due to the random thermal motion of the air molecules and is responsible for the phenomenon of Brownian motion. Depending on the nature of the particle and the conditions of the environment, it may be acted upon also by electrical, thermal, and sonic forces, and by photophoretic and diffusiophoretic forces. The conditions under which these forces occur and their relative magnitudes are considered below.

2-1.1 The Resistance Force

The fluid properties of density and viscosity are responsible for the resistance that acts on a particle having motion relative to the fluid. It is possible to separate the inertial forces (associated with density) and the viscous forces only in special cases, one of which applies to most of the particle sizes of interest to us. A consideration of the special cases is useful in that it provides a fairly simple physical picture of the fluid–particle interactions that are involved.

Figure 2.1(A) represents a sphere that has motion relative to the fluid in which it is immersed. It does not matter whether the sphere is fixed and the fluid is flowing around it or the fluid is stationary and the sphere is moving through it; the fluid elements in front of the sphere have to be given an acceleration normal to the direction of relative motion of the sphere. If the sphere's diameter is D and its relative velocity is U, then per unit of time it will have to push aside a column of fluid of length U and cross-sectional area $\pi D^2/4$. If the density of the fluid is ρ_f, then the mass of fluid that must be accelerated per unit of time is $(\pi/4)\cdot\rho_f D^2 U$. The acceleration of the fluid can be expected to be proportional to the relative velocity between sphere and fluid, that is, the faster the sphere moves through the fluid, the faster the fluid must be moved aside to permit its passage. Therefore, the momentum imparted to the

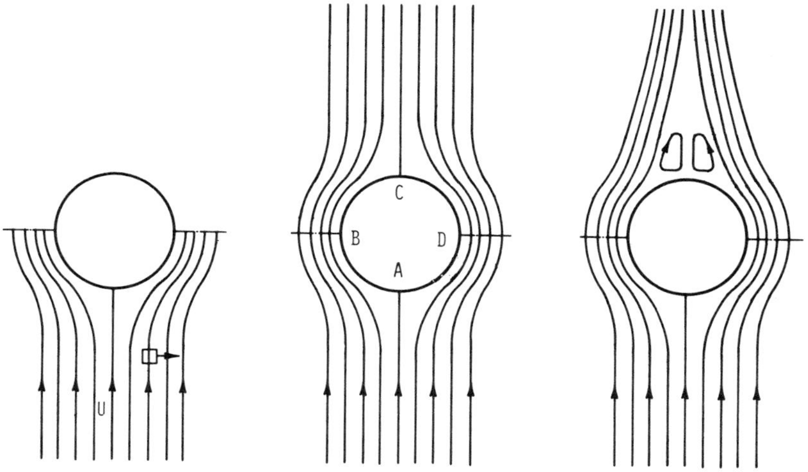

FIG. 2.1. Concepts of fluid resistance. (A) Inertial resistance due to acceleration of fluid elements to one side. (B) Ideal fluid. (C) Real fluid at velocity giving rise to turbulent wake.

cylinder of fluid per unit of time is proportional to the square of the sphere's velocity:

$$\text{momentum/unit time} \propto \rho_f U^2 D^2.$$

By definition, the rate of change of momentum of the fluid is the force required to displace it. Inertial resistance, given by the equal and opposite force which the fluid exerts on the sphere is

$$F_R = C \cdot \rho_f U^2 D^2.$$

This method of looking at fluid resistance, which dates back to Newton, requires that C be a constant. While it provides a simple physical picture of what happens and is accurate under certain conditions, it ignores the fluid motion at the back of the sphere, which is the dominant factor in inertial resistance. It is more realistic to relate inertial resistance to the pressure difference across the sphere. In the case of an ideal fluid, that is, one having zero viscosity, the streamlines of flow about the sphere will be symmetrical as shown in Fig. 2.1(B). According to Bernoulli's theorem, the pressure and velocity at any point on the streamlines are related by

$$P + \rho_f U^2/2 = H,$$

where H is a constant. At A, the velocity is zero and the pressure at that point is $P_A = H$. As the fluid passes around the sphere, the velocity increases and the pressure decreases, reaching a minimum value at B and D. As the fluid closes in behind the sphere, its velocity decreases again and its pressure increases until point C is reached where the velocity is again zero and the pressure is $P_C = H$. Because of the symmetrical flow pattern, any force acting on a small surface element on the front of the sphere, due to the pressure at that element, is offset by an equal and opposite force on an equivalent element diametrically opposite it on the back of the sphere. Thus, in an ideal fluid there is no resistance to the sphere's motion.

In a real (viscous) fluid, however, friction between the fluid and sphere and between adjacent layers of fluid causes a fluid element, which has been accelerated to one side, to lag behind on the downstream side. This causes a zone of stagnant fluid to form behind the sphere (Fig. 2.1C). Now the symmetries of pressure and velocity about the sphere are lost, and the excess of pressure in the front gives rise to a net force acting to retard the motion of the sphere. This force is related to the product of the velocity head $\frac{1}{2}\rho_f U^2$ and the cross-sectional area A of the sphere and is written as

$$F_R = C \cdot \tfrac{1}{2}\rho_f U^2 A, \tag{2.2}$$

where the proportionality factor C depends on the degree of turbulence on the downstream side of the sphere.

On the other hand, if the relative velocity between fluid and object is very low, or if the sphere is very small, or if the fluid is very viscous, then the zone of stagnant fluid may be negligibly small and the resistance offered to the motion of the sphere is due predominantly to the viscous friction between fluid and sphere. In this case, inertial effects are neglected and the resistance is calculated from viscous forces alone. A detailed derivation of the law governing viscous resistance is beyond the scope of this discussion, but it is possible to present a physical picture that brings out the roles of viscosity, relative velocity, and sphere diameter in viscous resistance. In Fig. 2.2, the flow pattern is symmetrical and the shearing stress on two elements of area, dA, in symmetrical positions on either side of the flow axis, will be identical. The shearing stress τ is the force per unit area acting parallel to the element of area under consideration. It is given by

$$\tau = f/dA = \eta\, dU/dr,$$

where η is the cofficient of viscosity of the fluid and dU/dr is the velocity gradient normal to the sphere's surface. Therefore, the force acting on dA in the tangential direction is

$$f = \eta (dU/dr)\, dA.$$

Combining the two forces on the two equal areas yields f_η, which is the force

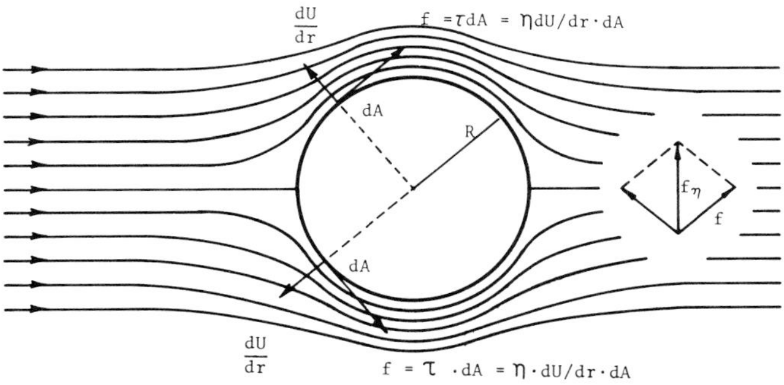

FIG. 2.2. Viscous forces on a particle.

resisting the motion of the sphere due to viscous stresses on the two elements of area. The absolute value of f_η will depend on the angle between the two f's, but it will be porportional to f:

$$f_\eta \propto \eta(dU/dr)\, dA.$$

To find the total resistance it is necessary to sum over the whole area of the sphere. After summation the total force will be proportional to both η and A, the total area. Also, at the surface of the sphere, $dU/dr \propto U/R$, and the resistance due to friction is proportional to the product of these three quantities:

$$F_R \propto \eta(U/R)A \propto \eta UR.$$

In addition, the energy loss due to the viscous drag on the sphere leads to a pressure drop across it, which contributes an additional resistance force also proportional to ηUR. In the particular case of a sphere, Stokes found that the constant of proportionality is 6π and

$$F_R = 6\pi\eta UR = 3\pi\eta UD. \tag{2.3}$$

The relationship between the resistance force and velocity for these two special cases is

Newton's law: $F_R \propto U^2$,

Stokes's law: $F_R \propto U$.

Stokes's law applies when viscous resistance is sufficiently large that flow about the sphere shows front and back symmetry; Newton's law applies when inertial effects are very much greater than viscous resistance. In between, there is a wide range in which F_R is proportional to U^n, where $1 < n < 2$. This variation of F_R with velocity requires some criterion for comparing the significance of each effect under a given set of conditions. The relevant conditions are fluid density ρ_f, fluid viscosity η, the relative velocity U between sphere and fluid, and the sphere diameter D. They enter into each force in the following way:

$$\text{inertial resistance} \propto \rho_f U^2 D^2,$$

$$\text{viscous resistance} \propto \eta U D.$$

The appropriate criterion, called the Reynolds number, is simply the ratio of these quantities

$$\text{Re} = \rho_f U^2 D^2 / \eta U D = \rho_f U D / \eta = U D / \nu, \tag{2.4}$$

where $\nu = \eta / \rho_f$, is the kinematic viscosity of the fluid. For nonspherical particles, other than cylinders, D is replaced with the hydraulic diameter, $4A/P$, where A is the particle's cross-sectional area in the direction of flow and P is the wetted perimeter of that area.

The fluid resistance can be put into a general form covering the full range of values of Reynolds numbers by applying a dimensional analysis attributed to Rayleigh [3]. He reasoned that the resistance force was related to the properties of the fluid and the particle in the following way:

$$F_R = \text{constant} \cdot \eta^a \rho_f{}^b U^c D^d.$$

Putting both sides in terms of the dimensions mass, length, and time:

$$MLT^{-2} = M^a \cdot L^{-a} \cdot T^{-a} \cdot M^b \cdot L^{-3b} \cdot L^c \cdot T^{-c} \cdot L^d.$$

Equating exponents and solving for b, c, and d in terms of a, gives

$$F_R = \text{constant} \cdot \text{Re}^{-a} \cdot \tfrac{1}{2}\rho_f U^2 A = C_R \cdot \tfrac{1}{2}\rho_f U^2 A. \tag{2.5}$$

C_R is called the drag coefficient and is plainly a function of Re. The index "a" varies from 1 to 0 as the Reynolds number goes from essentially 0 to 10^3 or so. The exact manner in which C_R varies with Reynolds number for spheres is shown in Fig. 2.3.

For the two special cases described above, the general resistance equation must be consistent with the corresponding special equations, which require that the drag coefficient have the following values:

$$\text{Newton's law:} \quad C_R = \text{constant} = 0.4,$$

$$\text{Stokes's law:} \quad C_R = 3\pi\eta U D / (\tfrac{1}{2}\rho_f U^2 A) = 24/\text{Re}.$$

The range of applicability of Stokes's law can be judged by comparing values of 24/Re with empirical values of C_R as shown in Table 2.1. The percent error

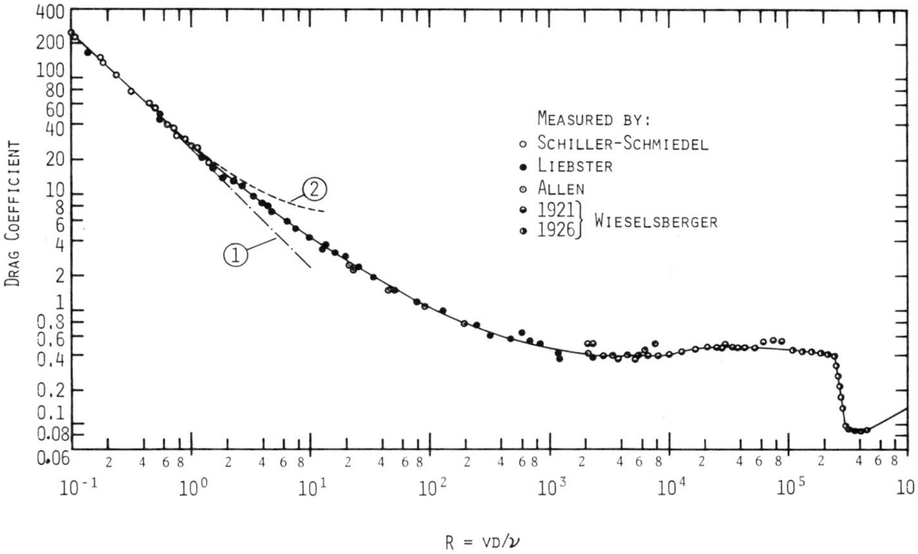

FIG. 2.3. Drag coefficient as a function of Reynolds number for spheres [2]. (1) Stokes's law; (2) Oseen's approximation, $C_R = (24/Re)\ [1+(0.187/Re)]$. Courtesy of McGraw-Hill Book Company.

equals $100(1-24/C_R \text{ Re})$. The tabulated values of C_R a e averages from experimental data [4]. For higher values of Re, Klyachko's formula [1]

$$C_R = (24/\text{Re})(1+\text{Re}^{2/3}/6) \tag{2.6}$$

is accurate within a few percent for $1 < \text{Re} < 400$. Below $\text{Re} = 1$, it reaches a maximum error of about 8% when $\text{Re} \simeq 0.3$. Some values of C_R according to Eq. (2.6) are included in Table 2.1, together with their errors relative to the experimental values.

TABLE 2.1

COMPARISON OF 24/Re WITH EMPIRICAL VALUES OF C_R

Re	24/Re	C_R	Percent error	Klyachko	Percent error
0.3	80.0	80.0	negligible	86.0	7.5
0.5	48.0	49.5	3.0	53.0	7.1
0.7	34.3	36.5	6.0	38.8	6.3
1.0	24.0	26.5	9.4	28.0	5.7
2.0	12.0	14.4	16.7	15.2	5.6

For $0.6 < \mathrm{Re} < 3.0$ the following formula, derived by Hänel [5] from the work of Pettyjohn and Christiansen [6], is more accurate than either the Stokes or Klyachko formulas:

$$C_R = (24/\mathrm{Re})(1 + 0.13\ \mathrm{Re}^{0.85}). \tag{2.7}$$

The relationship between C_R and Re brings out forcefully the importance of "dynamical similarity." When the pattern of flow depends only on fluid density and viscosity, results obtained using one fluid in a particular geometric arrangement are applicable to another fluid in a similar geometric arrangement provided the values of Re are the same in both cases. The curve of Fig. 2.3 spans four orders of magnitude for the drag coefficient and more than six for Reynolds number, yet the relationship has proved to be valid for a wide variety of fluids. In this way, phenomena that are difficult to handle in a particular fluid may be studied with a fluid more adaptable to experimentation.

(a) *Slip Correction Factors.* The derivation of Stokes's law included the boundary condition that the relative velocity between sphere and fluid must go to zero at the surface of the sphere. This means that a thin layer of fluid adheres to the sphere, rather than moving with the fluid. Applying this boundary condition to a gas, such as air, implies that the gas molecules, which must, on the average, have a net velocity in the direction of flow just before they strike the sphere, rebound in such a way that there is no net velocity of molecules adjacent to the sphere surface. This cannot occur in reality and the gas at the boundary retains some velocity in the direction of flow. As a result, the resistance force acting on a particle in air is less than that calculated from Stokes's law.

The effect of slip is negligible for large particles; it only starts to be of concern at a diameter of about 1 μm, where the resistance force is reduced about 16% from the Stokes's law value. Slip was first treated theoretically by Cunningham [7]; the correction factor that is applied to Stokes's formula now bears his name, although the values assigned to it come entirely from experimental data. The slip effect was studied extensively by Millikan [8] and his students because of its importance in making an accurate determination of the electronic charge. They measured the terminal settling velocities of paraffin and oil spheres over a wide range of pressures and expressed the slip correction factor as the ratio of the observed velocity to that calculated using Stokes's law. McKeehan [9], for instance, measured slip factors having values from 1.003 to 197, with a relative probable error of 2–3%. After Knudsen and Weber [10], the slip factor K_s was usually expressed in the form

$$K_s = 1 + (\lambda/a)[A + B \cdot \exp(-Ca/\lambda)], \tag{2.8}$$

where a is the particle's radius and λ is the mean free path of air molecules.

In 1945, Davies [11] reviewed the available data concerning the slip effect (see Fig. 2.4) and, taking into account the latest information on mean free path, recommended the following values for the constants:

$$A = 1.257, \qquad B = 0.400, \qquad C = 1.10.$$

Fuchs [1], relying entirely on Millikan's data, assigned to these three constants the values of 1.246, 0.42, and 0.87, respectively. Either set gives essentially the same values of K_s.

Making use of the inverse proportionality between mean free path and pressure, Davies put his results into the more useful form

$$K_s = 1 + (2/PD)[6.32 + 2.01 \exp(-0.1095PD)], \qquad (2.9)$$

where P is the air pressure in centimeters of Hg and D is the diameter in micrometers.

(b) *Fluid Resistance for Nonspherical Bodies.* Theoretical calculations of the resistance force have been carried out for ellipsoids, discs, and cylinders [12]. Information concerning the fluid resistance acting on bodies of other nonspherical shapes, even regular, isometric solids, comes from experimental studies. Fortunately, the data indicate that certain useful generalizations can

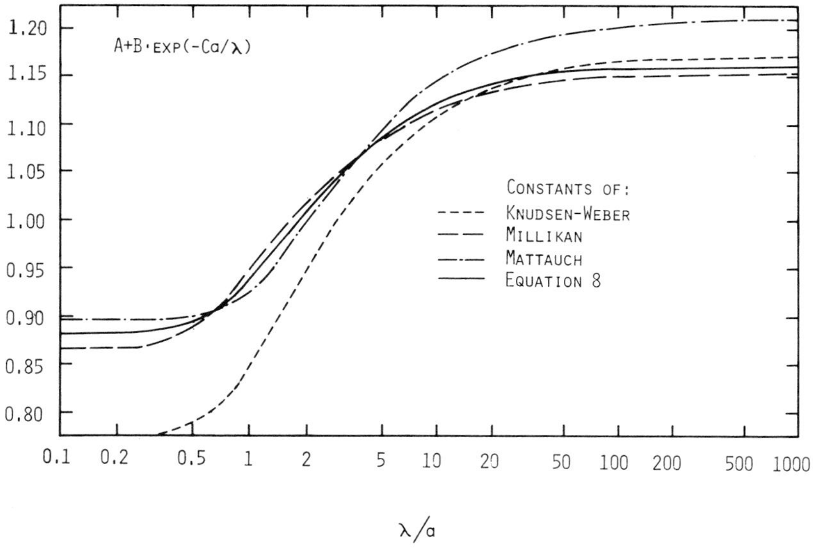

FIG. 2.4. Slip correction data as a function of mean free path/particle radius [11] (for old value of $\lambda = 0.091\ \mu$m). Courtesy of Institute of Physics and The Physical Society.

TABLE 2.2

RESISTANCE FORCE RELATIONSHIPS FOR ISOMETRIC PARTICLES

Shape	K_R[a]	$C_R \cdot Re$[b]	Re	Ref.
sphere	1.00	24.0	$\leqslant 0.07$	[13]
cube-octahedron	0.99	22.9	$\leqslant 0.07$	[13]
cube	0.98	25.6	$\leqslant 0.07$	[13]
cube	0.94	24.3	$\leqslant 0.1$	[14]
cube	0.97	25.4	$\leqslant 0.1$	[15]
cube-octahedron	0.98	22.8	$\leqslant 0.01$	[6]
octahedron	0.98	19.4	$\leqslant 0.01$	[6]
cube	0.97	25.1	$\leqslant 0.01$	[6]
tetrahedron	0.96	17.0	$\leqslant 0.01$	[6]

[a] $K_R = F_R/3\eta U(\pi S)^{1/2}$; F_R is the experimental resistance force.
[b] $C_R \cdot R_e = 8F_R/\eta PU$ [Eq. (2.12)].

be made to simplify the estimation of the fluid resistance for nonspherical particles.

Several workers [13–15] have pointed out that when inertial effects can be neglected, the resistance force acting on a nonspherical particle moving at a velocity U with respect to a fluid, can be put in a form analogous to Stokes's law:

$$F_R = 3\pi\eta DU \cdot K_R, \tag{2.10}$$

where D is some characteristic "diameter" of the particle and K_R is a resistance shape factor. Gurel *et al.* [13] showed that by setting $D = (S/\pi)^{1/2}$, where S is the surface area of the particle, and $K_R = (3.0/\pi)^{1/2}$, the resistance, expressed as

$$F_R = 3\eta U(3S)^{1/2}, \tag{2.11}$$

agreed within a few percent with experimental values for a variety of isometric and compact shapes.*
Following their suggestion, Eq. (2.10) with $D = (S/\pi)^{1/2}$, has been used to calculate K_R for a variety of isometric shapes for which data are available in the literature. The results are shown in Table 2.2. Values of K_R near unity indicate that the drag on the nonspherical particle is close to that on a sphere of the same surface area.

For certain shapes, other forms of the resistance equation may be useful.

* Fuchs [1] terms their results "unsuitable," apparently because of an unfortunate printing error which consistently replaced $3 \cdot \sqrt{3}$ with $\sqrt[3]{3}$.

Writing the Reynolds number in terms of the hydraulic diameter,

$$\mathrm{Re} = \rho_f (4A/P)\, U/\eta,$$

the general form of the resistance equation can be expressed as

$$F_R = (C_R \cdot \mathrm{Re})\, \eta P U/8. \tag{2.12}$$

Squires and Squires [16] have shown that for very thin discs moving with their flat face perpendicular to the direction of motion,

$$C_R \cdot \mathrm{Re} = 20.34,$$

and when moving edgewise,

$$C_R \cdot \mathrm{Re} = 21.36.$$

For comparison, spheres have a value of 24. Values of $C_R \cdot \mathrm{Re}$, which were calculated from the experimental data referred to above using appropriate values of the perimeter P, are also tabulated in Table 2.2.

Data for particles of more extreme shapes are shown in Table 2.3. Since they each have three axes of symmetry, they should show no preferred orientation while falling [17]. McNown and Malaika [14] point out, however, that for $\mathrm{Re} \gtrsim 0.1$ particles tend to orient themselves in a single stable position, causing an increase in F_R of about 20% for $\mathrm{Re} = 1$.

Stöber and his associates [18, 19] studied the settling velocities of aggregates of similar spheres. Some of their results are shown in Table 2.4, with some earlier data from Kunkel [20]. The latter refer to Reynolds numbers much larger than those at which Stöber's measurements were made. For the cluster aggregates, some of the particle surface is inaccessible to the fluid, so that K_R diminishes continuously as the number of particles increases and the appropriate surface area to use in Eq. (2.10) becomes that of the aggregate's envelope. Assuming the n particles are packed in cubic configuration, the use of the envelope area provides an estimate of the resistance shape factor

$$K_R' = 0.806 \cdot n^{1/6} K_R,$$

which is also included in Table 2.4. For the chains, Stöber obtained a relationship that is equivalent to

$$K_R = 0.862 \cdot n^{1/6}.$$

Values calculated with this equation agree with the experimental values within about 3%.

At Re less than 0.5, the resistance force per unit length for cylinders having

TABLE 2.3

RESISTANCE FORCE RELATIONSHIPS FOR SOME EXTREME SHAPES

Shape	Axial ratio[a]	Orientation	K_R[b]	$C_R \cdot Re$[c]	Re	Ref.
Prism	1:4	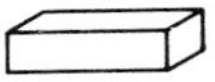 ↓	1.07	19.4	<0.1	[14]
	4:1	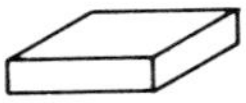 ↓	1.12	20.7	<0.1	[14]
	Very large	,,	1.15	17.3	0.21	[20]
	Very large	,,	1.28	19.3	0.74	[20]
	Very large	,,	1.37	20.6	0.99	[20]
Rod	4:1	 ↓	1.11	17.8	<0.1	[14]
	1:1	,,	0.94	21.6	<0.1	[14]
Disc	1:4	 ↓	1.12	23.4	<0.1	[14]
	Very small	,,	1.17	19.8	0.23	[20]
	Very small	,,	1.28	21.6	0.85	[20]
	Very small	,,	1.34	22.7	1.15	[20]
Double cone	1:1	 ↓	1.01	20.4	<0.1	[14]
	1:4	,,	1.03	17.8	<0.1	[14]
	4:1	 ↓	1.12	14.7	<0.1	[14]

[a] For prism: side of square face/third dimension. For rod or cylinder: length/diameter. For double cone: cone height/radius of base.
[b] $K_R = F_R/3\eta U(\pi S)^{1/2}$; F_R is the experimental resistance force.
[c] $C_R \cdot \mathrm{Re} = 8F_R/\eta PU$ [Eq. (2.12)].

a length much greater than their diameter is given by [21]

$$F_R/L = 4\pi\eta U/(2.00 - \ln \mathrm{Re}). \tag{2.13}$$

This can be put in the form

$$F_R/L = (4\pi\eta v/D) \cdot \mathrm{Re}/(2.00 - \ln \mathrm{Re}).$$

For $10^{-9} < \mathrm{Re} < 10^{-3}$, which covers the range of Reynolds numbers applicable to settling velocities of cylindrical fibers of respirable size,

$$\mathrm{Re}/(2.00 - \ln \mathrm{Re}) \simeq 0.175 \cdot \mathrm{Re}^{1.07},$$

with an error generally less than 6% ,and

$$F_R/L = 0.70\pi\eta U(UD/v)^{0.07}. \tag{2.14}$$

In this range, the resistance force is nearly proportional to the velocity and is not much affected by cylinder diameter.

At low Reynolds numbers, the experimental data cited above indicate that the resistance force is proportional to velocity for particles of both spherical and nonspherical shape. The reciprocal of the constant of proportionality is called the particle's mechanical mobility, Z:

$$\begin{aligned} Z &= U/F_R \\ &= K_s/(3\pi\eta D), \qquad \text{for spheres.} \end{aligned} \tag{2.15}$$

TABLE 2.4

RESISTANCE SHAPE FACTORS FOR AGGREGATES OF SIMILAR SPHERES[a]

	Clusters						Chains	
	Latex spheres [19] Re $\ll 0.1$		Steel balls [18] Re < 0.005		Glass spheres [20]		Latex spheres	Glass spheres
n	K_R	K_R'	K_R	K_R'	K_R	K_R'	K_R	K_R
2	1.00	0.92	0.98	0.90	1.068 (0.48)[b]	0.97	1.00	1.06 (0.48)
3	0.96	0.93	0.97	0.94	1.07 (0.60)	1.04	1.06	1.10 (0.50)
4	0.93	0.94	0.93	0.95			1.05	1.30[c] (0.55)
5	0.91	0.96	0.92	0.97			1.11	
6	0.87	0.95	0.92	1.00	0.97 (0.88)	1.05	1.16	
7	0.87	0.97	0.90	1.00			1.20	
8	0.85	0.97	0.88	1.00			1.22	1.56 (0.63)
9	0.84	0.98	0.85	0.99				
10	0.84	1.00	0.86	1.02				
12	0.83	1.02						
14	0.81	1.02						
16	0.79	1.01						
18	0.79	1.03						
20	0.76	1.02						
22	0.73	0.98						

[a] The following terms are used: $K_R = F_R/3\eta U(\pi S)^{1/2}$; F_R is the experimental resistance force, $K_R' = 0.806 n^{1/6} K_R$, n is the number of unit spheres.

[b] Reynolds numbers are in parentheses.

[c] Sphere centers in common plane.

This quantity, which has units of velocity/unit force and is analogous to electrical mobility, is useful in discussing the motion of particles because its application is not restricted to spheres.

2-1.2 The Gravitational Force

Airborne particles are subject at all times to the force of gravity, which gives rise to a velocity in the direction of the center of the earth. The force is equal in magnitude to the product of the particle's mass and the acceleration due to gravity:

$$\bar{F}_G = m\bar{g}.$$

A particle of density ρ_p starting to fall at $t = 0$ will accelerate according to

$$m \cdot dU/dt = F_G(1-\rho_f/\rho_p) - U/Z, \tag{2.16}$$

where the first term on the right allows for the buoyancy of the fluid and U/Z is the fluid resistance at low Reynolds numbers. Since all the forces act along the same line, the vector notation has been eliminated. Integration of this equation yields the particle's velocity as a function of time:

$$U = F_G(1-\rho_f/\rho_p) \cdot Z \cdot [1-\exp(-t/mZ)]. \tag{2.17}$$

For the time being, the term allowing for buoyancy will be left in, although it can be neglected when the fluid is air. For $t \gg mZ$, the settling velocity U approaches a terminal value

$$U_G = F_G(1-\rho_f/\rho_p) \cdot Z = mZg(1-\rho_f/\rho_p). \tag{2.18}$$

The combination of gravity and buoyancy represents a constant force applied to the particle. Therefore, any constant force, or constant resultant of several forces, in cooperation with the fluid resistance would give the particle a velocity that could be described by an equation similar to (2.17):

$$U = F \cdot Z \cdot [1-\exp(-t/mZ)].$$

This also would lead to a terminal velocity equal to $F \cdot Z$. The product of mass and mobility is called the relaxation time of the particle:

$$\tau = mZ.$$

In time τ, the particle reaches a fraction, $1-1/e$, of its terminal velocity. Note that τ depends only on the mass and mobility of the particle and not at all on the magnitude or nature of the forces bringing about the terminal velocity. For spheres of diameter D micrometers, and density ρ_p gm/cm^3, at 20°C,

$$\tau = mZ = 3.0 \times 10^{-6} \rho_p D^2 K_s \qquad \text{sec.}$$

Even quite large particles very rapidly reach their terminal velocity under the action of the resistance force and a constant resultant of applied forces.

Because of the importance of sedimentation in aerosol stability and particle dynamics, U_G is used to define two "equivalent diameters." For any particle having a given terminal settling velocity:

(a) The *Stokes diameter* D_S is the diameter of a sphere having the same bulk density and the same terminal settling velocity; and

(b) The *aerodynamic diameter* D_A is the diameter of a sphere of unit density having the same terminal settling velocity.

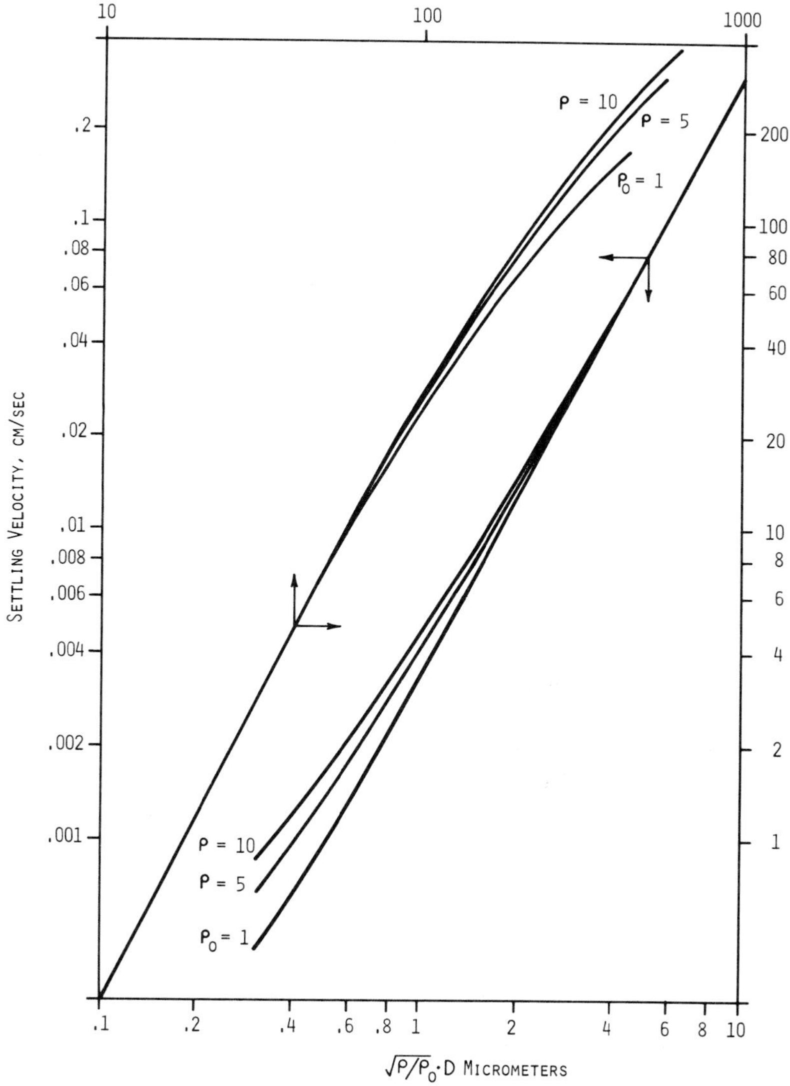

FIG. 2.5. Terminal settling velocity of spheres as a function of size and density.

The diameters are related as follows:

Terminal settling velocity U_G equals

$$mZg = \rho_p D_S{}^2 K_{sS} g/18\eta = \rho_0 D_A{}^2 K_{sA} g/18\eta.$$

any particle	sphere having particle's bulk density	sphere of unit density

The second subscript on the slip factor K_s identifies the diameter to which it refers, and ρ_0 represents unit density.

This is the defining relationship for the Stokes diameter D_S and the aerodynamic diameter D_A. It has certain shortcomings at small values of U_G because the quantity K_s, which is a function of the diameter being defined, is included in its definition.

As the Reynolds number increases, the terminal settling velocity given by Eq. (2.18) becomes increasingly inaccurate and the general resistance equation (2.5) must be used to calculate accurate values of U_G. After a constant velocity is reached,

$$m \cdot g = \tfrac{1}{2} C_R \rho_f U^2 A.$$

Using Klyachko's formula (2.6) for C_R,

$$\rho_p D^3 = (18\eta v/g) \cdot \mathrm{Re} \cdot [1 + (\mathrm{Re}^{2/3}/6)]. \tag{2.19}$$

In arriving at this equation, the slip factor and the buoyant effect of the air have been neglected. For a given particle density, the value of D can be calculated for a selected value of Re. Knowing the diameter for that value of Re, the corresponding velocity, U, can be calculated using Eq. (2.4). U is the terminal settling velocity of a sphere of diameter D and density ρ_p. In this way, a graph such as that shown in Fig. 2.5 can be constructed, relating U to aerodynamic diameter. The portion of the curve at small values of U was calculated using Eqs. (2.15) and (2.18).

2-1.3 The Brownian Motion Force

The particles of an aerosol come into thermal equilibrium with the air molecules and, like them, acquire an average thermal energy

$$\tfrac{1}{2} m U_B{}^2 = 3kT/2, \tag{2.20}$$

where k is Boltzmann's constant, T is absolute temperature, m is the particle's mass, and U_B is its root-mean-square thermal velocity given by

$$U_B = (18kT/\pi\rho_p D^3)^{1/2}. \tag{2.20a}$$

For unit density particles, 1 μm in diameter, $U_B = 0.48$ cm/sec at 20°C, the average velocity of the particle due to its thermal energy is

$$\bar{U}_B = (8kT/\pi m)^{1/2}. \tag{2.20b}$$

Its instantaneous velocity is constantly changing in magnitude and direction, however, so the thermal motion of the particles, like that of gas molecules, must be treated as a statistical phenomenon. Although single collisions between a molecule and a particle have little effect on the latter, it is possible to define a mean free path for particles that is useful in understanding Brownian motion phenomena [1]. Assume the particle follows a curved path as shown in Fig. 2.6. At any point, its velocity is tangential to the curve. At point *A*, for instance,

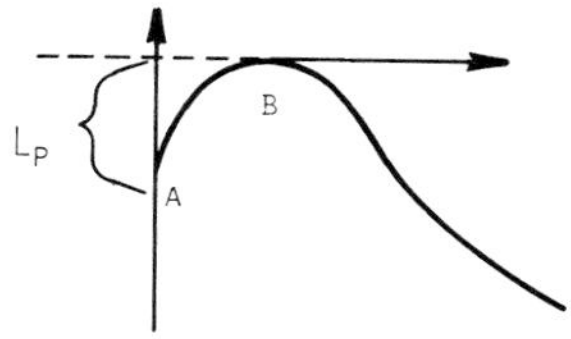

FIG. 2.6. Particle mean free path.

its velocity is as shown by the arrow. The second arrow shows its velocity at point *B*, where it no longer has a component in the direction of the velocity at *A*. The mean free path L_p of a particle is then defined [22] as the average distance a particle will travel in a given direction before its velocity has turned through 90°. This is simply the stopping distance (see below) of the particle at its *average* thermal velocity:

$$L_p = K_s m\bar{U}_B/3\pi\eta D. \tag{2.21}$$

Using Eq. (2.20b), the mean free path becomes

$$L_p = 1.37 \times 10^{-4} \cdot K_s \cdot (\rho_p D)^{1/2} \quad \text{cm} \tag{2.22}$$

for D in centimeters and ρ_p in g/cm^3. Some representative values at room temperature and unit density are tabulated below.

Diameter (μm)	L_p (cm)	Δ (cm^2/sec)
1	1.58×10^{-6}	2.75×10^{-7}
0.1	1.19×10^{-6}	6.81×10^{-6}
0.01	3.01×10^{-6}	5.23×10^{-4}

The diffusion coefficient of the particle,

$$\Delta = (kT/3\pi\eta D)\,K_s, \tag{2.23}$$

can be put in terms of its mean free path:

$$\Delta = \pi L_p \bar{U}_B/8,$$

a form that is analogous to the relationship between Δ and mean free path for molecules. L_p is of the same order of magnitude as the molecular mean free path, but $\bar{U}_B$ is very much smaller for particles than molecules, so that the former have quite small diffusion coefficients, as indicated in the table.

2-1.4 Other Forces

(a) *Electrostatic.* A particle carrying a charge q in a region of an applied electric field $\bar{E}$ experiences a force

$$\bar{F}_E = \bar{E}q \tag{2.24}$$

the direction of which depends on the polarity of the charge. The particle need not carry a net charge, since an applied electric field will polarize the particle, forming an electric dipole. If the electric field is not homogeneous, there will be a net force [23]

$$\bar{F}_E = [(\xi-1)/(\xi+1)]\,(D^3/16)\ \text{grad}\ E^2 \tag{2.25}$$

acting on the particle to move it in the direction of increasing field strength. In this equation, ξ is the dielectric constant of the particle and grad E^2 is the rate of change of E^2 in the direction of increasing E. This force may be encountered in concentric cylinder condensers.

On the other hand, a charged particle may create an electric field in the form of an image force when near a polarizable object. Then it will be acted on by a force that depends on the magnitude of its charge and the dielectric and geometric nature of the object. The same effect can be produced when a neutral particle approaches a charged object. Some values for geometric arrangements encountered in aerosol work are given in Table 2.5.

Electrostatic forces are very important because they can be used to impose a large precipitating velocity on airborne particles, making it possible to clean or sample air efficiently. Since particles are likely to carry a charge naturally, they may move in an unpredictable manner; this is particularly undesirable in laboratory studies.

(b) *Thermal Forces.* The force F_T acting on an airborne particle of diameter D in a temperature gradient ∇T, at a point where the temperature is T, is

$$F_T = f(H_a, H_p, P, T, D)\,(D/T)\cdot\nabla T, \tag{2.26}$$

where P is the pressure and H_a and H_p are, respectively, the thermal conductivities of air and particle. For a given material, the force is approximately proportional to the particle diameter when that is several times larger than the mean free path of air molecules. As diameter decreases, the function f gradually changes until it is proportional to D and the particles experience a force proportional to the square of the diameter [26]. Like electrostatic forces, thermal forces can occur naturally and should be considered in any practical situation. Their chief application is in the collection of air samples for particle size analysis.

TABLE 2.5.

IMAGE FORCES ON AIRBORNE PARTICLES[a]

Particle charge	Object	Object charge	Force on particle	Ref.
q_p	plane	0	$\frac{q_p^2}{4\rho^2} \cdot \frac{\xi_2 - 1}{\xi_2 + 1}$	[24]
q_p	cylinder	0		
0	cylinder	Q_c (per unit length)	$\frac{Q_c^2 D_p^3}{2\rho^3} \cdot \frac{\xi_1 - 1}{\xi_1 + 2}$	[25]
q_p	sphere	0	$q_p^2 \cdot \frac{R^3}{\rho_1^3} \cdot \frac{2\rho_1^2 - R^2}{(\rho_1^2 - R^2)^2} \cdot \frac{\xi_2 - 1}{\xi_2 + 1}$	[23]

[a] The terms used are:

ρ	is the distance from particle center to surface of object,
ρ_1	$\rho + R$,
R	is the object's radius,
D_p	is the particle diameter, and
ξ_1, ξ_2	are the dielectric constants of particle and object.

The magnitude of the thermal force relative to the force of gravity is shown in Fig. 2.7 [27], with curves for some other forces of interest.

(c) *Miscellaneous Forces.* Airborne particles may be acted upon also by photophoretic [28], diffusiophoretic [26], and sonic [29] forces. The latter have been of some interest in industrial hygiene because of their use to promote the coagulation and sedimentation of particles in off-gases [30]. Diffusiophoresis may be of some significance for the deposition of particles in the respiratory tract [31]. Photophoretic forces do not seem to have much significance for the problems of concern here.

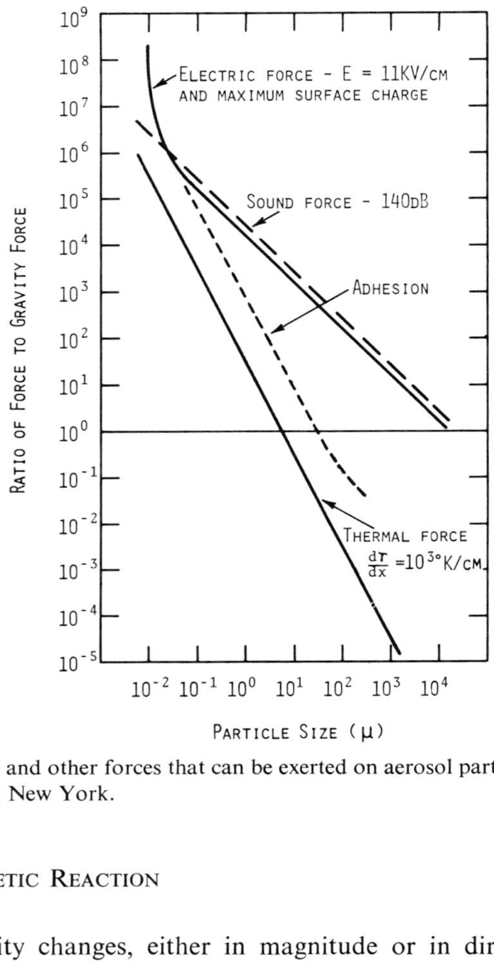

FIG. 2.7. Electrical and other forces that can be exerted on aerosol particles [27]. Courtesy of Academic Press, New York.

2-1.5 THE KINETIC REACTION

When air velocity changes, either in magnitude or in direction, particles suspended in it will lag behind owing to their inertia, giving rise to the kinetic reaction, $-mdU/dt$, where U is the relative velocity between air and particle. If air initially at rest is suddenly set in motion at a constant velocity U_0, suspended particles will approach that velocity according to

$$-m \cdot dU/dt = U/Z, \tag{2.27}$$

so that their velocity relative to the air diminishes exponentially:

$$U = U_0 \exp(-t/mZ). \tag{2.28}$$

Conversely, a particle injected into still air at the initial velocity U_0 is slowed down according to the same equation. The distance S_0 that the particle will travel before coming to rest is of considerable interest and can be found readily by dividing both sides of Eq. (2.27) by $U = dx/dt$ and integrating:

$$m \cdot dU/dx = -1/Z,$$

$$\int_{U_0}^{0} dU = -(1/mZ)\int_{0}^{S_0} dx, \tag{2.29}$$

$$S_0 = mZU_0.$$

S_0 is called the stopping distance of the particle and is the product of its mobility and its initial momentum. It is significant in equations describing the motion of particles in devices which effect collection by causing the particle-laden air to be deflected about an obstacle. For initial conditions outside the Stokes region, Klyachko's formula (2.6) gives

$$S_0 = (\rho_p/\rho_a) \cdot D[\mathrm{Re}_0^{1/3} - \sqrt{6}\tan^{-1}(\mathrm{Re}_0^{1/3}/\sqrt{6})], \tag{2.30}$$

where $\mathrm{Re}_0 = DU_0/\nu$.

2-2 Properties Affecting Aerosol Stability

A number of factors are constantly at work to alter the mass and number concentrations and particle size distribution of an aerosol. Particles are removed from the air primarily by sedimentation, although electrostatic effects are often significant, as is diffusion at small particle sizes. Evaporation and condensation are important for particles that have appreciable vapor pressures at ordinary temperatures or that are hygroscopic. Coagulation is a primary cause of changes in number concentration and particle size distribution. In the following discussion, the nature of electrostatic charges and the various processes responsible for the unstable character of an aerosol are considered.

2-2.1 Electrical Properties

Aerosol particles may carry charges acquired in the process by which they became airborne or they may pick up charges from collisions with atmospheric ions. The latter process may also cause the discharge of a particle initially carrying a charge.

(a) *Sources of Electric Charge.* When dry, nonmetallic particles are dispersed into the air by elutriation, for instance, they may carry charges acquired by the so-called triboelectric effect [32]. If, before dispersion, the particles are in contact with other particles of the same material or with a surface of the same material, then in the airborne state they will have a symmetrical charge distribution with equal numbers of positively and negatively charged particles

[33]. The average number of charges per particle appears to be related to the particle diameter [34]. Very roughly, for the particle diameter in micrometers, the absolute value of the charge in units of the electronic charge is

$$|\bar{q}| = 25D.$$

If the particles are in contact with a dissimilar material before dispersion, then the number of positively and negatively charged particles may or may not be equal. In general, if one material is an insulator and the other a metal, asymmetry will be quite marked, with the insulator carrying a negative charge and the metal carrying a positive charge.

Aerosols formed by the condensation of vapors produced by burning or by an electric arc are usually highly charged, although to an equal amount with respect to polarity. In this case, the vapors have been raised to a temperature sufficiently high that some of the molecules are thermally ionized. On the other hand, if vapor formation occurs at relatively low temperatures, subsequent condensation produces particles that are generally uncharged.

One of the most important processes by which aerosols are produced for laboratory studies is the atomization of solutions, colloidal suspensions, or suspensions of insoluble particles. When a droplet is formed by mechanical disruption of a body of conducting liquid, it may carry away a net charge as a result of the purely statistical fluctuation of positive and negative charges within the small volume element destined to become the droplet. If the concentration of ions within the liquid is sufficiently low, the fluctuations are due almost entirely to the random thermal motion of the ions. von Smoluchowski [35] showed that, under these circumstances, the average absolute value of the excess charge occurring in a spherical volume of diameter D micrometers, expressed in units of the electronic charge, is

$$|\bar{q}| = 8.2 \times 10^{-7} \cdot D^{3/2} N^{1/2}, \tag{2.31}$$

where N is the number of ions of one sign per cubic centimeter.

This formula refers to the droplet volume while still in the bulk liquid. As it is being formed into a droplet, however, there may be time for it to discharge partially. As the ionic strength increases, the discharge rate increases. At the same time, interionic forces tend to reduce the net charge in a given element of volume before it is detached, so that the charge carried away by the droplet is less than that given by Eq. (2.31).

Natanson [36] extended the theoretical work to include interionic forces and carried out an extensive experimental test of the theory. He found that the charge on the droplets agreed with that predicted in Eq. 2.31 even at values of N at which significant interionic effects should have occurred. Up to an ionic concentration of about 10^{15} cm^{-3}, there was essentially no discharge of the droplet during its formation. As the ionic strength further increased, the droplet charge diminished continuously until $N \simeq 10^{18}$ cm^{-3}, where it leveled off at a value found experimentally by Natanson to be

$$|\bar{q}| = 5.6D^{1/2} \qquad \text{electronic charges,}$$

where D is, again, in micrometers. Recent measurements in this region of ionic concentrations [37] indicate that this formula is valid up to about $N = 10^{20}$, but beyond that the charge distributions become asymmetrical and have more charges per particle than the formula predicts.

Several aspects of charging by atomization are of particular significance in laboratory work. Organic liquids and aqueous suspensions of colloidal or insoluble particles may well have ionic concentrations in the region of maximum charging effect. Changing the size of residual particles by diluting the solution to be atomized not only favors a larger charge on the initial droplet but leaves a smaller particle with a correspondingly larger electrical mobility.

(b) *Charge Distributions of Airborne Particles.* Whatever the charge on an aerosol at the time of its formation, it will subsequently tend toward an equilibrium condition that will depend only on the electrical conductivity of the air around it. The electrical conductivity of the air is due to the presence of ions produced primarily by the radioactive gases, radon and thoron, and their decay products in air, by radioactivity in soil, and by cosmic radiation. The electrical conductivity λ of air is the current per unit electric field strength that would flow across each square centimeter of a plane perpendicular to the direction of the electric field. It is the sum of the net number of positive ions passing through each unit of area each second in the direction of the electric field and the net number of negative ions crossing the same area each second in the opposite direction:

$$\lambda = \lambda_+ + \lambda_- = N_+ \varepsilon\mu_+ + N_- \varepsilon\mu_-, \tag{2.32}$$

where N_+ is the number of positive ions per cubic centimeter, N_- is the number of negative ions per cubic centimeter, ε is the electronic charge, and μ_+, μ_- are the electrical mobility of positive ions and negative ions in cm^2/volt-sec. At formation, both types of ions have mobilities of about 2.0 cm^2/volt-sec, but the mobility of the positive ion diminishes with time, averaging about 1.6 cm^2/volt-sec [38].

Under equilibrium conditions, the rate p at which ions are being formed equals the sum of the rate at which they are being lost by recombination and the rate at which they become attached to particles or nuclei:

$$p = \alpha N_+ N_- + \beta N,$$

where β is a function of the aerosol concentration and composition, $N = N_+ + N_-$, and α is a recombination coefficient having a value of 1.6×10^{-6}

cm^3/sec. Values of p usually fall between 7 and 15 ion pairs/cm^3-sec with an average of about 10 [39]. If there are no particles or nuclei in the air, $\beta = 0$, and the concentration of the ions, assuming equal values for both polarities, will be about 3500 cm^{-3}. Moderate aerosol concentrations can reduce the ion concentration to a few hundred per cubic centimeter.

In a bipolar ion field, charging or discharging of particles continues until an equilibrium condition is reached in which the frequency function of elementary charges q on particles of diameter D is approximately

$$f(q) = [\varepsilon/(\pi DkT)^{1/2}] \exp[-(q-q_0)^2/2\sigma^2], \tag{2.33}$$

where k is Boltzmann's constant, T is temperature,

$$\sigma^2 = DkT/2\varepsilon^2, \qquad \text{and} \qquad q_0 = \sigma^2 \cdot \ln(\lambda_+/\lambda_-).$$

If the negative and positive conductivities are equal, then q_0 is zero and the charges are normally distributed about a mean of zero, as shown in Fig. 2.8

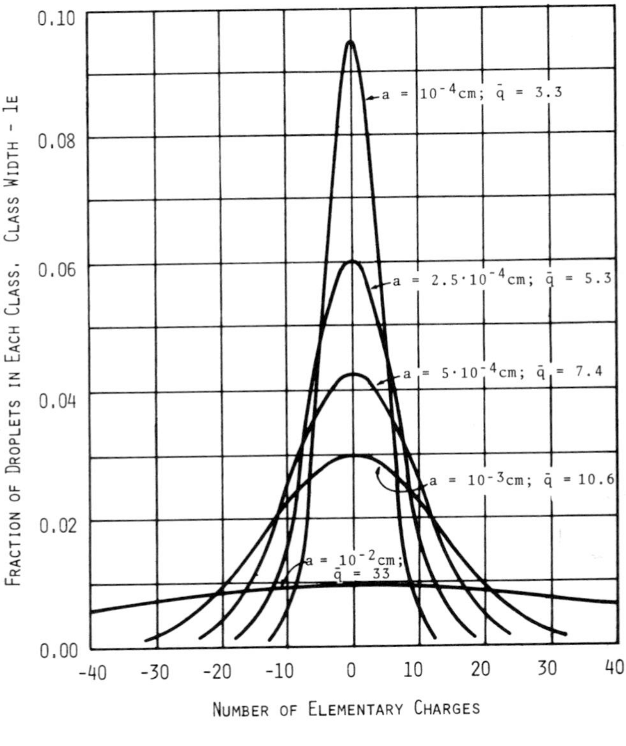

FIG. 2.8. Equilibrium distribution of charges on particles [40] (a = radius). Courtesy of Academic Press, New York.

[40]. In this case, the average number of charges, sign ignored, on a particle D micrometers in diameter is

$$\bar{q} = 2.37 \cdot D^{1/2}.$$

Gunn ignored image forces in deriving the charge distribution equation above and treated charge as a continuous quantity. The errors introduced into his analysis tended to cancel each other and Eq. (2.33) is quite accurate down to $D = 0.2\ \mu\text{m}$ [41]. If the conductivities are not equal, the curves in Fig. 2.8 are shifted to the left or right as λ_+ is less than or greater than λ_-, and the average absolute value of the charge is increased.

At equilibrium, the magnitude of the air's conductivity does not affect either the charge distribution or the average absolute value of the charge. It does affect, however, the rate at which equilibrium is approached. An object carrying an excess charge in air will approach equilibrium according to

$$q = q_0 \cdot \exp(-4\pi\lambda t),$$

with a half-life of discharge equal to

$$T_{1/2} = 0.055/\lambda \qquad \text{sec}.$$

A sparsely charged object will approach equilibrium with the same half-life. Gunn [40] estimates that in ordinary air the charging time is of the order of 400 sec. This is based on the assumption that the conductivity does not change as a result of the charging or discharging process. In the case of a highly charged aerosol, the ion concentration would be depleted rapidly and the subsequent approach to equilibrium would depend on the rate of ion formation. At the normal rate of 10 $\text{cm}^{-3}\ \text{sec}^{-1}$, equilibrium could not be attained before serious changes occurred in the aerosol. For laboratory studies, it is necessary to expose the aerosol to a very high bipolar concentration of ions to bring about a rapid equilibration of particle charge. This can be achieved with radioactive sources [42] or with a corona discharge in a suitably designed jet orifice [43].

2-2.2 Other Physical Properties

(a) *Sedimentation.* The manner in which settling affects aerosol concentration can be considered for two extreme cases: tranquil settling and thoroughly stirred settling. In the former case, a cloud of monodisperse particles of initial concentration C_0 in an enclosure of height H and horizontal projected area A settles out at a rate

$$dN/dt = -C_0 U_G A, \tag{2.34}$$

where N is the total number of particles having terminal settling velocity U_G that are airborne in the enclosure. The number remaining at time t is

$$N(t) = C_0 AH[1-(U_G t/H)]. \tag{2.35}$$

and the concentration, averaged over the whole volume, is

$$\bar{C} = C_0[1-(U_G t/H)]. \tag{2.36}$$

The nature of tranquil settling is such, however, that the concentration in the chamber is either C_0 or zero. Sedimentation proceeds so that the concentration at time t is C_0 up to a height, $H-U_G t$, and zero above that height. If the aerosol initially contains a range of particle sizes, then the size distribution, as observed at a height h will change with time as shown in Fig. 2.9.

In thoroughly stirred settling, a uniform concentration is maintained to within a very small distance, x, of the bottom of the enclosure, where still air conditions prevail. The number of particles of a given size falling into the still air region per unit time is CAU_G and the rate at which the concentration of particles of that size changes is

$$\frac{dC}{dt} = -\frac{CAU_G}{A(H-x)} = -\frac{CU_G}{H-x}. \tag{2.37a}$$

In terms of the initial concentration C_0, the concentration at time t is

$$C = C_0 \cdot \exp[-U_G t/(H-x)] \simeq C_0 \cdot \exp(-U_G t/H). \tag{2.37b}$$

For a range of particle sizes, the rate at which concentrations change with time is difficult to describe quantitatively. Sinclair [44] has obtained an approximate relationship, valid when $t \ll H/U_G$, for stirred settling of an aerosol characterized initially by a lognormal distribution of particle sizes. He found that the particulate surface concentration C_s and the particulate mass concentration C_m diminished with time according to

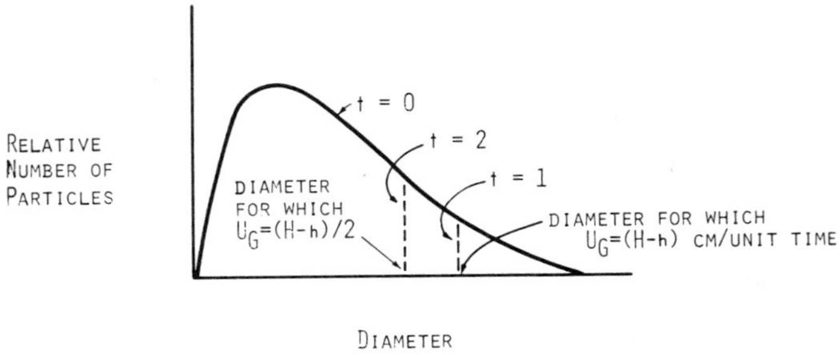

FIG. 2.9. Particle size distribution change with time at height h.

$$dC_s/dt = -3.0 \times 10^{-3} \rho_p D_3{}^2 C_s/H = -U_{G_3} C_s/H, \tag{2.38a}$$

and

$$dC_m/dt = -3.0 \times 10^{-3} \rho_p D_3{}^2 C_m \exp(2 \ln^2 \sigma_g)/H = -U_{G_3} C_m \exp(2 \ln^2 \sigma_g)/H. \tag{2.38b}$$

D_3 and σ_g are, respectively, the mass median diameter in micrometers and the geometric standard deviation of the initial size distribution. U_{G_3} is the terminal settling velocity in centimeters per second corresponding to D_3. Both stirred settling [45] and tranquil settling [44] have been used to estimate size distribution parameters.

(b) *Coagulation.* Coagulation, which arises from relative velocities between particles, causes a continuous change in the number concentration and size distribution of an aerosol. The relative velocities may be due to Brownian motion, differences in particle sedimentation rates, electrostatic effects between particles or imposed on particles by an external field, velocity gradients in laminar or turbulent flow, or ultrasonic forces.

Brownian motion, or thermal, coagulation was studied theoretically by Smoluchowski [46]. He showed that for a monodisperse aerosol the rate at which n, the number of particles per unit volume, should change with time is

$$dn/dt = -K_C n^2, \tag{2.39}$$

where K_C is a coagulation coefficient given by

$$K_C = (4kT/3\eta) K_s = 3.0 \times 10^{-10} K_s \quad \text{cm}^3/\text{sec} \quad (\text{at} \quad T = 296°\text{K}). \tag{2.40}$$

K_s is the slip factor for the particles and is the only quantity in the coagulation constant that depends on particle size. Integration of the rate equation gives

$$n = n_0/(1 + n_0 K_C t), \tag{2.41}$$

where n_0 is the initial concentration of particles. Despite the fact that particle size changes with coagulation, this equation has been found to hold quite well for aerosols that were, initially, not severely polydisperse [47]. Moreover, while polydispersity must increase the rate of thermal coagulation [48], its effect on ordinary size distributions does not appear to be very great [49].

When all particles carry charges of the same polarity, their mutual repulsion causes the coagulation rate to decrease. If the charge on the aerosol is symmetrical with respect to polarity, as is so frequently the case in practice, the rate of coagulation is increased, but the effect does not appear to be large [27]. If the charges are very large or there is an impressed electric field in the space confining the aerosol, then coagulation may be greatly enhanced, with the formation of long chain aggregates [27].

Coagulation due to different rates of settling is important in such problems as the scavenging of atmospheric aerosols by raindrops. As an effect between aerosol particles of two different sizes, it is not as great as ordinary thermal coagulation. Coagulation may be greatly increased by ultrasonic forces, but as pointed out above, the effect is primarily of interest for cleaning processes and its practical usefulness is still open to question. The ratio of the coagulation rate due to a velocity gradient to that due to thermal motion is [49]:

$$GD^2/6\pi\Delta \qquad \text{for laminar shear flow,}$$

and

$$\frac{bD^2}{64\pi\Delta}\left(\frac{\psi}{v}\right)^{1/2} \qquad \text{for turbulent shear flow.}$$

G is the velocity gradient, D is particle diameter, Δ is the particle diffusion coefficient, ψ is the rate of energy dissipation per unit mass of fluid, v is the kinematic viscosity, and b is either 4 or 25, depending on the source of the derivation [49]. For large particles, coagulation rates due to turbulent velocity gradients can be much larger than those due to thermal effects.

(c) *Evaporation and Condensation.* Evaporation and its inverse, condensation, are important factors in the stability of an aerosol if it is composed of liquid droplets or particles of a hygroscopic substance. The physical processes involved will be discussed here in terms of evaporation, but the methods and results are applicable to condensation, provided that the directions of thermal and concentration gradients are kept in mind.

An isolated droplet in still air evaporates at a rate [50]

$$dM/dt = -4\pi R\Delta C_0/[(4\Delta/\alpha RU) + R/(R+\delta)]. \tag{2.42}$$

C_0 is the mass concentration of molecules corresponding to a saturated vapor pressure at the temperature of the droplet; Δ is the diffusion coefficient of the molecules, and U is their average thermal velocity; R is the droplet radius and M is its mass; α is the condensation coefficient of the molecules, that is, the probability that a molecule of the vapor, striking a surface of its own liquid, will condense; and δ is a length related to the mean free path λ of air molecules:

$$\delta = \lambda \cdot [(m_1+m_2)/m_2]^{1/2}. \tag{2.43}$$

m_1 and m_2 are, respectively, the masses of air and vapor molecules. It is assumed that the vapor concentration goes to zero at large distances from the particle. For large values of R, the denominator in Eq. (2.42) approaches unity and the equation becomes that of Langmuir [51] for evaporation controlled by diffusion of molecules away from the droplet.

For droplets of water or aqueous solutions or suspensions, the vapor concentration at a distance from the droplet has a value of $\beta C(T_a)$, where β is the relative humidity expressed as a fraction, and $C(T_a)$ is the saturated vapor concentration at ambient temperature T_a. Evaporation of the water cools the droplet to a temperature T_0 at which the value of C_0 is reduced to $C(T_0)$. If the droplet is a dilute solution obeying Raoult's law, the concentration of vapor molecules at the surface is further reduced by the factor

$$f = (1-f')/[1+(n-1)f'], \tag{2.44}$$

where f' is the mole fraction of the solute in the solution and n is the effective number of ions contributed by each molecule. The rate of evaporation is related to the temperature by

$$-L(dM/dt) = 4\pi RH_a(T_a-T_0). \tag{2.45}$$

L is the latent heat of vaporization of water and H_a is the thermal conductivity of air. Replacing dM/dt with Eq. (2.42), the temperature difference is

$$T_a - T_0 = \left[\frac{\Delta L}{H_a}\right][f\cdot C(T_0) - \beta C(T_a)]\Bigg/\left[\frac{4\Delta}{\alpha RU} + \frac{R}{R+\delta}\right]. \tag{2.46}$$

Handbook data [52] indicate that, for small temperature differences,

$$C(T_a) = C(T_0) + b(T_a-T_0), \tag{2.47}$$

where b depends on the range of termperature that is under consideration. This leads to the following expression for the rate of change of droplet mass:

$$\frac{dM}{dt} = 4\pi\Delta RC(T_a)(\beta-f)\Bigg/\left[\frac{R}{R+\delta} + \frac{4\Delta}{\alpha UR} + \frac{fbL\Delta}{H_a}\right]. \tag{2.48}$$

dM/dt is negative (evaporation) when $(\beta-f) < 0$ and positive (condensation) when $(\beta-f) > 0$. The equation must be integrated stepwise in increments of M and t.

Values reported for the condensation coefficient α range from 0.015 to 1.0. Unfortunately, it is an important quantity when estimating the rate of growth of a droplet or the time required for one to evaporate completely. Figure 2.10 shows the times required for sodium chloride particles to grow to various masses by the accretion of water. The calculations with which curve 1 was obtained referred to particles that were large enough to make it possible to

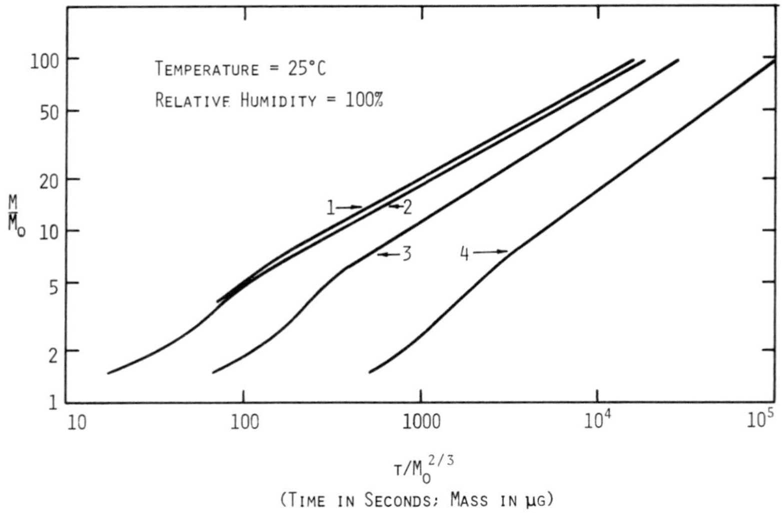

FIG. 2.10. Growth of sodium chloride particles in a humid atmosphere. (1) $\alpha = 1$, all values of m_0 [53]. From Eq. (2.48), (2) $\alpha = 0.034$, $m_0 > 1$ μg; (3) $\alpha = 0.034$, $m_0 = 10^{-6}$ μg; and (4) $\alpha = 0.034$, $m_0 = 10^{-9}$ μg.

ignore the limitation that is placed on evaporation rate due to the finite rate at which molecules can leave a liquid surface. The other curves were derived from Eq. (2.48), assuming a condensation coefficient equal to 0.034 [50]. Recently, some workers claimed its value should be 0.015 [54]; however, others argue that values below unity are due to experimental difficulties [55].

The hygroscopic nature of a particle is an important factor in the deposition of particles in the lung. In the humid atmosphere of the lung soluble particles become droplets of solution and grow rapidly. Figure 2.11 shows how droplets of sodium chloride solution increase in diameter as they move through the respiratory tract [56].

Usually, the hygroscopic crystal is considerably more dense than water, and the increase in its diameter as it takes up water is less significant to lung deposition than the ratios of the diameters would suggest. Dense substances of high molecular weight show a considerable increase in diameter before there is any significant increase in their aerodynamic diameter. Figure 2.12 shows this effect for cesium chloride, which has a relatively high molecular weight and has been used in studies of inhalation toxicity.

Large droplets will have a relative motion with respect to the air that, in effect, moves them out of their own vapor, thus lowering the concentration of molecules in the vicinity of the droplet. This "ventilation" effect was studied by Frössling [57], who found that the rate of evaporation dM/dt as given for

Both components are scattered symmetrically in front and in back of the particle, as shown in Fig. 2.14(A). The ratio i_1/i_2 varies with the scattering angle, but is not a function of the particle size. The total light scattered, however, varies rapidly with size, being proportional to the square of the particle's volume V.

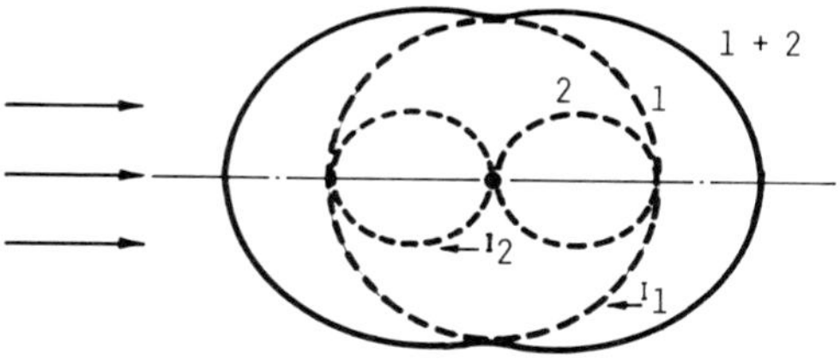

A

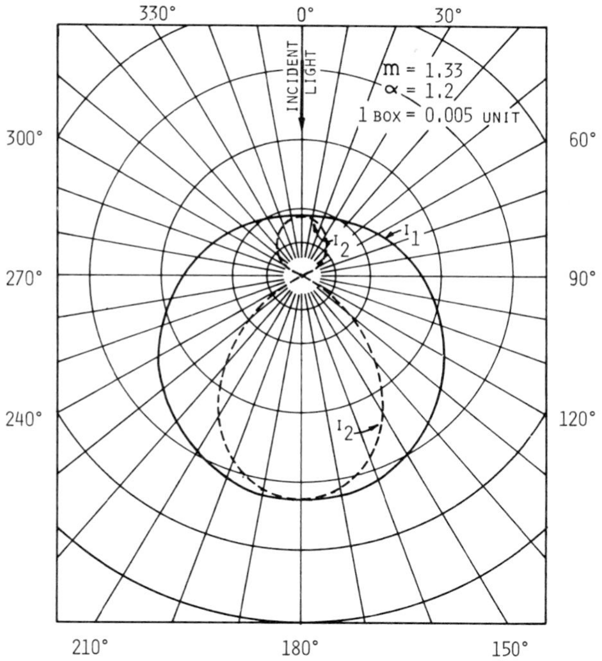

B

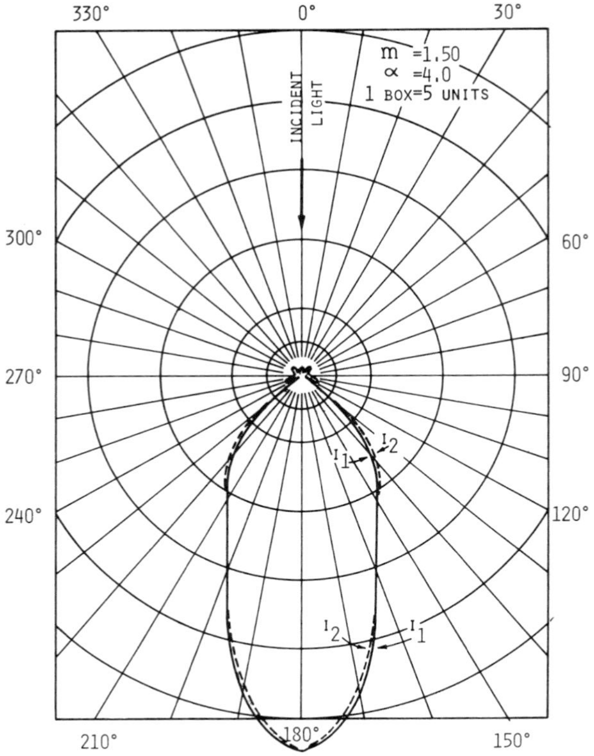

FIG. 2.14. Angular distribution of scattered light. (A) Rayleigh scatter [66], courtesy of H. C. van de Hulst. (B) and (C) Mie scatter [65].

2-3.2 RAYLEIGH–GANS SCATTERING

By assuming that each volume element of a particle scattered light independently according to Rayleigh scattering, it was possible to extend the former equations so that the criterion for its validity became $2\alpha(m-1) \ll 1.0$. If the right-hand side of Eq. (2.53) is multiplied by the factor [66]

$$(1/9)(m^2+2)^2 \cdot G^2(\theta, \alpha)$$

then $I(\theta)$ is the intensity of the scattered light at the angle θ, for Rayleigh–Gans scattering. The function $G(\theta, \alpha)$ has been tabulated by van de Hulst [66]. Again, the ratio i_1/i_2 is proportional to $1/\cos^2\theta$, but is not a function of particle size. Now, more light is scattered forward than backward. This modification of the Rayleigh equation extended the region in which calculations were possible to particles of about 0.1 μm diameters.

2-3.3 Mie Scattering

As the particle size increases beyond the region described by the Rayleigh–Gans approximation, the scattering relationship becomes very complex and, up to values of $2\alpha(m-1) \gg 1$, the exact Mie theory is necessary. As the value of this quantity increases, more and more of the scattered light is in the forward direction as shown in Fig. 2.14(B), for which the data were calculated by digital computer [65]. In this region, just beyond the range of the Rayleigh–Gans approximation, the degree of polarization begins to depend on the size of the particle and, over a very limited range, it is possible to make use of the ratio of i_2 to i_1 to measure the particle size of a monodisperse aerosol. An example of the narrowness of the size range over which this is possible is shown in Fig. 2.15 for water droplets in air. Above about 0.4 μm, a number of different diameters give the same value of i_2/i_1. The curve continues to oscillate at diameters above 0.7 μm, and a similar behavior is shown for other values of m.

As the particle size increases further, the intensity of scattering becomes more pronounced in the forward direction, and the curve begins to show maxima and minima (Fig. 2.14C) at positions that are functions of the wavelength of the incident light. There are approximately α minima between 0° and 180°.

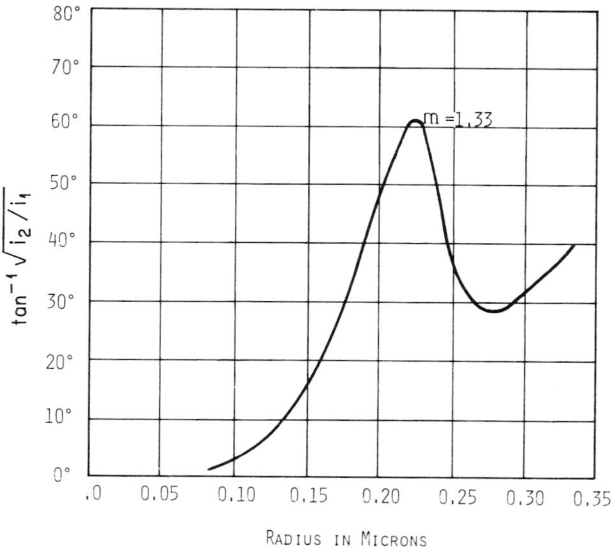

Fig. 2.15. Relative polarization ratios of light scattered at 90° [65].

When a particle is irradiated with white light, the scattered light has the appearance of a rainbow, and the number of times that a given color predominates in going from the front to the back of a particle is related to its size. Of particular significance in this respect is the number of reds that are seen. Figure 2.16(A) shows how this number varies with particle size and Fig. 2.16(B) shows the angular position of the corresponding maxima.

The extinction coefficient E varies with $\alpha = \pi D/\lambda$ as shown in Fig. 2.17. As m increases, the maxima and minima shift to the left and the values of E at the maxima increase. If E is plotted as a function of $\rho = 2\alpha(m-1)$, curves for

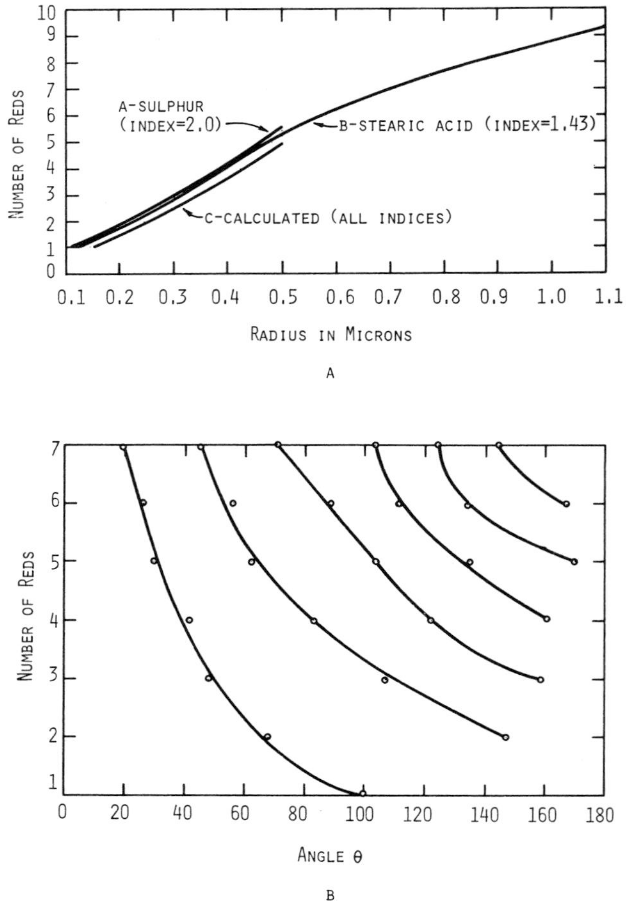

FIG. 2.16. Number and position of red bands [65]. (A) Number as a function of radius; (B) angular position of reds.

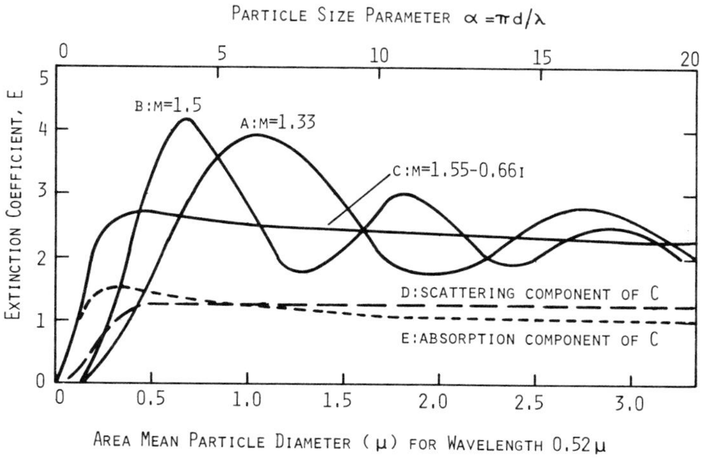

FIG. 2.17. Extinction coefficient as a function of size [67]. Courtesy of Academic Press, London.

different values of m oscillate in the same manner, with maxima at $\rho \simeq 4.2$, 10.8, 17.2, and 23.6 and minima at $\rho \simeq 7.6$, 14.1, and 20.3. As α continues to increase, the oscillations are gradually damped out and $E \rightarrow 2$.

2-3.4 MECKE'S APPROXIMATION

In this approximation, the light incident directly on the particle is treated according to geometric optics, and light passing near the particle is scattered according to Kirchhoff's theory of diffraction. The contribution of each to scatter at a given angle is calculated separately and then combined under the assumption that the two components do not interfere. For this to be approximately true, it is necessary that

$$2\alpha(m-1) \gg 1.$$

The angular distribution of the diffracted light is a function of wavelength, particle size, and scattering angle, but not of the relative refractive index. Most of the diffracted light falls within about 20° of the direction of propagation. The light scattered by diffraction is equal in quantity to that incident on the particle. It is not polarized. Light reflected or refracted by the particle is scattered through a larger angle, and the distribution of intensity is not affected much by particle size. The reflected and refracted light is plane polarized, but the effect is not useful for discriminating particle sizes. The amount of light scattered in this way is very nearly equal to the amount falling on the particle, so that the extinction coefficient in this region is $E \simeq 2.0$.

2-3.5 Scatter from Particles of Irregular Shape

The discussion above refers to individual transparent spherical particles. If the particles are irregular in shape, the scattering will be different because a particle of a given size will have a scattering effect that will depend on its orientation. A number of irregular particles of the same size, randomly oriented with respect to the incident beam, will scatter light in a somewhat more diffuse manner than the same concentrations of spheres. This tends to smooth out the fluctuations in the scattering coefficient curve.

Figure 2.18 shows some experimental results for irregularly shaped quartz particles [68]. The extinction coefficient rises steadily from $\alpha = 0$ until it reaches a constant value of 2 at $\alpha = 20$. All of the fluctuations predicted by the Mie theory have been damped out. This is partly due to particle orientation, but due also to the spread in particle sizes, since the diameters represent the averages of size increments obtained by repeated sedimentation.

For the larger particles, Fig. 2.19 shows that the scattering is almost entirely in the forward direction. Although the values of α are quite large, the value of m is so low that the scattering does not fall in the region of the Mecke approximation. This is emphasized by the polarization results in Fig. 2.19, which show that up to $\alpha = 20$, there is a definite effect of particle size on the

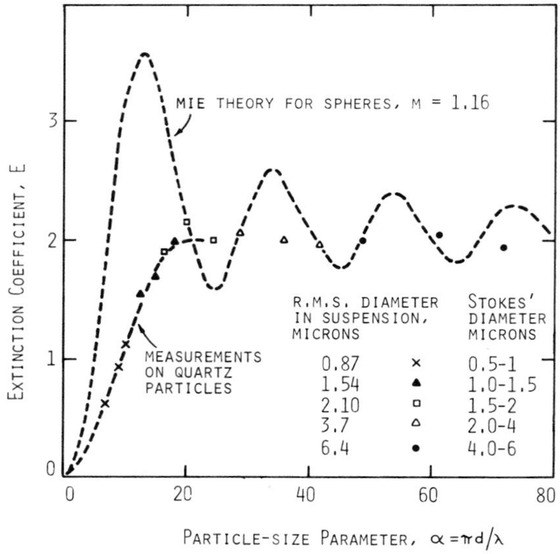

Fig. 2.18. Extinction coefficients for transparent particles of irregular shape [68]. Courtesy of Pergamon Press, Book Division.

value of i_1/i_2. By the time the Mecke region is reached, however, the polarization effect is practically independent of size.

Breuer *et al.* [69] have measured the intensity of scattered light from individual particles suspended in an electric field. Both spheres and particles of irregular shape were used and scattering of both visible and infrared radiations was observed. Their arrangement permitted them to capture a test particle for observation in the light microscope after its scattering characteristics had been recorded. Their work was related to hazard evaluation in German coal mines, where much of the monitoring work is done using a tyndalloscope. In this work, they could find no significant differences between the scattering characteristics of coal particles and waste rock particles.

Unlike most of the particles referred to above, coal absorbs some of the incident radiation, as do many other materials. Absorption is included in the index of refraction as an imaginary term, which has a marked effect on scattering properties (see Fig. 2.17).

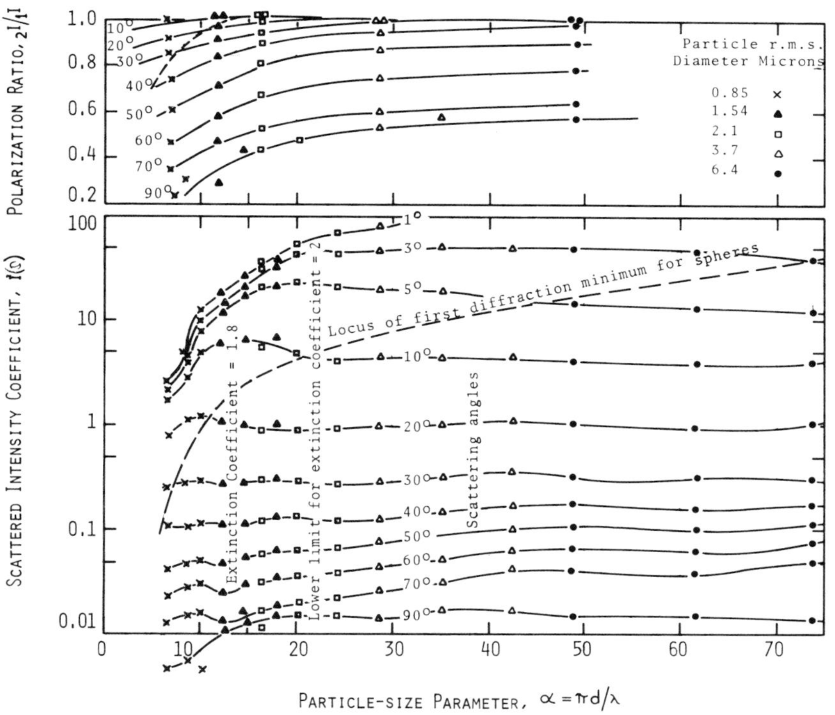

FIG. 2.19. Angular contours of scattering and polarization by quartz particles in water at wavelengths 0.365, 0.436, and 0.546 μm [68]. Courtesy of Pergamon Press, Book Division.

References

1. N. A. Fuchs, *The Mechanics of Aerosols*, Pergamon, Oxford, 1964
2. H. Schlichting, *Boundary Layer Theory*, p. 16, McGraw-Hill, New York, 1960.
3. Lord Rayleigh, in S. Goldstein (Ed.), *Modern Developments of Fluid Mechanics*, Oxford Univ. Press (Clarendon), London and New York, 1938.
4. C. E. Lapple and C.B. Shepherd, Calculation of Particle Trajectories, *Ind. Eng. Chem.*, *32:* 605–617 (1940).
5. G. Hänel, Bemerkungen zur Theorie der Düsen-Impaktoren, *Atmos. Environ.*, *3:* 69–83 (1969).
6. E. S. Pettyjohn and E. B. Christiansen, Effect of Particle Shape on Free Settling Rates of Isometric Particles, *Chem. Eng. Progr.*, *44:* 157–172 (1948).
7. E. Cunningham, On the Velocity of Steady Fall of Spherical Particles Through a Fluid Medium, *Proc. Roy. Soc.*, *A–83:* 357–365 (1910).
8. R. A. Millikan, Coefficients of Slip in Gases and the Law of Reflection of Molecules from the Surfaces of Solids and Liquids, *Phys. Rev.*, *Ser. 2*, *21:* 217–238 (1923).
9. L. W. McKeehan, The Terminal Velocity of Fall of Small Spheres in Air at Reduced Pressures, *Phys. Rev.*, *33:* 153–172 (1911).
10. M. Knudsen and S. Weber, Luftwiderstand gegcn die langsame Bewegung kleiner Kugeln, *Ann. Phys.*, *Leipzig*, *36:* 981–984 (1911).
11. C. N. Davies, Definitive Equations for the Fluid Resistance of Spheres, *Proc. Phys. Soc.*, *57:* 259–270 (1945).
12. R. Gans, Wie fallen Stäbe und Scheiben in einer reibenden Flüssigkeit?, *Sitzungsb. Math-Phys. Klasse Akad. Wiss. München*, *41:* 198–203 (1911).
13. S. Gurel, S. G. Ward, and R. L. Whitmore, Studies of the Viscosity and Sedimentation of Suspensions, *Brit. J. Appl. Phys.*, *6:* 83–87 (1955).
14. J. A. McNown and J. Malaika, Effects of Particle Shape on Settling Velocity at Low Reynolds Number, *Trans. Amer. Geophys. Un.*, *31:* 74–82 (1950).
15. J. F. Heiss and J. Coull, The Effect of Orientation and Shape on the Settling Velocity of Non-Isometric Particles in a Viscous Medium, *Chem. Eng. Progr.*, *48:* 133–140 (1952).
16. L. Squires and W. Squires, The Sedimentation of Thin Discs, *Trans. Amer. Inst. Chem. Eng.*, *33:* 1–12 (1937).
17. C. N. Davies, The Sedimentation of Small Suspended Particles, *Trans. Inst. Chem. Eng. Suppl.*, *25:* 25–39 (1947).
18. W. Stöber, A. Berner, and R. Blaschke, The Aerodynamic Diameter of Aggregates of Uniform Spheres, *J. Colloid Interface Sci.*, *29:* 710–719 (1969).
19. W. Stöber, H. Flachsbart, and D. Hochrainer, Der aerodynamische Durchmesser von Latexaggregaten und Asbestfasern, *Staub*, *30:* 277–285 (1970).
20. W. B. Kunkel, Magnitude and Character of Errors Produced by Shape Factors in Stokes's Law Estimates of Particle Radius, *J. Appl. Phys.*, *19:* 1056–1058 (1948).
21. H. Lamb, *Hydrodynamics*, 6th ed., p. 617, Dover, New York, 1945.
22. N. A. Fuchs, On the Theory of Coagulation, *Z. Phys. Chem.*, *A–171:* 199–208 (1934).
23. M. Abraham and R. Becker, *Electricity and Magnetism*, Hafner, New York, 1932.
24. D. R. Corson and P. Lorrain, *Introduction to Electromagnetic Fields and Waves*, Freeman, San Francisco, California, 1962.
25. J. Pich, The Theory of Filtration, in C. N. Davies, (Ed.), *Aerosol Science*, p. 235, Academic Press, London and New York, 1966.
26. L. Waldmann and K. H. Schmitt, Thermophoresis and Diffusiophoresis of Aerosols, in C. N. Davies, (Ed.), *Aerosol Science*, Academic Press, London and New York, 1966.

27. K. T. Whitby and B. Y. H. Liu, The Electrical Behavior of Aerosols, in C. N. Davies (Ed.), *Aerosol Science*, Academic Press, London and New York, 1966.
28. O. Preining, Photophoresis, in C. N. Davies (Ed.), *Aerosol Science*, Academic Press, London and New York, 1966.
29. O. Brandt, H. Freund, and E. Heidemann, Schwebstoffe im Schallfeld, *Z. Phys.*, *104:* 511–533 (1937).
30. H. W. St. Clair, Agglomeration of Smoke, Fog, or Dust Particles by Sonic Waves, *Ind. Eng. Chem.*, *41:* 2434 (1949).
31. G. M. Hidy and J. R. Brock, Lung Deposition of Aerosols—A Footnote on the Role of Diffusiophoresis, *Environ. Sci. Tech.*, *3:* 563–567 (1969).
32. L. B. Loeb, The Basic Mechanisms of Static Electrification, *Science*, *102:* 573–576 (1945).
33. H. Sachsse, Uber die elektrischen Eigenschaften von Staub und Nebel, *Ann. Phys.*, *14:* 396–412 (1932).
34. W. B. Kunkel, The Static Electrification of Dust Particles on Dispersion into a Cloud, *J. Appl. Phys.*, *21:* 820–832 (1950).
35. M. von Smoluchowski, Experimentally Demonstrable Molecular Phenomena Which Contradict Conventional Thermodynamics, *Phys. Z.*, *13:* 1069–1080 (1912).
36. G. L. Natanson, The Electrification of Drops During Atomization of Liquids as a Result of Fluctuations in the Ion Distribution, *Zh. Fiz. Khim.*, *23:* 304–314 (1949).
37. H. Y. Chow and T. T. Mercer, Charges on Droplets Produced by Atomization of Solutions, *Amer. Ind. Hyg. Ass. J.*, *32:* 247–255, (1971).
38. L. B. Loeb, *Fundamentals of Electricity and Magnetism*, 3rd ed., p. 489, Dover, New York, 1947.
39. J. Bricard, Electric Charge and Radioactivity of Naturally Occurring Aerosols, in C. N. Davies (Ed.), *Aerosol Science*, Academic Press, London and New York, 1966.
40. R. Gunn, The Statistical Electrification of Aerosols by Ionic Diffusion, *J. Colloid Sci.*, *10:* 107–119 (1955).
41. D. Keefe, P. J. Nolan, and T. A. Rich, Charge Equilibrium in Aerosols According to the Boltzmann Law, *Proc. Roy. Irish Acad.*, *60A:* 27–45 (1959).
42. T. T. Mercer, Aerosol Production and Characterization: Some Considerations for Improving Correlation of Field and Laboratory Derived Data, *Health Phys.*, *10:* 873–887 (1964).
43. K. T. Whitby, *Homogeneous Aerosol Generators*, Tech. Rep. No. 13, Univ. of Minnesota, NP–10020, 1961.
44. D. Sinclair, *Handbook on Aerosols*, p. 67, USAEC, Washington, D.C., 1950.
45. R. L. Dimmick, M. T. Hatch, and J. Ng, A Particle-Sizing Method for Aerosols and Fine Powders, *AMA Arch. Ind. Health*, *18:* 23–29 (1958).
46. M. von Smoluchowski, A Mathematical Theory of the Kinetics of Coagulation of Colloidal Solutions, *Z. Phys. Chem.*, *92:* 129–168 (1917).
47. H. S. Patterson and W. Cawood, The Reproducibility and Rate of Coagulation of Stearic Acid Smokes, *Proc. Roy. Soc. London*, *136A:* 538–548 (1932).
48. J. Pich, A Mathematical Study of the Wiegner Effect in Colloid Coagulation, in T. T. Mercer, P. E. Morrow, and W. Stöber (Eds.), *Assessment of Airborne Particles*, Thomas, Springfield, Illinois, 1972.
49. G. Zebel, Coagulation of Aerosols, in C. H. Davies (Ed.), *Aerosol Science*, Academic Press, London and New York, 1966.
50. R. S. Bradley, The Rate of Evaporation of Microdrops in the Presence of Insoluble Monolayers, *J. Colloid Sci.*, *10:* 571–575 (1955).
51. I. Langmuir, The Evaporation of Small Spheres, *Phys. Rev.*, *12:* 368–370 (1918).

52. *Handbook of Chemistry and Physics*, 41st ed., p. 2326, Chemical Rubber Publ., Cleveland, Ohio, 1959.
53. W. L. Crider, R. H. Milburn, and S. D. Morton, The Evaporation and Rehydration of Aqueous Solutions, *J. Meteorol., 13:* 540–547 (1956).
54. H. Wakeshima and K. Takata, Growth of Droplets and Condensation Coefficients of Some Liquids, *Japan J. Appl. Phys., 2:* 792–797 (1963).
55. A. F. Mills and R. A. Seban, The Condensation Coefficient of Water, *Int. J. Heat Mass Transfer, 10:* 1815–1827 (1967).
56. R. H. Milburn, W. L. Crider, and S. D. Morton, The Retention of Hygroscopic Dusts in the Human Lungs, *AMA Arch. Ind. Health, 15:* 59–62 (1957).
57. N. Frössling, The Evaporation of Falling Drops, *Gerlands Beitr. Geophys., 52:* 170–216 (1938).
58. W. E. Ranz and W. R. Marshall, Evaporation from Drops, *Chem. Eng. Progr., 48:* 141–146 (1952).
59. G. D. Kinzer and R. Gunn, The Evaporation, Temperature, and Thermal Relaxation Time of Freely Falling Water Drops, *J. Meterol., 8:* 71–83 (1951).
60. R. H. Milburn, Theory of Evaporating Water Clouds, *J. Colloid Sci., 12:* 378–388 (1957).
61. N. H. Fletcher, *The Physics of Rainclouds*, Cambridge Univ. Press, London and New York, 1966.
62. C. Orr, F. K. Hurd, and W. J. Corbett, Aerosol Size and Relative Humidity, *J. Colloid Sci., 13:* 472–482 (1958).
63. A. E. H. Love, The Scattering of Electric Waves by a Dielectric Sphere, *Proc. London Math. Soc., 30:* 308 (1899).
64. G. V. Rozenberg, Optical Investigations of Atmospheric Aerosol, *Usp. Fiz. Nauk, 95:* 159–208 (1968).
65. D. Sinclair, Optical Properties of Aerosols, in *Handbook of Aerosols*, p. 81, USAEC, Washington, D. C., 1950.
66. H. C. van de Hulst, *Light Scattering by Small Particles*, Wiley, New York, 1957.
67. J. R. Hodkinson, The Optical Measurement of Aerosols, in C. N. Davies (Ed.), *Aerosol Science*, p. 287, Academic Press, London and New York, 1966.
68. J. R. Hodkinson, Light Scattering and Extinction by Irregular Particles Larger Than the Wavelength, in M. Kerker (Ed.), *Proc. ICES*, p. 87, Pergamon, Oxford, 1962.
69. H. Breuer, J. Gebhart, and K. Robock, Zur Bestimmung von Staub-konzentrationen in Steinkohlenbergbau auf der Basis der Lichtstreuung, *Staub, 30:* 426–431 (1970).

3

Particle Size and Size Distributions

A knowledge of the size distribution of airborne particles is essential to anyone wishing to predict, even approximately, the fate of a particular aerosol. In general, one's interest centers on some specific property of the aerosol as a whole and on the manner in which that property changes with time and position. A health physicist, for instance, concerned with the spread of radioactive particles from a stack, would like to know how the activity is distributed with respect to the deposition velocities of the particles, since this would be the most appropriate information for predicting the fate of the contaminant after its introduction into the atmosphere. A meteorologist, on the other hand, might be interested in the light-scattering properties of particles rather than their toxic properties, while someone studying chemical reactions between particles and other atmospheric contaminants might be interested in the particulate surface area. In each case, an attempt must be made to determine the manner in which the property of interest is distributed with respect to a diameter that characterizes the dynamic nature of the aerosol. If the particles were nonporous spheres of known density and chemical composition, a knowledge of the frequency distribution of their geometric diameters would be sufficient to permit calculation of the various physical properties and dynamic characteristics of interest.

Unfortunately, the particles encountered in practice generally are irregular in shape and the diameter assigned to them is an arbitrary quantity, the significance of which depends on the manner in which it is defined. The most common method of measurement requires the use of an optical or electron microscope to provide a particle image of readily visible size. A particle's diameter is then defined in terms of some geometric characteristic of its silhouette that can be measured conveniently. Another method requires the measurement of some physical property of the particle. Its diameter in this case is defined as equal to that of a hypothetical sphere that possesses the same property to the same extent.

Measurement of geometric diameters of individuals in a sample provides an estimate of the relative frequency distribution of that diameter among the population of particles from which the sample was taken. Such a distribution is seldom of interest in itself and its usefulness depends on the accuracy and convenience with which the distributions of real interest can be calculated from it. The latter are best estimated directly, whenever possible, since their estimation from the size frequency distribution requires shape factors that seldom are known accurately and often are unavailable. In addition, there are factors of a purely statistical nature which increase the errors of estimation when extrapolating from one distribution to another. Despite these shortcomings, the determination of size frequency distributions has been, and continues to be, an important method of aerosol technology, if only because our instrumentation does not always permit us to make direct measurement of the significant distributions. This chapter, therefore, will deal with the measurement and significance of geometric diameters, with the shape factors relating them to various particulate properties, with the mathematical description of size distributions, and with the statistics of sampling from a lognormal distribution of particle sizes. Measurement of property-related diameters will be considered later.

3-1 The Geometric Diameters of Nonspherical Particles

Figure 3.1 shows a silhouette of a particle of irregular shape on which three of the better-known "statistical" geometric diameters have been designated.

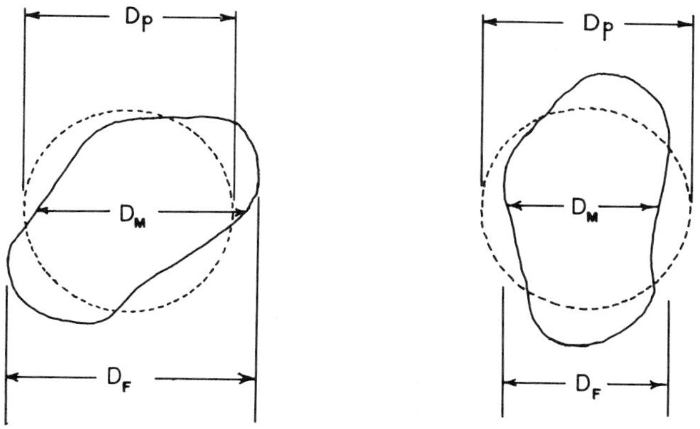

FIG. 3.1. Geometric diameters of an irregularly shaped particle. D_p is the diameter of dashed circle.

These diameters are statistical in the sense that they have quantitative significance only when averaged over a large number of measurements. The first to be defined was Martin's diameter (D_M) [1]. Martin was seeking a simple relationship between the number of particles and their diameters when grinding crushed sand. The particles with which he was concerned were quite irregular in shape and some arbitrary dimension had to be chosen to describe a particle's size before he could make an organized compilation of measurements. From analogy with the diameter of a circle, he chose to define the diameter of an irregularly shaped particle as the length of a chord, parallel to a given reference line, which divided the particle into two equal areas. Classifying all particles of the same chord length as having the same "statistical" size, he was able to organize his results into meaningful distributions. Subsequently, he showed that the size frequency distribution of finely ground particles followed the compound interest law, rather than the Gaussian error function. His research established the statistical diameter as a rational definition of particle size.

The later definition of Feret's [2] diameter (D_F) was probably a straight forward matter of convenience. It was defined as the projection of the particle on an arbitrary fixed reference line. It made particle size measurement more rapid and less subject to errors by the microscopist, who no longer had to estimate what constituted two equal areas of a particle.

The projected area diameter (D_P) is the diameter of a circle having the same area as the particle's silhouette. More formal definitions [3] sometimes include the phrase "when viewed in a direction perpendicular to the plane of greatest stability," a qualification that invalidates the primary theoretical advantage in the use of this diameter but that is, fortunately, seldom satisfied in aerosol sampling. As the definition of the size of a particle silhouette, the projected area diameter has the considerable advantage that it is independent of the orientation of the silhouette, i.e., it is not a two-dimensional statistical diameter.

Projected area diameter was not a very useful definition when measurements were made on individual particles because it was necessary to obtain a highly magnified image of the particle and measure its area with a planimeter. However, with the introduction of the Patterson–Cawood graticule [4] (Fig. 3.2A) it became possible to sort particle profiles rapidly into size intervals defined by two successive circles which, for the modified graticule shown in Fig. 3.2(B), differ in area by a factor of 2. The increase in speed with which particle size distributions could be determined more than compensated for any loss in accuracy, and the projected area diameter became the measurement of choice in Great Britain. Its popularity was not so great in the U.S., perhaps because of the rapidly expanding application of electron microscopy to particle size analysis, but its use has increased greatly since the introduction of the Endter–Gebauer Particle Size Analyzer (see Fig. 3.3). With this instrument, a circle of light of variable diameter can be projected onto a particle silhouette on an optical or electron photomicrograph and its size adjusted until its area appears equal to that of the particle. By depressing a foot-switch, the operator can automatically record the particle in the size interval within which the diameter of the circle of light falls. The instrument includes 48 size intervals, with a choice of either logarithmic or linear increments, each in two different size ranges. The particle sizes can be recorded either as a differential or an integral distribution.

While these three are the best-known of the statistical geometric diameters, a variety of other possible measurements could serve a similar purpose; e.g., the maximum and minimum chords, the circle of equal perimeter, the circumscribed circle, etc. Some of these are being studied for application with computer-interfaced systems of measurement [6].

3-1.1 The Geometric Significance of D_P, D_F, and D_M

Cauchy [7] showed that the surface area of an irregularly shaped, convex solid is equal to four times its projected area averaged over all possible orientations. (Because the original papers on this work are not readily accessible, Vouk [8] provided another proof of the relationship.) By virtue of this theorem, the *root-mean-square* projected area diameter becomes a three-dimensional statistical diameter with which surface area can be estimated. For particle size work, the validity of the relationship requires that particles have no reentrant surfaces and show no preferred orientation with respect to the direction of observation. According to Walton [9], the former condition is usually satisfied with respect to particle profiles. Evidence concerning particle orientation is equivocal. It is clear from observations of deposits in horizontal elutriators [10], centrifuges [11], and thermal [12, 13] and electrostatic precipitators (see

A

B

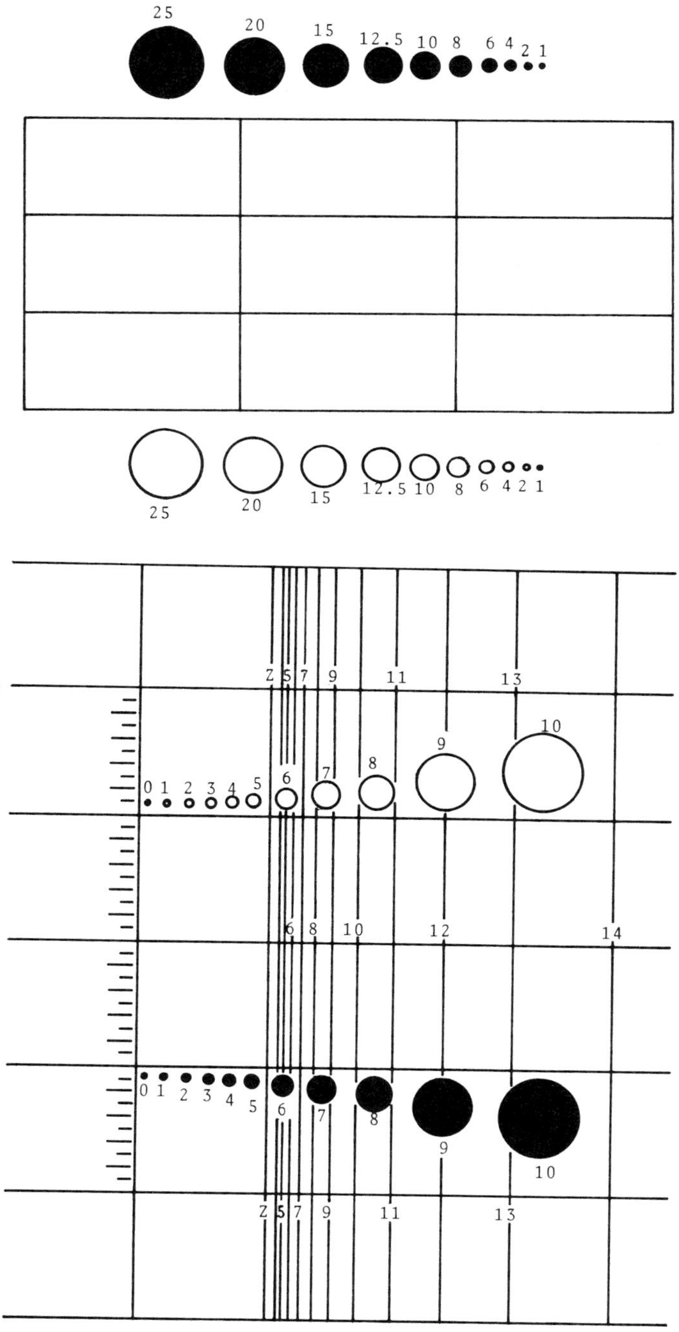

FIG. 3.2. Eyepiece graticules. (A) Original by Patterson–Cawood; (B) recent modification by May [5]. Courtesy of Institute of Physics and The Physical Society.

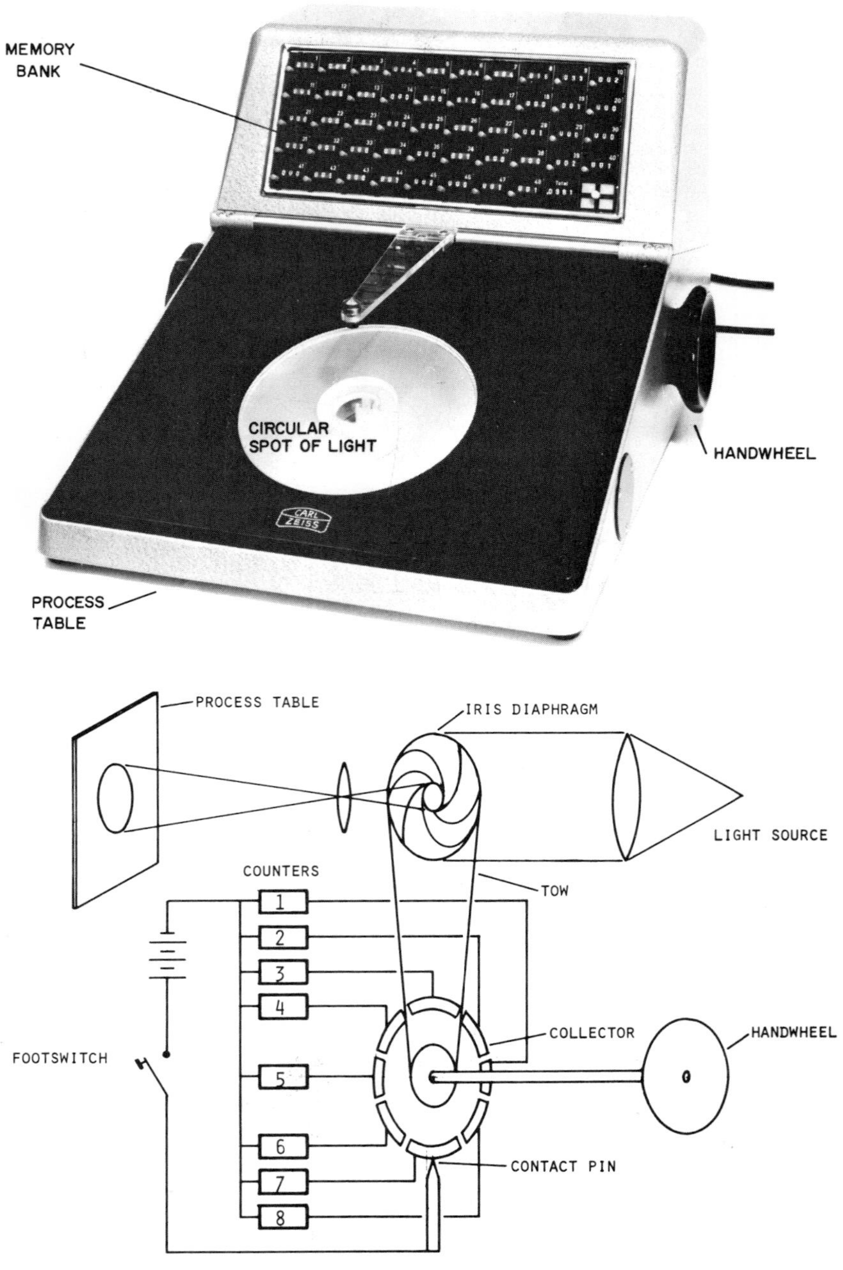

FIG. 3.3. The Endter–Gebauer Particle Size Analyzer. Above: The model manufactured by Zeiss; below: schematic representation of measurement procedure [6]. Courtesy of Charles C. Thomas, Publisher.

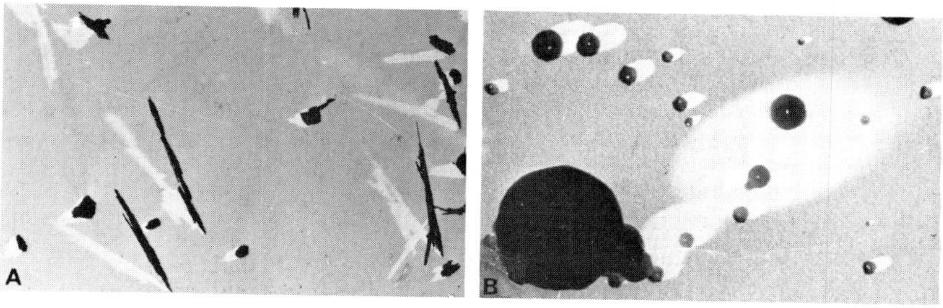

FIG. 3.4. Some examples of deposition with marked instability. (A) Asbestos fiber collected in centrifuge; (B) uranine-sodium chloride particles collected in thermal precipitator.

Fig. 3.4) that particles of irregular shape are not oriented in the position that is most stable in the usual mechanical sense. On the other hand, completely random orientation has not been demonstrated. Both Corn [14] and Hodkinson [15] found that particles, initially oriented at random in aqueous suspension, showed marked evidence of orientation after settling out of suspension. Neither author, however, felt that the results could be applied to similar particles in air. It seems likely that particles deposited from air retain whatever orientation they have at the moment they come into contact with a surface.

Feret's diameter had been widely used [16], and occasionally abused [17], before Walton [9] gave it respectability by the application of another of Cauchy's theorems. According to this relationship, the mean projected length of an irregularly shaped convex particle profile is equal to the particle's perimeter divided by π. An example of the correlation between Feret's diameter and the perimeter of irregularly shaped particles is shown in Fig. 3.5.

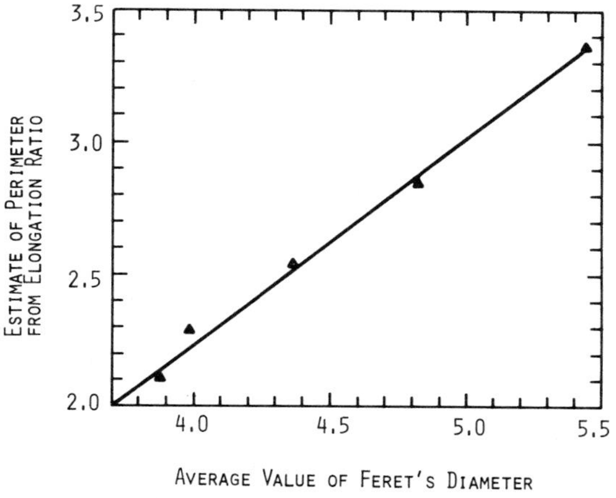

FIG. 3.5. Experimental relationship between Feret's diameter and profile perimeter [18]. Courtesy of Elsevier Sequoia S.A.

Martin's diameter has been defined as the "mean chord" of a particle profile [17] and as the "mean of chords through the centroids of a group of particles" [19]. It is apparent at once that the first definition is incorrect, since Martin's diameter for spherical particles is equal to the largest chord of the particle profile. Kaye [18] has pointed out that the second definition is not true for profiles of irregular shape, although the difference between the centroid diameter and Martin's diameter may be too small to be significant in practice. In the following table the relative values of the three diameters are shown for several ellipses and rectangles. For these profiles, the mean value of Martin's diameter is, in fact, the centroid diameter, but it is not equal to the projected area diameter. The root mean square value of Martin's diameter, however, does equal the projected area diameter for profiles of ellipses and parallelograms. It should be more reliable, in general, than the arithmetic mean value.

	Ellipses [20][a]				Rectangles[a]			
	$\beta =$				$\beta =$			
	1	2	5	10	1	2	5	10
$D_M : D_F$	1.00	0.89	0.57	0.40	0.89	0.81	0.55	0.36
$D_M : D_p$	1.00	0.97	0.86	0.73	1.00	0.96	0.85	0.71

[a] $\beta =$ long axis (side)/short axis (side).

The definitions of Martin's and Feret's diameters imply that a large number of measurements will be made on the same profile at random orientations with respect to a fixed line. In practice, however, it is customary to make a single measurement on each profile under the tacit assumption that that particular profile will recur often enough in the sample, at random orientation, so that the measurements will yield a satisfactory average diameter. Davies [19] has shown that such an assumption is not necessary; that the average of a number of measurements made on a sample of particles oriented at random has the same significance for the group of particles as does the average of a number of measurements of the same profile taken at random. His derivation requires that the particle profiles have essentially the same shape, differing only in size. A similar argument relating profile area to particle surface applies to the projected area diameter for particles of similar shape.

The number of particles that must be measured before the statistical averaging process is valid has not been established. Clearly, it will be related to the particle shape and the spread in sizes. Some information on this point has been provided by Kaye [18], incidentally to a comparison between the standard method of measuring Feret's diameter and the so-called "antithetic variates" technique in which Feret's diameter at each orientation is taken as one-half the sum of the particle's projection on a fixed reference line and its projection on a second reference line at right angles to the first. The variations of the two measurements about the average value are shown in Fig. 3.6(A) and the relative efficiency of the two methods is shown in Fig. 3.6(B). The latter figure shows that for the usual method of measurement, 20 random determinations of the diameter will yield an average value that is within 5% of the true mean

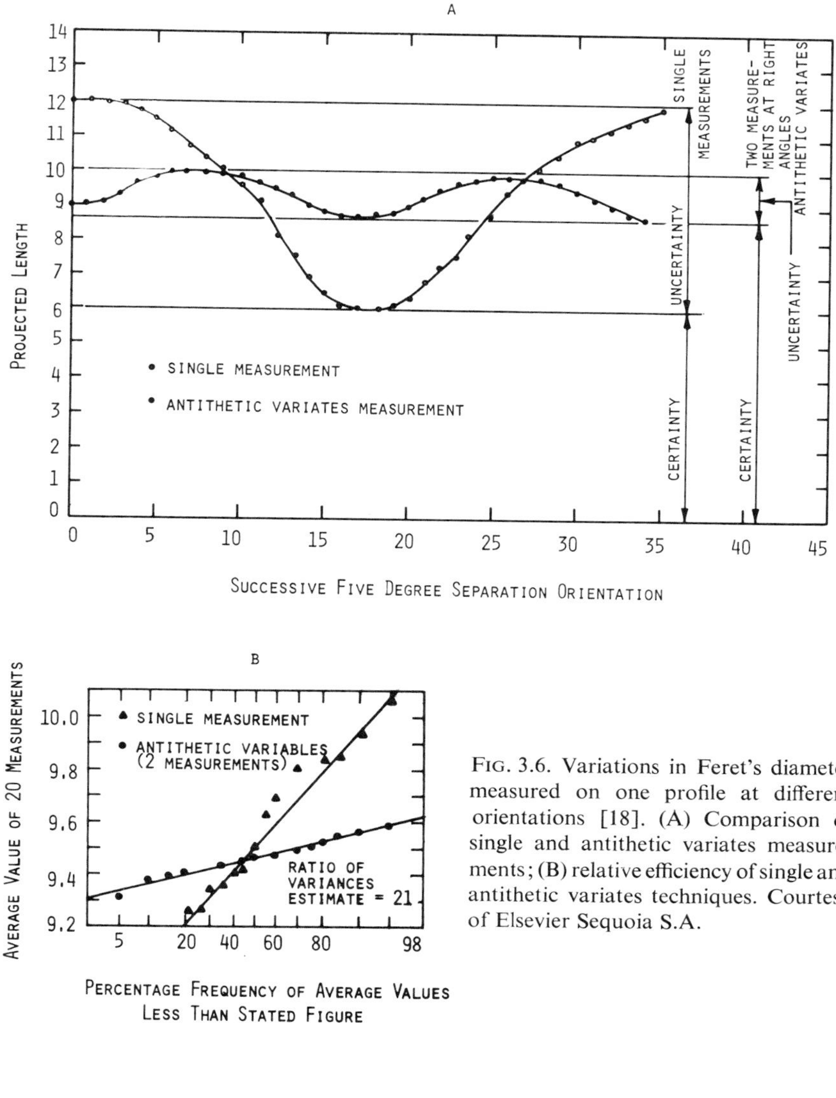

FIG. 3.6. Variations in Feret's diameter measured on one profile at different orientations [18]. (A) Comparison of single and antithetic variates measurements; (B) relative efficiency of single and antithetic variates techniques. Courtesy of Elsevier Sequoia S.A.

about 75% of the time. Using the antithetic variables method, the same amount of work yields an average value that is within 2% of the true mean 90% of the time. These results also bring out the fact that, while the overall average may be a valid estimate of the true mean diameter, the estimate in the spread of the statistical diameters will almost certainly be too high. This is a problem that does not seem to have received much attention.

3-1.2 THE "SPLITTING IMAGE" TECHNIQUE

The diffraction of light as it passes near a particle introduces some uncertainty into the measurement of particle size with the light microscope. Charman [21, 22] has studied this effect experimentally, and some of his results are shown in Fig. 3.7. The line marked "no diffraction" indicates how the light intensity would increase from zero over an opaque particle to the intensity of the incident light at the particle's edge if diffraction did not occur. The other curves show how diffraction makes the position of the particle's edge a matter of conjecture. Charman concluded that measurements smaller than several times the limit of resolution were subject to serious error [23]. In the past few years, optical microscopes have been described [24, 25] which minimize the effects of diffraction in particle size measurements. In one of these microscopes, the light from the specimen is split into two beams, each following a similar path through mirrors and prisms to reach the eyepiece. A prism in each

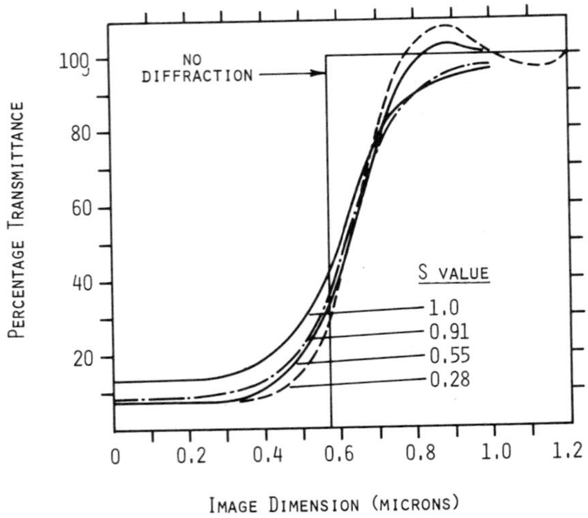

FIG. 3.7. Effect of diffraction on the definition of a particle's boundary [22]. S is the ratio of condenser numerical aperture to objective numerical aperture. Particle center at origin. Courtesy American Institute of Physics.

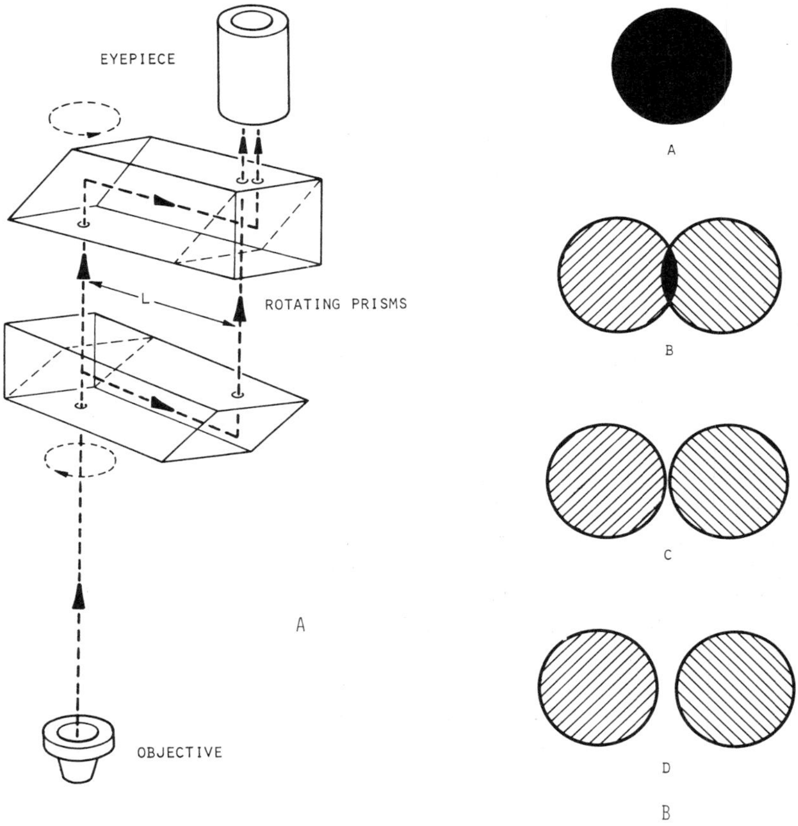

FIG. 3.8. The image-splitting microscope [24]. (A) Eyepiece optical system; (B) Example of field of view—A, unsheared image; B, shear < image diameter; C, shear = image diameter; D, shear > image diameter. Courtesy Vickers Instrument, Inc.

path can be rotated by means of a calibrated vernier, the two prisms synchronized to rotate the same amount, but in opposite directions. In the reference position of the prisms, the light paths are identical and the images produced by each beam exactly coincide in the eyepiece. When the prisms are rotated, the two images diverge and the angular rotation of the prisms, read from the verniers, can be calibrated in terms of the linear dimension representing the separation of corresponding points of the images. The method of measurement minimizes the diffraction edge of the particle and permits the determination of a variety of chord lengths. A schematic diagram of the microscope and an example of a field of view are shown in Fig. 3.8.

3-2 Geometric Shape Factors

3-2.1 Volume Shape Factor

Several particulate properties are proportional to the diameter of the particle raised to an appropriate power:

$$Q_r(D) = \alpha_r D^r. \tag{3.1}$$

Q_r is the amount of the property associated with a particle of diameter D and α_r is a shape factor that is valid only for a specific value of r, a specific method of measuring the diameter, a specific material and, perhaps, a specific method of preparation. For most nonspherical particles, the shape factors must be determined empirically. While Eq. (3.1) is valid for a number of different values of r, experimental investigations have been limited to α_3, the volume shape factor, and α_2, the surface shape factor, and even these have been studied for only a few materials.

Shape factors, like the diameters with which they are used, are statistical quantities and are not intended to be applied to individual particles. It is important, however, that on the average they should not vary with particle size. Robins [26] determined the volume shape factor as a function of projected area diameter for coal particles in stable orientation over a size range of 3 to 76 μm. He obtained a distribution of values which give a good fit to a normal distribution having a mean, $\alpha_3 = 0.21$, and a standard deviation $= 0.07$. He did not find that the distribution was affected by size. Hatch and Choate [27] demonstrated that the average value of α_3 for silica, granite, and calcite particles was constant over a wide range of sizes. They measured the number of particles per gram for each of 23 samples for which the particle size distributions were lognormal, having median diameters between 1.1 and 72 μm and geometric standard deviations between 1.26 and 2.17. Figure 3.9 shows the

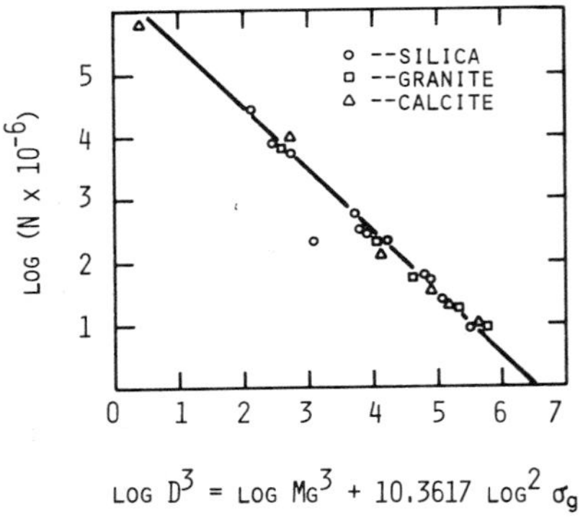

FIG. 3.9. Number of particles per gram as a function of particle diameter [27]. D is the diameter of average mass, M_g is the count median diameter, and σ_g is the geometric standard deviation. Courtesy of The Franklin Institute.

relationship they obtained between the number of particles per gram (the reciprocal of the average particle mass) and the cubed diameter of the particle of average mass according to the size distribution data. In the figure, the intercept on the y axis is equal to 6 plus $\log(1/\rho\alpha_3)$, where ρ is the particle density. It is of some importance that they measured Feret's diameter. More recently, Kotrappa [28, 29] determined volume shape factors, related to projected area diameter, for coal, uranium dioxide, thorium oxide, and quartz particles in the respirable size range and found they were unaffected by size.

Given that the volume shape factor is constant, its value does not have to be known to convert a number distribution obtained from microscope measurements to a mass distribution. It must be known, however, to convert one of the statistical diameters to an aerodynamic diameter. Some of the values reported, or calculated from data, in the literature are shown in Table 3.1.

TABLE 3.1

VOLUME SHAPE FACTORS FOR SEVERAL SUBSTANCES

	Volume shape factor based on				
Substance	Projected area diameter	Martin's diameter	Feret's diameter	Orientation	Ref.
Portland cement	0.48[a]	0.48	0.28	stable	[30]
Glass	0.35[a]	0.35	0.16	stable	[30]
Quartz	0.31[a]	0.31	0.15	stable	[30]
Silica	0.27[b]	—	0.14	—	[27]
Calcite	0.27[b]	—	0.14	—	[27]
Granite	0.27[b]	—	0.14	—	[27]
Quartz	0.31[b]	—	0.16	—	[31]
Coal	0.21	—	—	stable	[26]
Coal	0.25	—	—	?	[32]
Quartz	0.21	—	—	?	[32]
Quartz	0.29	—	—	not quite random	[15]
Quartz	0.34	—	—	random	[28]
Coal (Pittsburgh)	0.29[c]	—	—	?	[33]
Silica	0.26[c]	—	—	?	[33]
Mica	0.10[c]	—	—	?	[33]
Fly ash	0.61[c]	—	—	?	[33]

[a] Assuming $D_M \approx D_P$.
[b] Estimated from D_F value using $D_F/D_P = 1.2$.
[c] Volume measured by Coulter counter.

Except for Kotrappa's work, the particles for inspection were deposited from liquid suspension, which, in general, seems to favor some degree of orientation. While the results for similar materials vary considerably, the differences may well be real, since there is no fixed method of sample preparation involved. In fact, Steinherz' [30] data show that α_3 changed significantly when the method of grinding the quartz was changed. For comparison with the real values shown, theoretical volume shape factors related to the projected area diameter are given in Table 3.2. It appears that real particles have volume shape

TABLE 3.2.

THEORETICAL VOLUME SHAPE FACTORS

Particle shape	Volume shape factors related to D_P at random orientation for: Isometric particles	Regular shapes of axial ratio 2	5	10
Sphere	0.52			
Cube octahedron	0.45			
Octahedron	0.41			
Cube	0.38			
Tetrahedron	0.29			
Parallelepiped (rect × sq)		0.35	0.27	0.20
Oblate spheroid		0.46	0.26	0.14
Prolate spheroid		0.47	0.33	0.24
Cylinder		0.40	0.30	0.23

factors very similar to those of particles of regular shape having an axial ratio of about 5.

The shape factors given in Table 3.1 were determined, presumably, for individual particles. For aggregates of smaller particles, a volume shape factor alone is inadequate because the porosity of the aggregate reduces its density. If ρ is the apparent density of the aggregate and ρ_0 is the true density of the bulk material, then

$$\rho/\rho_0 = (V_0/V) = (V-V_v)/V = 1-p. \tag{3.2}$$

V is the volume (of the envelope) of the aggregate, V_0 is the volume of solid material in the aggregate, V_v is the volume of void space, and p is called the "voids" or porosity when expressed in percent. For compact aggregates of uniform spheres, the porosity and density ratios depend on the manner in which the spheres are oriented. Some examples are tabulated as follows.

Packing	Porosity (%)	ρ/ρ_0
Rhombohedral	25.9	0.74
Random	39.0	0.61
Cubic	47.6	0.52

Aggregates formed in the airborne state may show much larger porosities. Stein *et al.* [34] found that the ratio between projected area diameter and aerodynamic diameter was sometimes as great as 6.5 for particulates in urban air. If the equivalent volume diameter is assumed to be approximately equal to the projected area diameter, their observation indicates aggregate densities as low as 0.024 gm/cm^3 and porosities close to 0.99. Equally high porosities have been observed in laboratory aerosols, which often exhibit values >0.9 [35].

3-2.2 Surface Shape Factor

For individual, nonporous particles, the surface shape factor, α_2, should be equal to π when related to the root-mean-square projected area diameter measured on particles in random orientation. For Martin's and Feret's diameters, α_2 is reduced by a factor equal to the square of the ratio D_p to D_M or D_F. Like the volume shape factor, α_2 is a statistical concept that cannot be applied reliably to single particles. That it is unaffected by particle size is apparent from the determinations made by Hatch and Choate [27], which are summarized in Fig. 3.10. DallaValle [36] measured the pressure drop across a packed column of large (0.05–0.38 cm) quartz particles when subjected to a viscous flow of air and calculated their total surface area. For seven samples

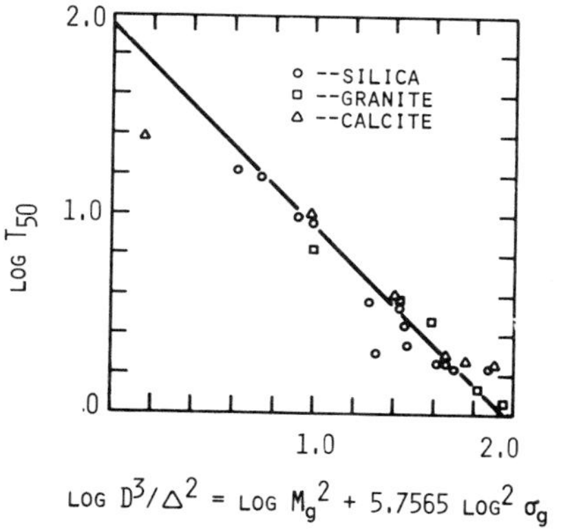

FIG. 3.10. Relative scattered light intensity ($\propto$ area) as a function of particle diameter [27]. D is the diameter of particle of average mass, Δ is the diameter of particle of average surface, M_g is the count median diameter, and σ_g is the geometric standard deviation. Courtesy of The Franklin Institute.

of different median diameters, he found an average value of $\alpha_2 = 2.6$, with no apparent trend related to particle size. Since he measured Feret's diameter, his results indicate a value of 1.1 for D_F/D_p. Some other values of α_2 are given in Table 3.3.

TABLE 3.3

PROJECTED AREA DIAMETER SURFACE SHAPE FACTORS

Substance	Orientation	Size range (μm)	Method of surface area measurement	α_2	Ref.
Quartz	stable	2–6	Light extinction	2.2	[14]
Quartz	?	0.2–10	N_2 adsorption	2.9–3.8	[32]
Quartz	stable	0.6–1.8	N_2 adsorption	5.2	[62]
Quartz	random	0.6–1.8	N_2 adsorption	5.2	[28]
Silica (vitreous)	stable	1.2–4.3	N_2 adsorption	4.7	[62]
Silica (tridymite)	stable	0·7–5.2	N_2 adsorption	4.8	[62]
Diamond	stable	3–6	Light extinction	1.8	[14]
Coal (bituminous)	stable	3.5–9	Light extinction	1.9	[14]
Coal (bituminous)	random	0.6–4.3	N_2 adsorption	15.6	[37]
Coal (anthracite)	stable	2–10	Light extinction	2.4	[14]

For aggregates of nonporous particles, α_2 will be greater than π when calculated from measurements of D_p^2 in random orientation and when related to the total surface area. The latter can be measured, for instance, by permeability methods. If the constituent particles of the aggregates are porous, then their internal structure contributes to their total surface area, which can be measured by gas adsorption methods. If their surface area is determined by measurement of light extinction, which yields a value approximately equal to the surface of an envelope about the aggregate, however, α_2 will be close to π. Thus, aggregates of porous particles may have three different surface shape factors for a given geometric diameter, depending on the method of surface measurement.

The ratio α_2/α_3 can be determined from combined measurements of specific surface area and particle size distribution using the relationship

$$\alpha_2/\alpha_3 = D_{vs}\,\rho\, S/M,$$

where D_{vs} is the volume-surface mean (Sauter's) diameter for the size distribution according to the type of diameter measured, and S/M is the surface area per unit mass of particles of density ρ. Several results for quartz are shown in Table 3.3a.

TABLE 3.3a

SOME MEASURED VALUES OF α_2/α_3 FOR QUARTZ

Microscope measurement	Size range	Method of surface area measurement	α_2/α_3	Ref.
D_P	0.2–10 μm	N_2 adsorption	14–18	[32]
D_P	0.1–2 μm	N_2 adsorption	34.6	[42]
D_F	0.05–0.38 cm	Permeability	6.3	[36]

3-3 Aerodynamic Shape Factors

To estimate the aerodynamic diameter of a particle of known density ρ from a measured diameter D, it is necessary to have a volume shape factor α_3 and a resistance shape factor K_R both related to the measured diameter. At its terminal settling velocity U_G the resistance force acting on a particle is [Eq. (2.10)]

$$F_R = 3\pi\eta D U_G \cdot K_R, \tag{3.3}$$

and the gravitational force acting on it is

$$F_G = mg = \alpha_3 \rho D^3 g. \tag{3.4}$$

The particle's terminal settling velocity is

$$U_G = \alpha_3 \rho D^2 g / 3\pi\eta K_R, \tag{3.5}$$

and its aerodynamic diameter D_A is

$$D_A = (6/\pi \cdot \rho/\rho_0 \cdot \alpha_3/K_R)^{1/2} \cdot D, \tag{3.6}$$

where ρ_0 represents unit density. Experimentally, the ratio D_A/D (or Stokes diameter/D) is actually measured, permitting estimation of the quantity α_3/K_R. It is useful to be able to separate the two shape factors, however, because K_R enters into equations involving forces that are not related to particle mass.

After α_3/K_R has been determined experimentally for a specific measured diameter and for particles of a given material, it can be applied to similar measurements of diameter, made on particles of the same type, to estimate their aerodynamic diameters. It could also be applied to particles of the same shape but of different materials. Some caution must be observed in the use of α_3/K_R because K_R becomes a function of size as the slip effect becomes significant. Since only a limited amount of experimental and theoretical data is available, only those cases in which D is the projected area diameter, the equivalent volume diameter, or the diameter of a fiber are discussed below. Unless stated otherwise, the slip effect is assumed to be negligible. In the following discussion, a second subscript is added to the shape factor to identify the diameter to which it refers.

3-3.1 Projected Area Diameter Resistance Shape Factor K_{RP}

This is the aerodynamic shape factor of most practical value and the one to which most available experimental data refer. When the projected area diameter D_p is measured on profiles of randomly oriented convex particles, its root-mean-square value is, on the average, the diameter of a sphere having the same surface area as the particle, and K_{RP} is equivalent to K_R as tabulated in Chapter 2. Tables 2.2 and 2.4 show that when slip is not important K_{RP} is close to unity for isometric particles and for cubicly packed clusters of spheres when the surface area refers to the cluster envelope. Table 2.3 shows that K_{RP} remains fairly close to unity even for many extreme shapes, if $\mathrm{Re} \gtrsim 0.1$.

Values of α_{3P}/K_{RP} and K_{RP} are given in Table 3.4 for particles of a variety

TABLE 3.4.

Aerodynamic Shape Factors[a] for Particles of Regular Shape

Particle shape	Aspect ratio	$\frac{\alpha_{3P}}{K_{RP}}$	K_{RP}	$\varkappa$
Sphere		0.52	1.0	1.00
Cube octahedron		0.47	0.99	1.03
Octahedron		0.42	0.98	1.06
Tetrahedron		0.30	0.96	1.17
Parallelepiped[b]	0.25	0.27	1.07	1.30
Parallelepiped[b]	4.0	0.24	1.12	1.40
Oblate spheroid	2	0.46	1.00	1.05
Oblate spheroid	5	0.26	0.98	1.24
Oblate spheroid	10	0.15	0.97	1.50
Prolate spheroid	2	0.46	1.01	1.05
Prolate spheroid	5	0.30	1.09	1.27
Prolate spheroid	10	0.19	1.23	1.60

[a] For negligible slip.
[b] See Table 2.3 for orientation.

of regular shapes. The shape factors for spheroids are theoretical values for average orientation. Experimental data for particles of irregular shape are given in Table 3.4a. The results for α_{3P}/K_{RP} as determined by different investigators agree rather well. The value of 0.19 for coal applies to a sample

TABLE 3.4a

AERODYNAMIC SHAPE FACTORS OF VARIOUS SUBSTANCES

Substance	Range of D_P, (μm)	α_{3p}/K_{RP}	α_{3p}	K_{RP}	$\varkappa$	Ref.
Coal	0.56–4.27	0.26±0.02	0.38±0.03	1.50±0.17	1.88±0.16	[29]
	5–15	0.29±0.06[a]	—	—	—	[43]
	2.5–12.5	0.27±0.02[a]	—	—	—	[10]
(mines)	2.7–25.6	0.27±0.07	—	—	—	[37]
(lab)	3.0–29.6	0.19±0.06	—	—	—	[37]
	>4	0.28	0.25±0.01	0.9	1.15	[32]
Glass	2–10	0.24±0.02	—	—	—	[10]
Quartz	0.65–1.85	0.24±0.02	0.35±0.04	1.43±0.2	1.84±0.22	[28]
	>4	0.23	0.21±0.01	0.91	1.23	[32]
	2–8	0.19±0.06[a]	—	—	—	[43]
China clay	2–8	0.20±0.05[a]	—	—	—	[43]
Rock (mine)	3.5–12	0.21±0.05	—	—	—	[37]
UO_2	0.21–0.63	0.40±0.02	0.34±0.03	0.85±0.27	1.11±0.07	[29]
	0.63–1.68	0.27±0.03	0.34±0.06	1.23±0.17	1.60±0.12	[29]
	0.21–1.68	0.36±0.06	0.34±0.04	0.95±0.21	1.24±0.24	[29]
ThO_2	0.23–0.68	0.31±0.02	0.23±0.03	0.75±0.06	1.19±0.03	[29]
	0.68–3.38	0.21±0.05	0.23±0.05	1.14±0.27	1.70±0.32	[29]
	0.23–3.38	0.26±0.06	0.23±0.04	0.93±0.27	1.42±0.33	[29]
Cotton	2.4–19	0.18±0.02[b]	—	—	—	[37]
Asbestos	2.1–25.5	0.17±0.03[b]	—	—	—	[37]

[a] Aggregates were not included.
[b] Mixed particles and fibers.

showing a greater degree of aggregation than was seen in the mine sample [37]. Kotrappa's data were corrected for slip factor [29]; all other data refer to particle sizes for which the slip effect is negligible. Under those conditions, K_{RP} should be close to unity; marked deviations from that value are probably related to difficulties in the measurement of α_{3P} or to an orientation that is not completely random.

3-3.2 EQUIVALENT VOLUME DIAMETER RESISTANCE SHAPE FACTOR $K_{RV}(\varkappa)$

McNown and Malaika [38] interpreted their data on settling velocities at low Reynolds numbers according to Eq. (2.10). They took D for a given particle as the diameter D_V of a sphere having the same volume as the particle. They called K_{RV} the coefficient of resistance and showed that it was the ratio between the settling velocity of the equivalent volume sphere and that of the particle. Their experimental results led to the conclusion that K_{RV} was very similar for particles of a given axis ratio and orientation, regardless of shape. Their study included calculation of theoretical values of K_{RV} for ellipsoids at a number of different axis ratios.

Fuchs [39] used the same method for expressing the dependence of fluid resistance on particle shape, but called K_{RV} the "dynamic shape factor," for which he adopted the symbol $\varkappa$. Davies [40] showed how particles of irregular shape could be replaced with equivalent spheroids by matching values of $\varkappa$, calculated from measurements of particle diameter, to theoretical values for spheroids. In this way, a correlation could be obtained between the dynamics and geometry of irregularly shaped particles. Stöber [41] made an extensive investigation of methods for calculating $\varkappa$ for particles of isometric shape, fibers, and aggregates of monodisperse spheres.

Values of $\varkappa$ for particles of regular shape are included in Table 3.4. Those for spheroids are theoretical. Values for particles of irregular shape are included in Table 3.4a.

When the resistance force is related to D_V, α_3 in Eq. (3.6) becomes $\pi/6$ and $K_{RV} = \varkappa$. If $\varkappa$ has been determined for particles of a certain shape, then subsequent measurements of D_V for particles of the same shape can be used to calculate D_A, using Eq. (3.6). At present, D_V can be determined for particles of irregular shape only by measuring their volumes using the Coulter technique. (See Chapter 7.)

3-3.3 Fiber Diameter Resistance Shape Factor K_{Rf}

When a cylindrical fiber falls normally to its long axis, both the gravitational force and fluid resistance [Eq. (2.14)] acting on it are proportional to the fiber length, so that its terminal settling velocity should depend only on its diameter. Using Eq. (2.4), the ratio of aerodynamic diameter to fiber diameter, for negligible slip effect and air at 20°C, is

$$D_A/D_f = 2.53\rho^{1/2}\mathrm{Re}^{-0.035} = 3.65\rho^{0.467}D_f^{-0.099}, \qquad (3.7)$$

when the fiber diameter is expressed in micrometers and ρ in gm/cm^3. Experimental determinations of D_A/D_f for fibers have been made by Timbrell [44] and Stöber *et al.* [45]. Their results, with corresponding values of α_{3f}/K_{Rf} and K_{Rf} are given in Table 3.4b.

Because the fluid resistance acting on a long fiber is a function of Reynolds number, its apparent aerodynamic diameter, when measured at an acceleration exceeding that of gravity by a factor ϕ, will be less than its true aerodynamic diameter by $\phi^{-0.035}$, and the apparent value of D_A/D_f must be corrected accordingly. With this correction, Stöber's results ($\phi \simeq 10^3$) for $10 \leqslant \beta \leqslant 100$ give

$$3.6 \leqslant D_A/D_f \leqslant 4.7 \qquad \text{for amosite,}$$

TABLE 3.4b

AERODYNAMIC SHAPE FACTORS FOR FIBERS[a]

Fiber type	Diameter (μm)	β = Length/diameter	D_A/D_f	α_{3f}/K_{Rf}	K_{Rf}	Ref.
Glass	1.5–8	$\simeq$2–309	$3.12/\sqrt{\Psi}$	$2.04/\Psi$	$0.385\beta\Psi$	[44]
Asbestos						
Amosite	0.6–3	—	3.5	2.56	0.31β	[44]
Amosite	<1	4–200	$2.18\beta^{0.116}$	$0.99\beta^{0.232}$	$0.79\beta^{0.768}$	[45][b]
Crocidolite	0.8–4	—	3.0	1.88	0.42β	[44]
Crocidolite	<1	4–150	$2.19\beta^{0.17}$	$\beta^{0.342}$	$0.78\beta^{0.658}$	[45][b]
Chrysotile	0.8–4	—	2.5	1.31	0.61β	[44]

[a] $\Psi = [1 + (0.5/\beta)]^{4.4}$.
[b] Values not corrected for Reynolds numbers.

and

$$4.1 \leqslant D_A/D_f \leqslant 6.0 \qquad \text{for crocidolite.}$$

Theoretical values for $0.5\ \mu\text{m} \leqslant D_f \leqslant 1.0$ and $\rho = 2.5\ \text{g/cm}^3$ are

$$5.6 \leqslant D_A/D_f \leqslant 6.0.$$

Timbrell's data for glass fibers, which should provide the best comparison with theory because he made a special effort to obtain fibers that were good approximations of right cylinders, show $D_A/D_f \to 3.12$ at large values of β. For the same range of diameters and $\rho = 2.5\ \text{g/cm}^3$, theory predicts that

$$4.5 \leqslant D_A/D_f \leqslant 5.4.$$

The difference between theory and experiment cannot be explained as due to orientation of the fiber's long axis in other than a horizontal position. If allowance could be made for such orientation, the theoretical values of D_A/D_f would be even larger.

3-3.4 RELATIONSHIPS AMONG DIFFERENT AERODYNAMIC SHAPE FACTORS

It is apparent from Eq. (3.6) that $(\alpha_3/K_R)\, D^2$ is constant, whatever diameter is measured. For particles of any shape, then

$$\frac{\alpha_{3P}}{K_{RP}} \cdot D_P^2 = \frac{\alpha_{3V}}{K_{RV}} \cdot D_V^2 = \frac{\pi}{6 \cdot \varkappa} \cdot D_V^2,$$

and

$$\varkappa = \frac{\pi}{6\alpha_{3P}} \cdot \frac{D_V^2}{D_P^2} \cdot K_{RP} = \left(\frac{\pi}{6\alpha_{3P}}\right)^{1/3} \cdot K_{RP}.$$

In the case of fibers, $\alpha_{3P} = \pi\beta/4(\beta+0.5)^{3/2}$, $\alpha_{3f} = \pi\beta/4$, and

$$\varkappa = \left(\frac{2}{3\beta}\right)^{1/3} \cdot (\beta+0.5)^{1/2} \cdot K_{RP}$$

$$K_{Rf} = (\beta+0.5)^{1/2} \cdot K_{RP}.$$

Wadell [46] defined two shape factors with which to correlate the aerodynamic properties of particles of irregular shape:

$$\text{sphericity} = \Phi = \frac{\text{surface of equivalent volume sphere}}{\text{actual surface of particle}}$$

$$\text{circularity} = \Theta = \frac{\text{circumference of circle of area equal to particle's projected area}}{\text{actual perimeter of particle}}.$$

In terms of the shape factors described above,

$$\Phi = (K_{RP}/\varkappa)^2 = (6\alpha_{3P}/\pi)^{2/3},$$

and

$$\Theta = D_P/D_F,$$

where D_F is Feret's diameter. Wadell found that Θ, which can be measured easily, gave a close approximation to Φ.

3-4 Particle Size Distributions

The discussion of particle size thus far has dealt only with the measurement of individual diameters and the shape factors of average diameters. An aerosol normally contains a wide range of particle sizes and cannot be defined adequately by an average diameter. A knowledge of the size distribution and a mathematical expression to describe it are highly desirable, especially when it is necessary to estimate particulate characteristics that are not measured directly. Many mathematical relationships have been proposed to describe particle size distributions and a number of them will be discussed later. Of these, the most widely used in aerosol work is the lognormal distribution, which is merely the normal distribution applied to the logarithms of the quantities actually measured. The discussion of the lognormal distribution, therefore, will be preceded by a summary of the normal distribution and the problems of sampling from it.

3-4.1 The Normal Distribution

Assume an essentially infinite population of individuals, each of which possesses a certain characteristic in an amount that can be assigned a numerical value x which is not the same for all individuals. If the x-values are normally distributed among the population, then the relative frequency with which values will occur in the interval, $x \pm dx/2$, is

$$f(x)\,dx = \frac{1}{\sigma(2\pi)^{1/2}} \cdot \exp[-(x-\mu)^2/2\sigma^2] \cdot dx, \tag{3.8}$$

where μ and σ are, respectively, the mean and standard deviation of the distribution and $f(x)$ is the normal probability density function. The mean is given by

$$\mu = \int_{-\infty}^{\infty} x \cdot f(x)\,dx, \tag{3.9}$$

and the variance (the square of the standard deviation) is given by

$$\sigma^2 = \int_{-\infty}^{\infty} (x-\mu)^2 \cdot f(x)\,dx. \tag{3.10}$$

The Greek letters μ and σ will be used to identify the parameters of the distribution of the infinite population of individuals; that is, they will represent the true mean and standard deviation. If these two parameters are known, the distribution is completely defined.

Equation (3.8) forms a bell-shaped curve centered over $x = \mu$. It can be plotted conveniently by setting $x = \mu + a\sigma$ and making the ordinate

$$y = \frac{1}{\sigma(2\pi)^{1/2}} \cdot \exp(-a^2/2).$$

The curve for $\sigma = 0.396$ is shown in Fig. 3.11. The element of area (cross-hatched) under the curve at x represents the relative number of individuals having x-values within the increment dx. The total area under the curve must equal unity, since

$$\int_{-\infty}^{\infty} f(x)\,dx = 1. \tag{3.11}$$

The fraction of all individuals having x values less than the specific value x_a is represented by the shaded area and is given by

$$F(x_a) = \int_{-\infty}^{x_a} f(x)\,dx. \tag{3.12}$$

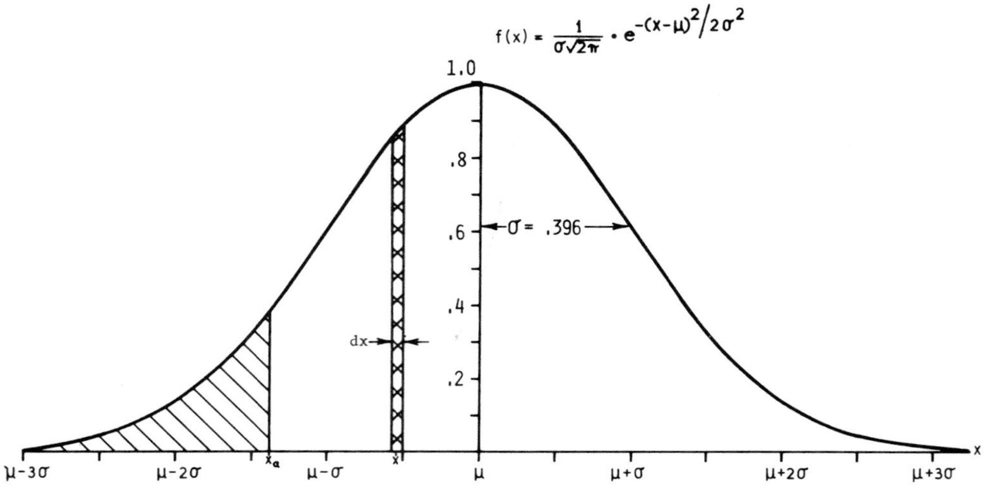

FIG. 3.11. Normal distribution curve.

$F(x_a)$ is the cumulative distribution function. For the normal distribution the integration must be carried out by approximation methods, but this distribution is so useful that tables of $F(x_a)$ are available in many handbooks (see below). The curve of $F(x_a)$, plotted against $(x_a-\mu)/\sigma$ to make it valid for all values of μ and σ, is shown in Fig. 3.12(A).

A more useful form of the cumulative distribution curve is obtained if the sigmoid curve is straightened by scaling the ordinate so that the distance between $F(x_a)-0.5$ and the origin is proportional to $(x-\mu)/\sigma$:

$F(x_a)$:	0.05	0.10	0.20	0.30	0.40	0.50
$(x_a-\mu)/\sigma$:	−1.65	−1.282	−0.842	−0.525	−0.255	0
$F(x_a)$:	0.60	0.70	0.80	0.90	0.95	
$(x_a-\mu)/\sigma$:	0.255	0.525	0.842	1.282	1.65	

The curve of Fig. 3.12(A) then becomes the straight line of Fig. 3.12(B). It is important to note that when $x_a = \mu+\sigma$, $F(x_a) = 0.841$, and when $x_a = \mu-\sigma$, $F(x_a) = 0.159$.

Tables of $F(x_a)$ are usually put in the form

$$F(x_a) = \frac{1}{(2\pi)^{1/2}} \int_{-\infty}^{z_a} \exp(-z^2/2)\, dz, \tag{3.13}$$

where $z = (x-\mu)/\sigma$, and z_a is the value of z corresponding to x_a. Since the distribution is symmetrical about μ, $F(x_a)$ [or $F(x_a)-0.5$] is tabulated only for positive values of z_a. z_a is the quantity that is linearly related to x_a in Fig. 3.12(B):

$$z_a = \alpha + \beta x_a, \tag{3.14}$$

where $\alpha = -\mu/\sigma$ and $\beta = 1/\sigma$.

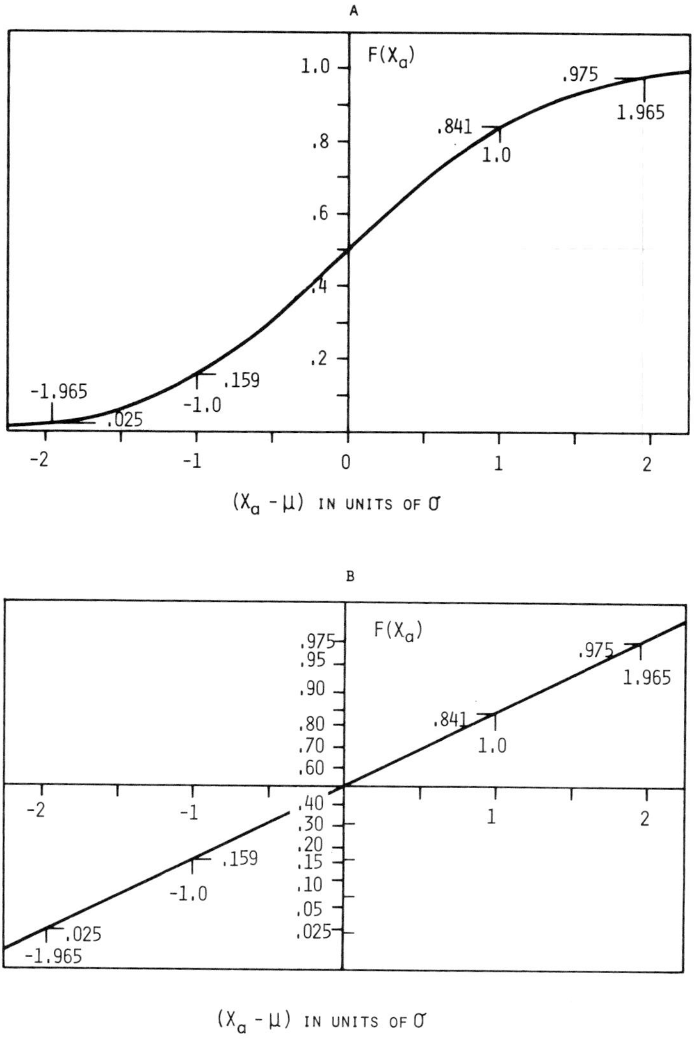

FIG. 3.12. (A) Cumulative distribution function in rectangular coordinates. (B) Cumulative distribution function in adjusted coordinates.

(a) *Estimating μ and σ.* The parameters, μ and σ, can never be determined exactly. Instead, they are estimated by calculating the statistics, m and s, from measurements of the x values of a number of individuals taken at random from the population. If each x value is measured essentially without error, the desired statistics are

$$m = \sum_N x/N, \tag{3.15}$$

and

$$s = \left[\sum_N (x-m)^2/(N-1)\right]^{1/2}, \tag{3.16}$$

where N is the number of individuals in the sample. In many cases, it is more practical to sort the individuals into size intervals and calculate the statistics as

$$m = \sum_{i=1}^{k} n_i x_i/N \tag{3.17}$$

and

$$s = \left[\sum_{i=1}^{k} n_i (x_i-m)^2/(N-1)\right]^{1/2}, \tag{3.18}$$

where n_i is the number of measurements that fall in the ith size interval and x_i is an average value assumed to be representative of x values in that interval. Formula (3.15) provides a maximum likelihood estimate of μ; s [Formula (3.16)] is the square root of the maximum likelihood estimate of σ^2, corrected for bias. The statistical relationships discussed below relate to them. If σ exceeds the width of the size intervals by a factor of 4 or more, however, formulas (3.17) and (3.18) are adequate [47].

The parameters, μ and σ, can also be estimated using the cumulative form of the distribution. The quantities

$$\begin{aligned} F_1 &= n_1/N, \\ F_2 &= F_1 + n_2/N, \\ &\vdots \\ F_i &= F_{i-1} + n_i/N, \\ &\vdots \\ F_{k-1} &= F_{k-2} + n_{k-1}/N, \end{aligned}$$

are plotted against the upper limit of the ith size increment, using the coordinate system of Fig. 3.12(B) (available commercially as arithmetic-probability paper). A straight line is drawn through the points, and the value of x at which the line's ordinate is 0.50, is an estimate of the mean. Letting $x_{0.84}$

represent the value of x at which the line's ordinate is 0.84, the standard deviation is taken as

$$s = x_{0.84} - m. \tag{3.19}$$

The uncertainties in drawing the straight line can be avoided by employing an algebraic method of calculation. From Eq. (3.14), the equation of the straight line will be

$$z(F) = a + bx, \tag{3.20}$$

where $z(F)$ values are obtained from tables for each of the quantities $F_1, F_2, \ldots, F_i, \ldots, F_{k-1}$, and x is the upper limit of the ith size increment. The constants, a and b, are calculated by the method of least squares, and the statistics are

$$m = -a/b,$$

and

$$s = 1/b.$$

This is essentially the method of probits [48] except that the probit was defined as

$$\text{Probit}(F) = 5 + z(F) = (5+a) + bx. \tag{3.20a}$$

(b) *Sampling Errors and Confidence Limits.* It is apparent that any inference about the population as a whole, based on the limited number of observations that are made, is subject to error. Unless some reasonable assertions can be made about the magnitude of that error, the inference is of doubtful value. It is here that statistical methods are applied that make it possible to place confidence limits on the observed statistics. These confidence limits bound a range of values, known as a confidence interval, about a sample statistic. For a given statistic, method of calculation, and sample size, there is a specific probability (confidence level) that a random sample will yield a confidence interval that includes the population parameter. A probability statement cannot be made for confidence limits calculated for any specific sample, because the parameter is either included within them or it is not. It can be said, however, that if similar limits are calculated for a number of similar samples, the parameter will be included in the confidence interval with a relative frequency very nearly equal to that defined by the confidence level.

To calculate confidence limits from the statistics, m and s, it is necessary to have a knowledge of their sampling distributions. Theoretically, if a great many samples of size N are examined, their means will be normally distributed about a mean value equal to μ, and the standard deviation of their distribution will be $\sigma/\sqrt{N}$. At one time, confidence limits were calculated with the aid of this

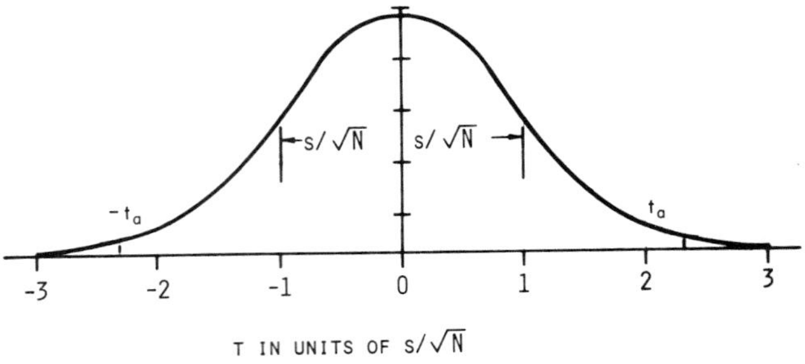

FIG. 3.13. Normal approximation to t distribution.

distribution, although it was necessary to assume $\sigma = s$. Now, they are based on the theoretical distributions of the quantity,

$$t = (m-\mu)/(s/\sqrt{N}),$$

known as "Student's t." When N is larger than about 30, the distribution of t is very nearly normal, with a mean equal to zero and a standard deviation equal to $s/\sqrt{N}$ as shown in Fig. 3.13. If $t = \pm t_a$ bound a fraction, $1-p$, of the total area under the curve, then there are $100(1-p)$ chances in 100 that

$$[m-t_a \cdot s/\sqrt{N}] < \mu < [m+t_a \cdot s/\sqrt{N}]. \qquad (3.21)$$

The terms in brackets are the upper and lower confidence limits on μ at the $100(1-p)$ percent level. At the 95% level for $N > 30$, $t_p \simeq 2$, and the confidence limits on μ are $m \pm 2s/\sqrt{N}$. For other levels of confidence and $N > 30$, t_a can be found from Eq. (3.13). At a given value of p, $t_a = z_a$ when $F(x_a) = 1-p/2$.

For small values of N, the distribution of t is no longer normal and it is necessary to use tables that relate p to t for various values of the number of degrees of freedom $(N-1)$ associated with m.

Similar limits can be put on σ^2, since the quantity

$$\chi^2 = (N-1)\, s^2/\sigma^2 \qquad (3.22)$$

has a known distribution for which the mean equals $N-1$ and the variance is $[2(N-1)]^{1/2}$. The distribution is not symmetrical, however, approaching normality only at quite large values of N. The $100(1-p)$-percent confidence limits on the variance are

$$\frac{(N-1)\, s^2}{\chi_u^2} < \sigma^2 < \frac{(N-1)\, s^2}{\chi_l^2}, \qquad (3.23)$$

where $u = p/2$ and $l = 1-p/2$. The appropriate values of χ^2, for $N-1$ degrees of freedom, are available in tables for values of $N-1$ up to about 30. To a very good approximation, the quantity $(2\chi^2)^{1/2}$ is distributed normally with mean equal to $(2N-3)^{1/2}$ and unit variance when $N-1 > 30$. The necessary values of χ^2 can then be calculated using Eq. (3.13) with

$$1 - u = F(x_a)$$

$$z_a = (2\chi_u{}^2)^{1/2} - (2N-3)^{1/2}. \tag{3.24}$$

The other limit is calculated by replacing u with l.

(c) *"Goodness-of-Fit" Tests.* The χ^2-distribution is also useful to test how well the data of a sample fit a normal distribution having parameters equal to the sample statistics, m and s. The observed range of sizes included in the sample is divided into M segments, each of which includes at least ten individuals [49]. Using the theoretical normal distribution based on m and s, the number of particles expected in each size increment,

$$\Delta N_t = [F(x_u) - F(x_l)] \cdot N,$$

is calculated. Equation (3.13) is used to determine the F's. x_u and x_l are the upper and lower boundaries of the size interval, respectively. For each size increment, the quantity

$$\Delta\chi^2 = (\Delta N_0 - \Delta N_t)^2 / \Delta N_t$$

is calculated, where ΔN_0 is the number of individuals observed in the increment. These are then summed to give

$$\chi^2 = \sum \Delta\chi^2.$$

The probability that a random sample from a normal distribution would give a larger value of χ^2 is taken from a table of χ^2, using $M-3$ degrees of freedom. If the probability is small, the assumption of normality is rejected; if it is large, the assumption is not rejected. The definitions of "large" and "small" are left to the discretion of the experimenter.

3-4.2 Properties of the Lognormal Distribution

If the quantity $x = \ln D$ is normally distributed, then the distribution of D is said to be lognormal [50]. The lognormal distribution is particularly useful in particle size analysis because of the characteristics described in the following.

Consider some property of a particle that can be defined quantitatively by

$$Q_r(D) = \alpha_r D^r, \tag{3.25}$$

where α_r is a constant (shape factor) for a given value of r. For a lognormal distribution, the relative number of particles having diameters whose logarithms fall in the interval $x \pm dx/2$ is given by Eq. (3.8), with $\mu_0 = \ln \delta_{0g}$, where δ_{0g} is the median diameter (or geometric mean diameter) of the population. (The double subscript is necessary to permit identification of median diameters of different distributions.) The value of Q averaged over all particles in the population is then

$$\bar{Q}_r = \int_{-\infty}^{\infty} Q_r(D) \cdot f_0(x) \cdot dx, \tag{3.26}$$

where

$$f_0(x) = \frac{1}{\sigma(2\pi)^{1/2}} \cdot \exp[-(\ln D - \mu_0)^2/2\sigma^2]. \tag{3.27}$$

Each interval, $dx = d(\ln D)$, contributes to $\bar{Q}_r$ an incremental amount given by

$$d\bar{Q}_r = Q_r(D) \cdot f_0(x) \cdot dx = \alpha_r \cdot e^{rx} \cdot f_0(x) \cdot dx,$$

since $D^r = e^{r \ln D} = e^{rx}$. Replacing $f_0(x)\, dx$ with its equivalent from Eq. (3.27) and making the algebraic adjustments necessary to separate the constant terms from the exponential leads to

$$d\bar{Q}_r = \alpha_r \delta_{0g}^r \cdot \exp(r^2\sigma^2/2) \cdot \frac{1}{\sigma \cdot (2\pi)^{1/2}} \cdot \exp[-(\ln D - \mu_0 - r\sigma^2)^2/2\sigma^2] \cdot dx. \tag{3.28}$$

Integration of this yields $\bar{Q}_r$, the average amount of the property per particle:

$$\bar{Q}_r = \alpha_r \delta_{0g}^r \cdot \exp(r^2\sigma^2/2). \tag{3.29}$$

The diameter of the particle for which $\bar{Q}_r = Q_r(D)$ is

$$\bar{\delta}_r = \delta_{0g} \cdot \exp(r\sigma^2/2). \tag{3.30}$$

Setting $\mu_r = \mu_0 + r\sigma^2$ in Eq. (3.28), we can designate the relative amount of Q_r associated with particles having diameters whose logarithms fall in the interval defined by $x \pm dx/2$ as

$$d\bar{Q}_r/\bar{Q}_r = f_r(x) \cdot dx = \frac{1}{\sigma(2\pi)^{1/2}} \cdot \exp -[(\ln D - \mu_r)^2/2\sigma^2]\, dx. \tag{3.31}$$

Comparing this with Eq. (3.27) reveals that the distribution of the property Q_r with respect to diameter is also lognormal having the same variance as the frequency distribution and a mean logarithm given by

$$\mu_r = \ln \delta_{rg} = \mu_0 + r\sigma^2. \tag{3.32}$$

The diameter

$$\delta_{rg} = \delta_{0g}\, e^{r\sigma^2} \tag{3.33}$$

is the median diameter of the Q_r distribution. Equation (3.31) is valid for any value of r. When $r = 0$, this equation reduces to Eq. (3.8).

To obtain the lognormal distributions described above, the range of diameters from $D = 0$ to $D = \infty$ was divided into equal logarithmic intervals and a fraction of Q_r, defined by Eq. (3.28), was allotted to each interval. For $r = 0$, we have the relative number, relative count, or relative frequency distribution; for $r = 2$, the relative surface distribution; and for $r = 3$, the relative volume or mass distribution. Another group of lognormal distributions is obtained if we divide the range of quantities from $Q = 0$ to $Q = \infty$ into equal logarithmic intervals and allot to each interval the relative number of particles having Q-values falling within that interval. From the definition of Q_r,

$$\ln Q_r = \ln \alpha_r + r \cdot \ln D = \ln \alpha_r + rx,$$

and

$$d(\ln Q_r) = r \cdot d(\ln D) = r \cdot dx.$$

Setting $\ln Q_{rg} = \ln \alpha_r + r\mu_0$ and substituting these quantities into the equation $f(\ln Q_r) \cdot d(\ln Q_r) = f_0(x)\,dx$, we find the relative number of particles having $\ln Q$ values in the interval $\ln Q \pm d(\ln Q)/2$ to be

$$f(\ln Q_r) \cdot d(\ln Q_r) = \frac{1}{r\sigma(2\pi)^{1/2}} \cdot \exp -[(\ln Q_r - \ln Q_{rg})^2/2(r\sigma)^2] \cdot d(\ln Q_r). \tag{3.34}$$

Hence, the number distribution of particles with respect to Q_r is also lognormal and has a variance $\sigma^2(Q_r) = r^2\sigma^2$, a geometric standard deviation, $\sigma_g(Q_r) = \sigma_g{}^r$, and a median value related to the count median diameter:

$$Q_{rg} = \alpha_r \delta_{0g}^r. \tag{3.35}$$

Distributions of this sort are seen in autoradiography [51], for instance, in which the distribution of particles with respect to radioactivity is lognormal. In the optical sizing of large particles from a lognormal distribution, the number of particles scattering light in a given logarithmic interval of intensity, should follow the same distribution.

3-4.3 Sampling from an Infinite Population of Particles Having a Lognormal Distribution of Diameters

The large amount of information available concerning the statistics of sampling from a normal distribution is directly applicable to particle size analysis when D is lognormally distributed. For a sample of N particles, the maximum likelihood statistics, m and s, are given by the following forms of Eqs. (3.15) and (3.16):

$$m = \ln D_{0g} = \sum_{1}^{N} \ln D_i/N, \tag{3.36}$$

and

$$s = \ln \sigma_g = \left[\sum_{1}^{N} (\ln D_i - \ln D_{0g})^2/(N-1)\right]^{1/2}. \tag{3.37}$$

In practice, D_{0g}, which estimates the population count median diameter δ_{0g}, and σ_g, which estimates the population geometric standard deviation e^{σ}, are reported rather than m and s. The mean logarithms of the D^r distributions are estimated using the Hatch–Choate [27] equation:

$$\ln D_{rg} = \ln D_{0g} + rs^2, \tag{3.38}$$

or

$$D_{rg} = D_{0g} \cdot \exp(rs^2).$$

The diameter of the particle having the average amount of Q_r is calculated from

$$\ln \bar{D}_r = \ln D_{0g} + rs^2/2, \tag{3.39}$$

or

$$\bar{D}_r = D_{0g} \cdot \exp(rs^2/2).$$

For samples of N particles, the sampling distribution of the mean logarithm of diameters m is normal and has a mean of μ_0 and a standard deviation equal to $\sigma/\sqrt{N}$. For the values of N of interest in particle size work, the sampling distribution of the variance s^2 is also normal, having a mean of σ^2 and a standard deviation equal to $\sigma^2(2/N)^{1/2}$. The sampling distribution of $\ln D_{rg}$, when estimated by means of Eq. (3.38), is normal, having a mean of μ_r and a standard deviation given by

$$\sigma(\ln D_{rg}) = (\sigma/\sqrt{N})(1+2r^2\sigma^2)^{1/2}. \tag{3.40}$$

The uncertainty in estimating μ_r (and, hence, D_{rg}) increases rapidly as $r\sigma$ increases. This is brought out in Fig. 3.14 [52], which compares the sampling distributions of the count and volume median diameters for samples of 200 particles taken from a population having $\sigma = 1.1$ ($\sigma_g = 3.00$). The histogram in the same figure is an experimental distribution of the volume median diameters of 300 samples from the same population.

The statistical relationships above are based on the assumption that each particle diameter is measured individually and accurately. In practice, however, it is customary to sort the particles into a series of size intervals. The proper

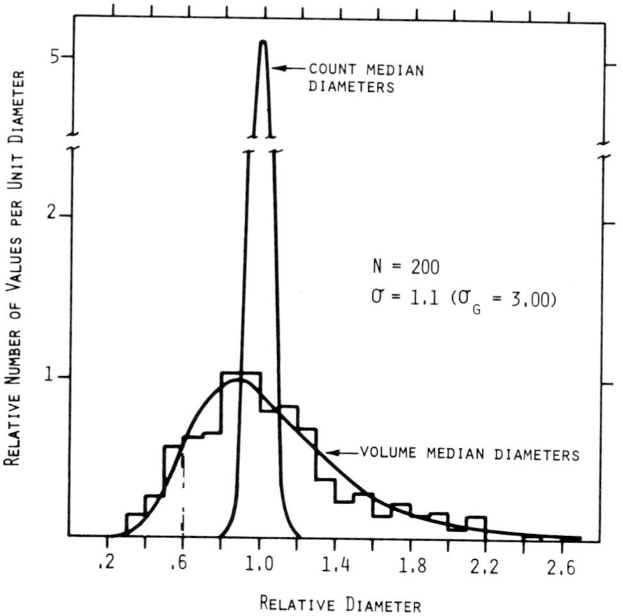

FIG. 3.14. Sampling distributions of count and volume median diameters. (For each distribution, diameters are scaled in units of the median diameter.)

estimates of μ_0 and σ then become quite complicated, and are not likely to be carried out without the services of a digital computer. The desired statistics can be approximated, however, by the following forms of Eqs. (3.17) and (3.18):

$$\ln D_{0g} = \left(\sum_{i=1}^{k} n_i \cdot \ln D_i\right) \Big/ \left(\sum_{i=1}^{k} n_i\right),$$

and (3.41)

$$s = \ln \sigma_g = \left[\sum_{i=1}^{k} n_i \cdot (\ln D_i - \ln D_{0g})^2 \Big/ \left(\sum_{i=1}^{k} n_i - 1\right)\right]^{1/2},$$

where n_i is the number of particles in the ith size interval, D_i is an average diameter for that interval, and k is the total number of size intervals. Alternatively, the statistics can be approximated by plotting, on logarithmic-probability paper, the cumulative percent

$$P_j = 100 \sum_{i=1}^{j} n_i \Big/ \sum_{i=1}^{k} n_i \qquad (j < k) \tag{3.41a}$$

against the upper limit of the jth size interval for a number of values of j and drawing the straight line which the resulting points appear to estimate. D_{0g} is the diameter at which the line has the coordinate $P = 50\%$. From Eq. (3.19),

$$\ln \sigma_g = \ln D_{84} - \ln D_{0g} = \ln(D_{84}/D_{0g}),$$

and (3.41b)

$$\sigma_g = D_{84}/D_{0g},$$

where D_{84} is the diameter at which the line has the coordinate $P = 84\%$. The statistics can also be calculated from the cumulative distribution using the method of probit analysis described in the previous section. In this case,

$$\text{Probit}(P_j/100) = a' + b \ln D_j.$$

For each value of $P_j/100$, the corresponding probit is calculated as described previously [Eq. (3.20a)] and a' and b are calculated by the method of least squares. The desired statistics are

$$m = \ln D_{0g} = (5 - a')/b,$$

and (3.42)

$$s = \ln \sigma_g = 1/b.$$

TABLE 3.5

ESTIMATION OF SAMPLE PARAMETERS[a] BY DIFFERENT MEANS

Size interval	Number of particles	Cumulative percent	Size interval	Number of particles	Cumulative percent
0–1	8	4.0	14–15	1	89.0
1–2	23	15.5	15–16	4	91.0
2–3	21	26.0	17–18	3	92.5
3–4	27	39.5	18–19	2	93.5
4–5	28	53.5	19–20	2	94.5
5–6	14	60.5	22–23	2	95.5
6–7	12	66.5	23–24	1	96.0
7–8	14	73.5	24–25	2	97.0
8–9	6	76.5	25–26	1	97.5
9–10	7	80.0	26–27	1	98.0
10–11	5	82.5	27–28	1	98.5
11–12	6	85.5	40–41	1	99.0
12–13	4	87.5	44–45	1	99.5
13–14	2	88.5	87.2	1	100

[a] The definition of the population parameters are:

$$\delta_{0g} = 5.0, \ \sigma_g = 2.46.$$

Table 3.5—*continued.*

Sample statistics:

	D_{og}	σ_g
True values	4.86	2.44
Estimates from group data using interval midpoint as average interval diameter	4.89	2.43
Estimates using probit analysis	4.85	2.40
Estimates from graph Fig. 3.15	4.80	2.40

An example of the results obtained by the different methods of calculation is given in Table 3.5. The data are a random sample of 200 particles from a population of particles having a count median diameter equal to five times the width of the class intervals, and a geometric standard deviation of 2.46. A graphical presentation of the data is shown in Fig. 3.15.

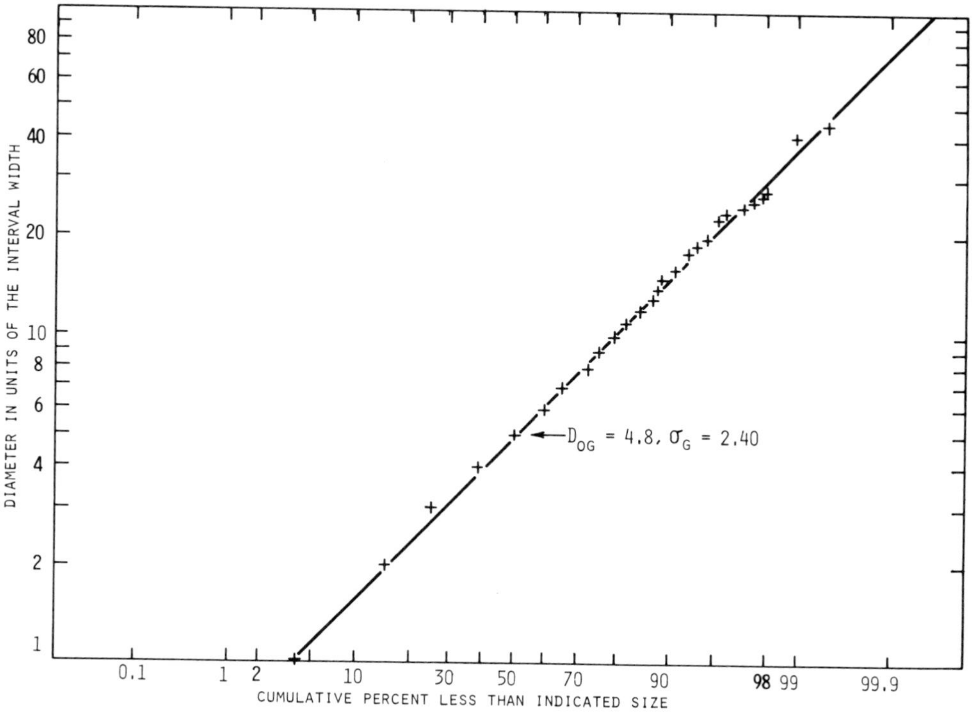

FIG. 3.15. Cumulative size distribution for data of Table 3.5.

Any estimates of population parameters can be adversely affected by inadequate magnification, which leaves images of some particles in a sample below the limit of observation. In addition, estimates based on grouped measurements are subject to error due to the use of an average diameter to represent all measurements in a given group. (The example above was chosen to minimize these errors.) The significance of the error depends on the ratio, δ_{0g}/w, where w is the interval width, and on the average diameter chosen to represent the individual class intervals. Figures 3.16 and 3.17 show how these factors affect the values of the sampling statistics [52]. The curves are theoretical results for an infinitely large population, truncated at $D = 0.2w$, and grouped in the manner used in practice. Experimental data show that the means of the sampling distributions of both statistics, for sample sizes of 100, 200, 500, and 1000, closely approximate the theoretical curves. The individual points in the two figures, which are theoretical values for truncation at $0.4w$, indicate that truncation is not a severe problem if D_0/w is greater than 3 or 4. Theoretical curves for the volume (mass) median diameters, calculated by applying the

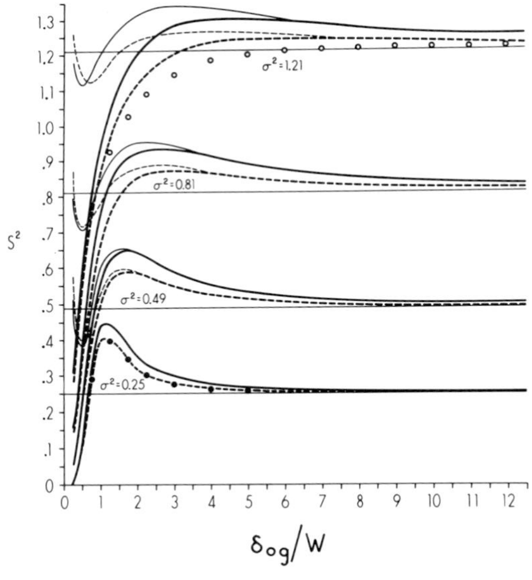

FIG. 3.16. Effect of magnification on class analysis estimates of the variance of infinitely large samples truncated at $0.2w$. Dashed lines: class interval represented by logarithm of its mean diameter; solid lines: class interval represented by average of logarithms of upper and lower interval boundaries. Lighter lines are nontruncated samples. Open and closed circles are respectively for $\sigma = 1.1$ and $\sigma = 0.5$, truncation at $0.4w$, analysis as for dashed lines [52]. Courtesy of American Industrial Hygiene Association.

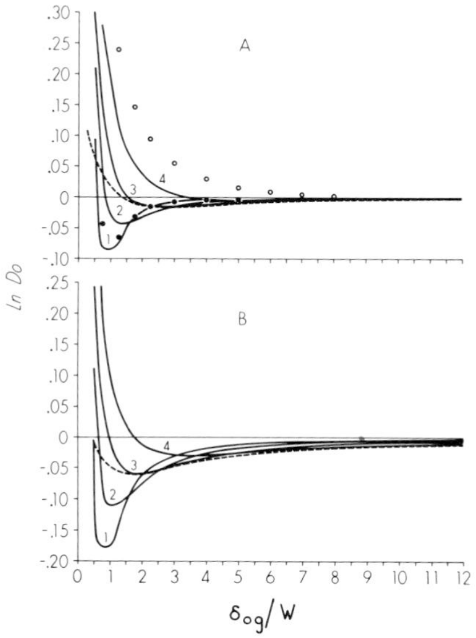

FIG. 3.17. Effect of magnification on class analysis estimates of the mean of infinitely large samples truncated at $0.2w$ [52]. A, class interval represented by logarithm of its mean diameter; B, class interval represented by average of logarithms of upper and lower interval boundaries. Curves 1, 2, 3, and 4 correspond to $\sigma = 0.5$, 0.7, 0.9 and 1.1. Dashed curves are nontruncated samples for $\sigma = 0.9$. The circles are for $\sigma = 1.1$ and truncation at $0.4w$. Courtesy of American Industrial Hygiene Association.

Hatch–Choate equation to the curves of Fig. 3.16 and 3.17, are shown in Fig. 3.18.

When the size measurements from a random sample of particles appear to be lognormally distributed, application of the sample statistics, m and s, to the Hatch–Choate equation is the method of choice for estimating the median

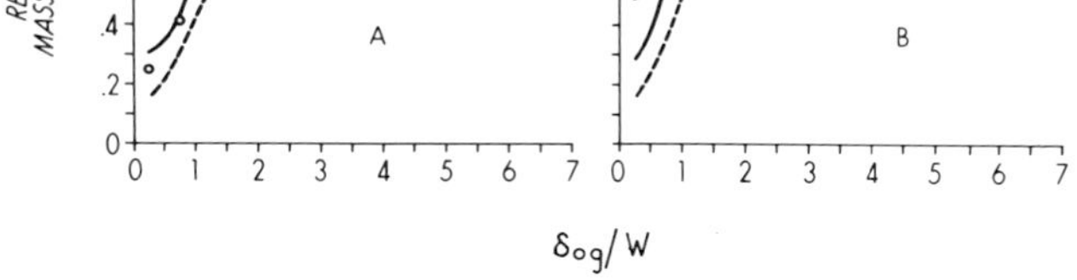

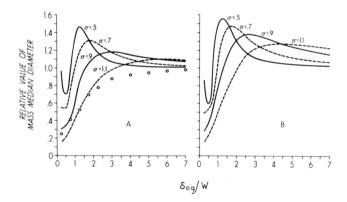

FIG. 3.18. Effect of magnification on class analysis estimates of volume (mass) median diameters for infinitely large samples truncated at $0.2w$ [52]. A, class interval represented by logarithm of its mean diameter; B, class interval represented by average of logarithms of upper and lower interval boundaries. Circles are for $\sigma = 1.1$, apparent maximum likelihood method. Courtesy of American Industrial Hygiene Association.

diameters of other distributions. Occasionally, an experimenter will attempt to calculate the latter diameters by weighting the observed frequency in each class interval with an average value of D^r for that interval. Although a graphical method of analysis is usually used, it is equivalent to the following calculation:

$$\ln D_{rg} = \sum n_i D_i^r \cdot \ln(D_i)/\sum n_i D_i^r. \tag{3.43}$$

Herdan [17] cites two sources of error associated with this method:

(1) The weighted proportions are too sensitive to sample fluctuations, since the method implies the raising of experimental errors to higher powers; and

(2) The grouping of observed dimensions into frequency intervals, which are in practice rather broad and sometimes nonuniform, makes an accurate calculation impossible.

Even if errors associated with these sources are avoided, however, there is a purely statistical problem which makes the use of the weighted-frequency method unreliable. The occurrence of large diameters in a random sample

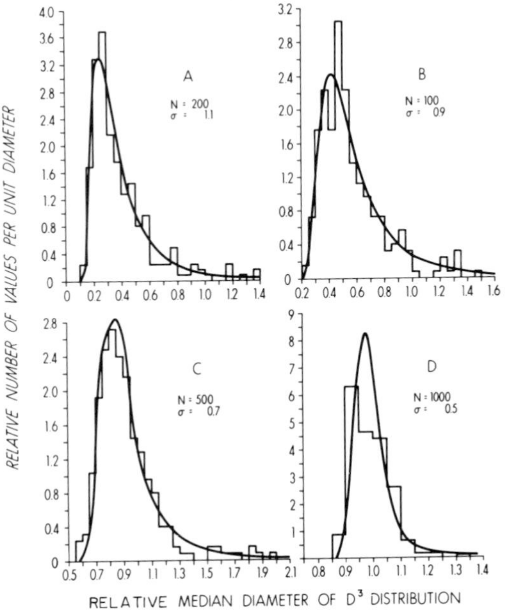

FIG. 3.19. Sampling distributions of volume (mass) median diameters when estimated by the method of weighted frequencies [53]. Smooth curves are theoretical; histograms are based on 250 computer-generated random samples of size N. Courtesy of Elsevier Sequoia SA.

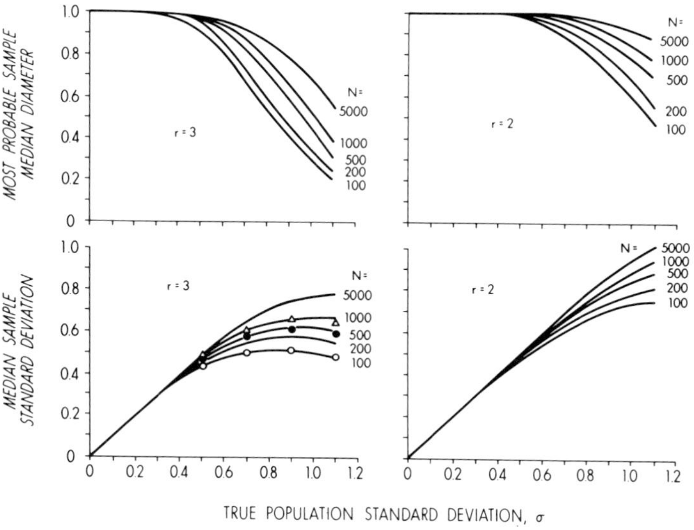

FIG. 3.20. Most probable median diameters and median standard deviations as functions of N and σ, when calculated according to method of weighted frequencies [53]. Points are experimental. Courtesy of Elsevier Sequoia SA.

depends on their relative number in the particle population rather than their relative contribution to the physical property under consideration. For this reason, the weighted-frequency method, which treats a random sample of a frequency distribution as if it were a random sample of a D^r distribution, shows a systematic bias in the estimation of the parameters of the latter distribution. The extent of the bias has been predicted theoretically and verified experimentally [53] under circumstances that eliminated the sources of error cited by Herdan. Figure 3.19 shows sampling distributions of the volume median diameter obtained in this way. A comparison of Fig. 3.19(A) with the sampling distributions of Fig. 3.14 brings out the marked bias introduced by this method. Figure 3.20 shows how the sampling statistics for the mass or volume ($r = 3$) distribution and the surface ($r = 2$) distribution vary with sample size and the logarithm of the geometric standard deviation.

(a) *Confidence Limits.* Because a large value of N is encountered in most particulate samples, the confidence limits on the various median diameters, calculated according to Equation (3.21), are given by

$$D_{rg}(\text{C.L.}) = D_{rg} \cdot \exp[\pm t_a \cdot \sigma(\ln D_{rg})], \qquad (3.44a)$$

where $\sigma(\ln D_{rg})$ is given by Eq. (3.40), using the sample estimate s in place of σ.

Using Eq. (3.23), the confidence limits on the population geometric standard deviation are found to be, approximately,

$$\sigma_g(\text{C.L.}) = \sigma_g{}^a, \tag{3.44b}$$

where $a = [2(N-1)]^{1/2}/[\pm t_a + (2N-3)^{1/2}]$ the positive value of t_a giving the lower confidence limit. t_a is used here in place of z_a to relate it to the limits on the mean at the same level of confidence.

The goodness-of-fit test can be carried out as described above if the transformation $x = \ln D$ is employed.

(b) *The Lognormal Distribution as an Artifact of the Sampling System.* The discussion above was based on the assumption that the samples were, in fact, taken from a population of particles whose diameters followed a lognormal distribution. Since there is no theoretical basis for the assumption, it must be justified on the grounds that the data, and the results calculated from them, are compatible with the overall experimental conditions. On occasion, however, the lognormal appearance of sample data is clearly an artifact of the method of measurement. Laskin [54] used the optical microscope to measure the diameters of uranium dioxide particles deposited on a glass slide and obtained a satisfactory lognormal size distribution. He then coated the sample with selenium, which greatly improved the resolution of his measurement technique, and again obtained a satisfactory lognormal distribution. However, the coated sample had a smaller median diameter and a larger geometric standard deviation than the uncoated sample. Fraser [55] reported similar results when the same sample was analyzed using both optical and electron microscopes. Cartwright [56] has encountered samples in which the diameter of peak frequency decreased continuously as he increased the magnification at which the sample was examined. He concluded that the sample had actually come from an exponential distribution (see below). Whatever the size distribution of the population from which the samples were taken, it is apparent that the appearance of lognormality in such cases was due to limitations in the measuring instruments.

3-4.4 Other Distributions

Although the lognormal distribution has received the most attention in aerosol work, a number of other distributions have been used, with varying degrees of success. Several of these are described below. In each case, examples of the cumulative form of the distribution, plotted on logarithmic-probability paper, are given to bring out any similarities to the lognormal distribution.

(a) *The Exponential Distribution.* This is the "compound interest law" distribution encountered by Martin [1] in his studies on fine grinding. In differential form, the number of particles per unit diameter is given by

$$dn/dD = a \cdot e^{-bD}. \tag{3.45}$$

The cumulative distribution giving the number of particles that have diameters smaller than D is

$$n(<D) = (a/b) \cdot (1 - e^{-bD}). \tag{3.46}$$

The total number of particles is $N = a/b$. The differential amount of the property $Q_r(D) = \alpha_r D^r$ contained in the increment of diameter dD is

$$dQ_r = \alpha_r D^r \cdot a \cdot e^{-bD} \cdot dD,$$

and the cumulative amount of the property is

$$\begin{aligned} \sum Q_r(D) &= a\alpha_r \cdot \int_0^D D^r \cdot e^{-bD}\, dD \\ &= (\alpha_r N/b^r)\{r! - e^{-bD}[(bD)^r + r(bD)^{r-1} + r(r-1)(bD)^{r-2} + \cdots + r!]\}. \end{aligned} \tag{3.47}$$

Setting $D = \infty$, the average diameters for the various properties can be calculated from

$$\bar{D}_r = (\sum Q_r/\alpha_r N)^{1/r} = \frac{1}{b}(r!)^{1/r}.$$

In particular, the arithmetic average diameter is the reciprocal of b:

$$\bar{D}_1 = 1/b.$$

Some of the diameters of interest are given below.

Property	Average diameter	Q_r-Distribution Median Diameter
Number	—	$D_{0g} = 0.69\ \bar{D}_1$
Surface	$\bar{D}_2 = 1.41\ \bar{D}_1$	$D_{2g} = 2.7\ \bar{D}_1$
Volume	$\bar{D}_3 = 1.82\bar{D}_1$	$D_{3g} = 3.7\ \bar{D}_1$

Cumulative distributions for the three properties tabulated are shown in Fig. 3.21. The diameters have been normalized by expressing them relative to the average diameter. In this form, all exponential distributions for a given value of r would fit the same curve. Since the number of particles per unit size increases continually as the particle size decreases, truncation may be a severe problem. As the degree of truncation increases, the cumulative number

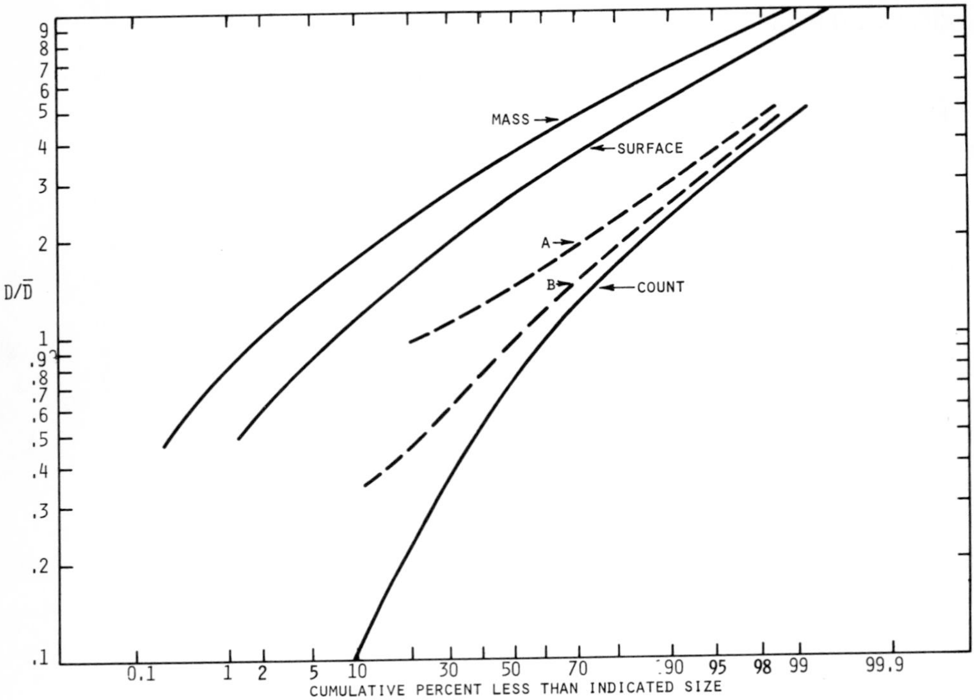

FIG. 3.21. Exponential size distribution. A, count distribution truncated at $D/\bar{D}_1 = 0.71$; B, count distribution truncated at $D/\bar{D}_1 = 0.22$.

distribution curve approaches a straight line on the logarithmic-probability grid. Some examples of this are included in Fig. 3.21.

The parameter b, which completely characterizes the distribution, can be estimated by plotting $\ln(\Delta n_i/\Delta D_i)$ against D_i, where Δn_i and ΔD_i are the number of particles and interval width, respectively, for the ith interval, centered on D_i. Except for the first, and possibly the second, class interval, the points should present a reasonably straight line of slope $-b$.

(b) *Power Function Distributions.* Power function distributions are usually expressed in terms of the number of particles per unit diameter:

$$dn/dD = KD^{-\gamma}.$$

This can be integrated to obtain the total number of particles N with which the relative number distribution, free of K, can be expressed as

$$\frac{dn}{N\,dD} = \frac{(\gamma-1)\,D^{-\gamma}}{[D_s^{1-\gamma} - D_1^{1-\gamma}]}. \tag{3.48}$$

The cumulative distribution becomes

$$\frac{n(<D)}{N} = \frac{D_s^{1-\gamma} - D^{1-\gamma}}{D_s^{1-\gamma} - D_l^{1-\gamma}}. \tag{3.49}$$

D_s and D_l are respectively the smallest and largest diameters in the distribution. The average value of any property of the particle defined by the diameter raised to the power r is

$$\bar{Q}_r = \alpha_r \cdot \frac{(\gamma-1)}{(r-\gamma+1)} \cdot D_s{}^r \cdot \frac{[(D_l/D_s)^{r-\gamma+1} - 1]}{[1-(D_l/D_s)^{1-\gamma}]}. \tag{3.50}$$

For the special cases of $\gamma = 1$, and $r = \gamma - 1$, the values of $\bar{Q}_r$ are

$$\bar{Q}_r(\gamma = 1) = \frac{\alpha_r D_s{}^r}{r} \cdot \frac{[(D_l/D_s)^r - 1]}{\ln(D_l/D_s)}, \tag{3.51}$$

and

$$\bar{Q}_r(r = \gamma - 1) = \frac{(\gamma-1) \cdot \alpha_r \cdot D_s{}^r}{[1-(D_l/D_s)^{1-\gamma}]} \cdot \ln\left(\frac{D_l}{D_s}\right). \tag{3.52}$$

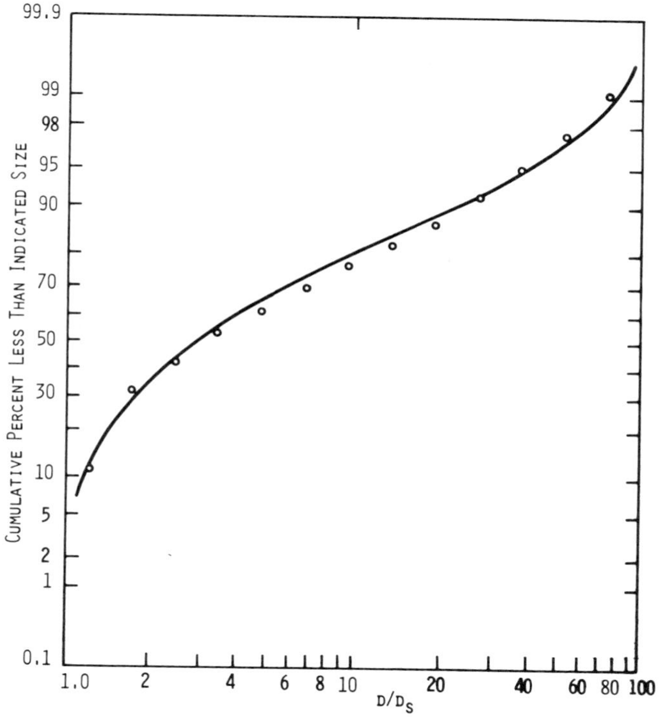

FIG. 3.22. Power law distribution. Circles, points from curve 1, Fig. 2 of Ref. [56]. Solid curve, theoretical, $\gamma = 1.6$, $D_l = 100 D_S$.

This is actually a three parameter distribution, requiring estimates of γ, D_s, and D_l. The two diameters are estimated from the appropriate values in the sample, and $-\gamma$ is the slope of the straight line obtained when $\log(dn/dD)$ is plotted against $\log(D)$. The distribution has been used extensively by members of the Safety in Mines Research Establishment in Sheffield, England. A typical example of their distributions (curve 1, Fig. 2, of Ref. 56) is shown in Fig. 3.22, in which the cumulative form of the distribution is plotted on a logarithmic-probability grid.

(c) *The Rosin–Rammler Distribution.* This distribution was derived from a theoretical consideration of the probability of obtaining particles of various sizes during the pulverization of coal. The differential and cumulative forms of the volume distribution, which was the one actually derived, are

$$(1/V_t) \cdot dV/dD = b \cdot m \cdot D^{m-1} \cdot \exp(-bD^m), \tag{3.53}$$

and

$$V(<D)/V_t = 1 - \exp(-bD^m), \tag{3.54}$$

where V_t is the total volume of particles. It is usually assumed that D can vary from 0 to ∞ [57, 58], as is apparent from the cumulative form of the distribution. The relative number of particles per unit diameter is given by

$$(1/N) \cdot dn/dD = b \cdot \bar{D}_3{}^3 \cdot m \cdot D^{m-4} \cdot \exp(-bD^m), \tag{3.55}$$

where $\bar{D}_3$ is the diameter of the particle of average volume. The cumulative form of this distribution is

$$\frac{n(<D)}{N} = b^{3/m} \cdot \bar{D}_3{}^3 \cdot \int_0^y y^{-3/m} \cdot e^{-y}\, dy, \tag{3.56}$$

where $y = bD^m$. The integral, which is the incomplete gamma function, is tabulated in handbooks of mathematics.

The diameter $\bar{D}_r$ of the particle having the average value of the property Q_r is

$$\bar{D}_r = b^{-1/m} \cdot \left\{\frac{\Gamma[1+(r-3/m)]}{\Gamma[1-(3/m)]}\right\}^{1/r}, \tag{3.57}$$

where Γ denotes the complete gamma function.

This equation brings out one of the anomalous consequences of the distribution. According to Fritz [58], m usually lies between 0.7 and 1.5. For all values of $m \leqslant 3$, the argument of the gamma function in the denominator of Eq. (3.57) is equal to or less than zero, the corresponding value of the function itself is ∞, and $\bar{D}_r$ goes to zero. The distribution should include a smallest diameter as a third parameter, but this would make estimation of the statistics very difficult.

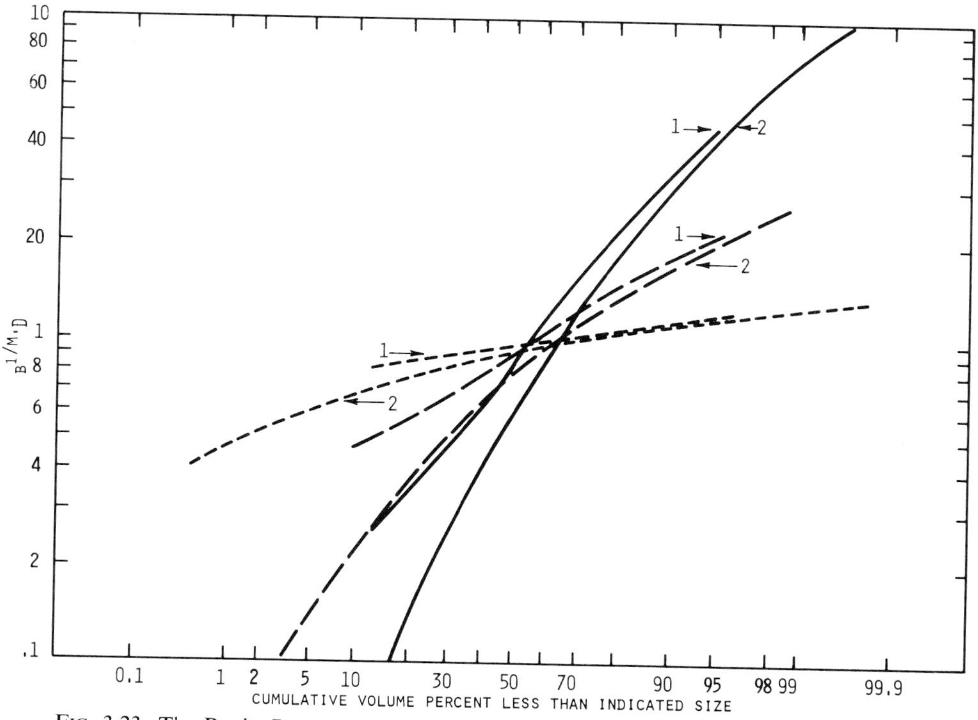

FIG. 3.23. The Rosin–Rammler distribution. Solid line, $m = 0.75$; long dashes, $m = 1.50$; short dashes, $m = 6.00$. Curve 1 = curve 2 truncated at 20%.

Experimentally, the volume distribution is used to estimate m and b. Equation (3.54) is expressed as

$$\ln \ln \{1 - [V(<D)/V_t]\}^{-1} = \ln b + m \ln D. \tag{3.58}$$

The term on the left is plotted against $\ln D$, a straight line is drawn through the data, and m and b are calculated from its slope and intercept. A plot of this sort tends to give disreputable data an air of respectability and may give the experimenter an unjustified feeling of confidence in his results. Kaye [6] calls it a "Procrustean device" with which "conformity is achieved by violent means."

Some examples of the distribution by volume are shown in Fig. 3.23 with adjusted plots showing the effects of truncation.

(d) *The Nukiyama–Tanasawa Distribution.* The two Japanese scientists for whom this distribution was named introduced it to describe the experimental data they obtained for droplet sizes in sprays produced by atomizers

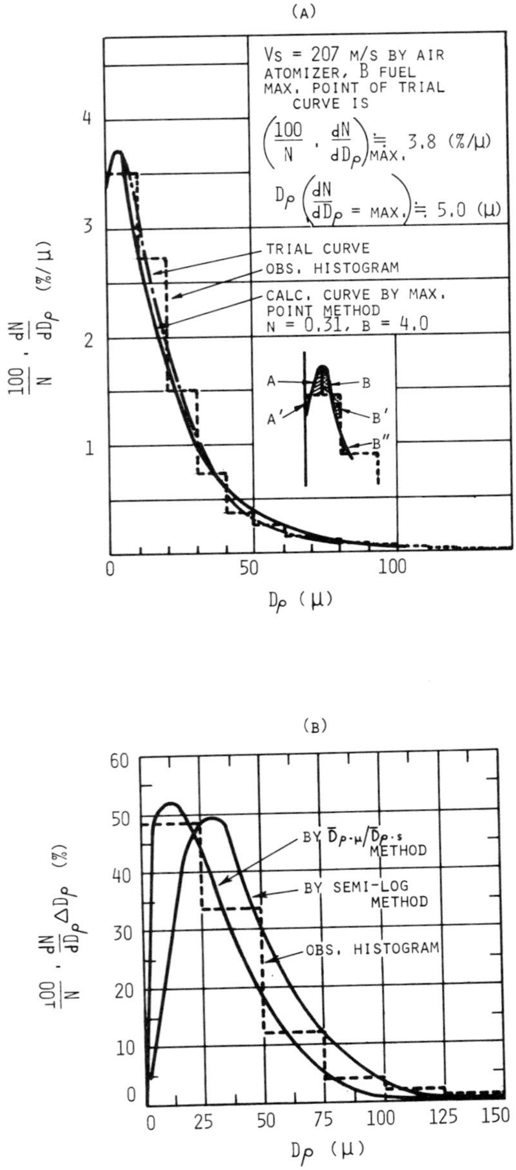

FIG. 3.24. Nukiyama–Tanasawa distributions for spray droplets [60]. Courtesy of the Society of Chemical Engineers, Japan.

operated under various conditions [59]. The relative number of particles per unit diameter is given by

$$\frac{1}{N}\cdot\frac{dn}{dD}=\frac{m\cdot b^{3/m}}{\Gamma(3/m)}\cdot D^2\cdot\exp(-bD^m), \tag{3.59}$$

where m and b are the distribution parameters that must be estimated. The diameter of the particle having the average value of the quantity Q_r is

$$\bar{D}_r=b^{-1/m}\cdot\left\{\frac{\Gamma[(r+3)/m]}{\Gamma(3/m)}\right\}^{1/r}. \tag{3.60}$$

Estimation of the distribution parameters requires a certain amount of trial and error. Equation (3.59) can be put in the form

$$\ln\left\{\frac{1}{ND^2}\cdot\frac{dn}{dD}\right\}=\ln\left\{\frac{m\cdot b^{3/m}}{\Gamma(3/m)}\right\}-bD^m. \tag{3.61}$$

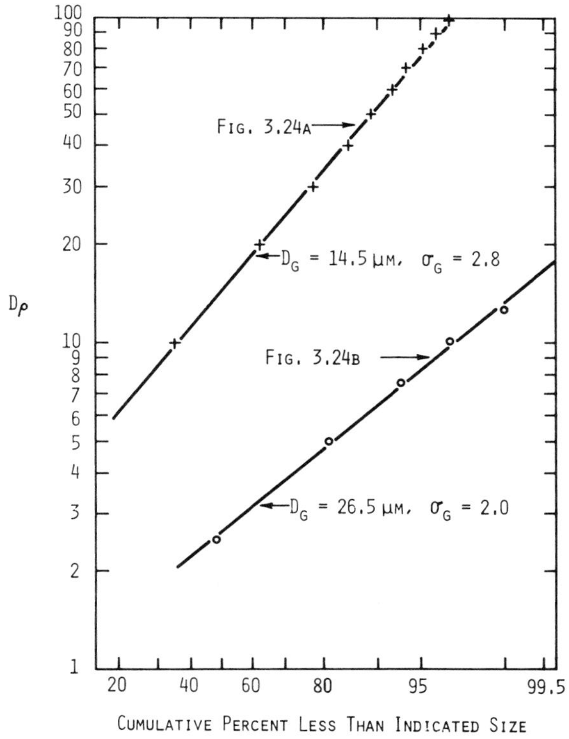

FIG. 3.25. Log-probability plot of data of Fig. 3.24.

The term on the left can be plotted against D^m to obtain a straight line of slope $-b$. However, it is necessary to assume a value of m, which can be adjusted later if the fit to a straight line is inadequate. Sakai and Sugiyama [60] developed an analytical method that requires an estimation of the arithmetic mean diameter and the Sauter mean diameter from the sample data. From the ratio of these two diameters, it is possible to estimate m. Thereafter, the graphical method described above can be used to estimate b, or it can be calculated with the aid of charts provided by the authors. Histograms for the two size distributions they considered are shown in Fig. 3.24. The continuous curves are based on Eq. (3.59) with m and b calculated by their methods. Their data have been used to plot cumulative distributions on logarithmic-probability coordinates. These are shown in Fig. 3.25. The data fit the lognormal distribution at least as well as they do the N–T distribution. A similar observation has been reported for droplets from a medical nebulizer [61].

References

1. G. Martin *et al.*, Researches on the Theory of Fine Grinding, *Trans. Brit. Ceram. Soc.*, *23:* 61–109 (1923–4).
2. L. R. Feret, La Grosseur des Grains des Matières Pulverisées, *Ass. Int. pour l'Essai des Mat.*, *2*, *group D:* 428–436 Zurich (1931).
3. H. Heywood, The Scope of Particle Size Analysis and Standardization, *Trans. Inst. Chem. Eng.*, *25:* 14–24 (1947).
4. H. S. Patterson and W. Cawood, The Determination of Size Distribution in Smokes, *Trans. Faraday Soc.*, *32:* 1084–1088 (1936).
5. K. R. May, A New Graticule for Particle Counting and Sizing, *J. Sci. Instrum.*, *42:* 500–501 (1965).
6. B. H. Kaye, Possible Automation Procedures for the Zeiss-Endter Particle Size Analyzer, in T. T. Mercer, P. E. Morrow, and W. Stöber (Eds.), *Assessment of Airborne Particles*, Thomas, Springfield, Illinois, 1972.
7. A. Cauchy, Notes on Various Theorems on the Lengths of Curves and the Areas of Surfaces, *C. R. Acad. Sci. Paris*, *13:* 1060 (1841).
8. V. Vouk, Projected Area of Convex Bodies, *Nature* (*London*), *162:* 330–331 (1948).
9. W. H. Walton, Feret's Statistical Diameter as a Measure of Particle Size, *Nature* (*London*), *162:* 329–330 (1948).
10. V. Timbrell, The Terminal Velocity and Size of Airborne Dust Particles, *Brit. J. Appl. Phys., Suppl.*, *3:* S 86-S 90 (1954).
11. H. H. Watson, *Brit. J. Appl. Phys., Suppl.*, *3:* S 94 (1954).
12. J. K. Donoghue, *Brit. J. Appl. Phys., Suppl.*, *3:* S 94 (1954).
13. A. T. Litvinov, Determination of the Dynamic Shape Factor and Apparent Density of Non-Spherical Aggregated Particles in Aerosols Containing a Solid Disperse phase, *Colloid J.* (*USSR*), *30:* 304–307 (1968).
14. M. Corn, Orientation of Dust Particles During and After Settling in Water, *Ann. Occup. Hyg.*, *6:* 251–266 (1963).
15. J. R. Hodkinson, The Effect of Particle Shape on Measures for the Size and Concentration of Suspended and Settled Particles, *Amer. Ind. Hyg. Ass., J.*, *26:* 64–71 (1965).

16. H. Heywood, A Comparison of Methods of Measuring Microscopic Particles, *Trans. Inst. Min. Met.*, *55:* 391–404 (1946).
17. G. Herdan, *Small Particle Statistics*, 2nd ed., Butterworths, London and Washington, D.C., 1960.
18. B. H. Kaye, Some Aspects of the Efficiency of Statistical Methods of Particle Size Analysis, *Powder Tech.*, *2:* 97–110 (1968).
19. C. N. Davies, Measurement of Particles, *Nature* (*London*), *195:* 768–770 (1962).
20. T. Church, Problems Associated with the Use of the Ratio of Martin's Diameter to Feret's Diameter as a Profile Shape Factor, *Powder Tech.*, *2:* 27–31 (1968/69).
21. W. N. Charman, Some Experimental Measurements of Diffraction Images in Low-Resolution Microscopy, *J. Opt. Soc. Amer.*, *53:* 410–414 (1963).
22. W. N. Charman, Diffraction Images of Circular Objects in High Resolution Microscopy, *J. Opt. Soc. Amer.*, *53:* 415–419 (1963).
23. W. N. Charman, The Production of Circular Discs of Known Diameter for Calibration Purposes in High Resolution Microscopy, *Appl. Opt.*, *1:* 249–251 (1962).
24. J. Dyson, The Precise Measurement of Small Objects, *AEI Eng.*, *1:* 1–5 (1961).
25. V. Timbrell, A Method for Measuring and Grading Microscopic Spherical Particles, *Nature* (*London*), *170:* 318–319 (1952).
26. W. H. M. Robins, The Significance and Application of Shape Factors in Particle Size Analysis, *Brit. J. Appl. Phys.*, *Suppl.*, *3:* S 82-S 85 (1954).
27. T. Hatch and S. Choate, Statistical Description of the Size Properties of Non-Uniform Particulate Substances, *J. Franklin Inst.*, *207:* 369–387 (1929).
28. P. Kotrappa, Shape Factors for Quartz Aerosol in Respirable Size Range, *Aerosol Sci.*, *2:* 353–359 (1971).
29. P. Kotrappa, Shape Factors for Aerosols of Coal, UO_2, and $Th0_2$ in Respirable Size Range, in T. T. Mercer, P. E. Morrow, and W. Stöber (Eds.), *Assessment of Airborne Particles*, Thomas, Springfield, Illinois, 1972.
30. A. R. Steinherz, The Shape of Particles in Finely Ground Powders, *Trans. Soc. Chem. Ind.*, *65:* 314–320 (1946).
31. J. M. DallaValle and F. H. Goldman, Volume-Shape Factor of Particulate Matter, *Ind. Eng. Chem. Anal. Ed.*, *11:* 545–546 (1939).
32. J. Cartwright, Particle Shape Factors, *Ann. Occup. Hyg.*, *5:* 163–171 (1962).
33. K. R. Schrag and M. Corn, Comparison of Particle Size Determined with the Coulter Counter and by Optical Microscopy, *Amer. Ind. Hyg. Ass. J.*, *31:* 446–453 (1970).
34. F. Stein, R. Quinlan, and M. Corn, The Ratio Between Projected Area Diameter and Equivalent Diameter of Particulates in Pittsburgh Air, *Amer. Ind. Hyg. Ass. J.*, *27:* 39–46 (1966).
35. J. M. Beeckmans, The Density of Aggregated Solid Aerosol Particles, *Ann. Occup. Hyg.*, *7:* 299–305 (1964).
36. J. M. DallaValle, Surface Area in Packed Columns, *Chem. Metal Eng.*, *45:* 688–691 (1938).
37. R. J. Hamilton, The Relation Between Free Falling Speed and Particle Size of Airborne Dusts, *Brit. J. Appl. Phys.*, *Suppl.*, *3:* S 90 – S 93 (1954).
38. J. S. McNown and J. Malaika, Effects of Particle Shape on Settling Velocity at Low Reynolds Number, *Trans. Amer. Geophys. Un.*, *31:* 74–82 (1950).
39. N. A. Fuchs, *The Mechanics of Aerosols*, p. 39, MacMillan, New York, 1964.
40. C. N. Davies, Shape of Small Particles, *Nature* (*London*), *201:* 905 (1964).
41. W. Stöber, Dynamic Shape Factors of Non-Spherical Aerosol Particles, in T. T. Mercer, P. E. Morrow, and W. Stöber, (Eds.), *Assessment of Airborne Particles*, Thomas, Springfield, Illinois, 1972.

42. W. Stöber, Statistical Size Distribution Analysis, *Laboratory Invest.*, *14:* 154–170 (1965).
43. H. H. Watson, Dust Sampling to Simulate the Human Lung, *Brit. J. Ind. Med.*, *10:* 93–100 (1953).
44. V. Timbrell, The Inhalation of Fibrous Dusts, *Ann. N.Y. Acad. Sci.*, *132:* 255–273 (1965).
45. W. Stöber, H. Flachsbart, and D. Hochrainer, Der aerodynamische Durchmesser von Latexaggregaten und Asbestfasern, *Staub*, *30:* 277–285 (1970).
46. H. Wadell, The Coefficient of Resistance as a Function of Reynolds Number for Solids of Various Shapes, *J. Franklin Inst.*, *217:* 459–490 (1934).
47. R. A. Fisher, *Statistical Methods for Research Workers*, 10th ed., Hafner, New York, 1948.
48. D. J. Finney, *Probit Analysis*, Cambridge Univ. Press, London and New York, 1947.
49. R. S. Burrington and D. C. May, *Handbook of Probability and Statistics with Tables*, p. 176, McGraw-Hill, New York, 1970.
50. J. H. Gaddum, Lognormal Distributions, *Nature* (*London*), *156:* 463–466 (1945).
51. R. J. Sherwood and D. C. Stevens, Some Observations on the Nature and Particle Size of Airborne Plutonium in the Radiochemical Laboratories, Harwell, *Ann. Occup. Hyg.*, *8:* 93–108 (1965).
52. T. T. Mercer, Effect of Magnification on the Estimation of the Parameters of a Lognormal Distribution, *Amer. Ind. Hyg. Ass. J.*, *31:* 552–564 (1970).
53. T. T. Mercer, Sampling Distributions of Surface and Mass Statistics of a Lognormal Distribution When Estimated by the Method of Weighted Frequencies, *Powder Tech.*, *3:* 65–71 (1970).
54. S. Laskin, in C. Voegtlin and H. C. Hodge, (Eds.), *Pharmacology and Toxicology of Uranium Compounds*, Vol I. pp. 463–505, McGraw-Hill, New York, 1949.
55. D. A. Fraser, Absolute Method of Sampling and Measurement of Solid Airborne Particulates, *A.M.A. Arch. Ind. Hyg. Occup. Med.*, *8:* 412–419 (1953).
56. J. Cartwright and J. W. Skidmore, The Size Distribution of Coal and Rock Dusts in the Electron and Optical Microscope Ranges, *Ann. Occup. Hyg.*, *3:* 33–57 (1961).
57. R. A. Mugele and H. D. Evans, Droplet Size Distribution in Sprays, *Ind. Eng. Chem.*, *43:* 1317–1324 (1951).
58. W. Fritz, Problems in the Experimental Determination of Particle Size and in the Rosin, Rammler and Sperling Graphic Representation of Particle Distribution, *Chem. Z. 83:* 819–823 (1959).
59. S. Nukiyama and Y. Tanasawa, An Experiment on the Atomization of Liquids, *Trans. Japan. Soc. Mech. Eng.*, *5:* 18, 62, 68, 131 (1939).
60. T. Sakai and S. Sugiyama, Application of Gamma Distribution Function on Fine Particles Methods of Determination of Coefficients, *Kagaku Kōgaku*, *3:* 133–134 (1965).
61. T. T. Mercer, M. I. Tillery, and H. Y. Chow, Operating Characteristics of Some Compressed-Air Nebulizers, *Amer. Ind. Hyg. Ass. J.*, *29:* 66–78 (1968).
62. I. Bergman and J. Cartwright, *Silica Powders of Respirable Size: Studies of Particle Shapes and Surfaces*, Safety in Mines Research Establishment Res. Rept. 225 (1964).

4

Measurement of Concentration

Except for certain methods based on light scattering, measurement of the concentration of an aerosol requires the collection of a portion of it for quantitative analysis. To separate particles from air, it is necessary to subject them to a force normal to some collecting surface; either a real applied force or a kinetic reaction in an accelerating flow field. Most of the forces discussed in Chapter 2 have been used for this purpose at one time or another, but only the few that have found fairly frequent application will be included in this discussion.

4-1 Filtration

A convenient and widely used method for collecting aerosol samples consists merely of drawing air through a suitable filter. A wide variety of filters are available, but in terms of physical structure they are either fibrous mats or porous membranes. The properties of fibrous filters have been studied extensively, both theoretically and experimentally, particularly with respect to collection efficiency and resistance to air flow. Membrane filters have been less thoroughly studied, perhaps because they have not been used so widely, their application as air filters having been restricted pretty much to air sampling.

Fibrous filters are made of cellulose (cotton) fibers, a mixture of cellulose and asbestos fibers, plastic fibers, or glass fibers. Their performance is closely related to fiber diameter, the smaller fibers having better collection properties. In the cellulose-asbestos filters, the cellulose fibers serve to give the filter strength, but the filtering action is due mainly to the small asbestos fibers. Filters of glass fibers having diameters of 1 μm or less are becoming increasingly popular for air sampling purposes.

Membrane filters are the dried residues of gels produced from colloidal suspensions of cellulose esters. They appear as sheets perforated with pores of irregular size and shape running irregular courses through them. The pores differ in diameter, but their size distribution is relatively narrow, clustering about a mean value that can be varied over a rather wide range during manufacture. Metal membrane filters, produced by an analogous process of powder metallurgy, are available also, but their application in air sampling, while promising, has been limited.

4-1.1 Fibrous Filters

(a) *Theoretical Considerations.* The theoretical study of fibrous filters begins with consideration of a single, isolated fiber, which is taken to be a straight cylinder of uniform cross section, set with its longitudinal axis perpendicular to the direction of air flow. It is assumed that all particles coming into contact with the fiber adhere to it. The particles themselves are assumed to be spheres. The efficiency with which the fiber removes particles from air flowing around it is then a function of the air flow pattern, the fiber diameter, the particle's density and diameter, and the nature of any charge on the particle, the fiber, or both.

Air flow patterns differ with the Reynolds number of flow about the fiber. The source of their effect on collection is apparent from Fig. 4.1, which shows streamlines of flow about a cylinder for Re = 0.2 and 10 and for an ideal (nonviscous) fluid. In that figure the coordinates are scaled in units of the fiber diameter and only one quadrant of the fiber cross section is shown. Each streamline passes through the point (2.0, 0.25). As Reynolds number increases, the flow pattern approaches that of ideal flow in which the streamlines are not significantly deflected until quite close to the cylinder. This flow favors collection of particles by impaction or interception.

If there are no electrostatic forces present, the collection of particles by a

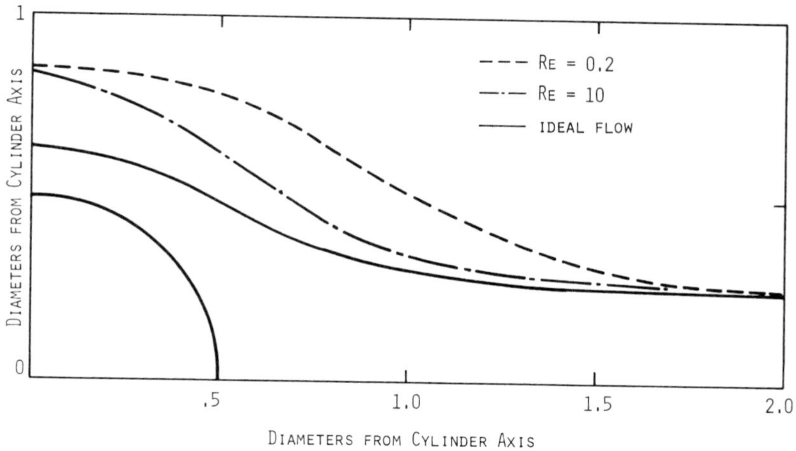

FIG. 4.1. Effect of Reynolds number on streamlines about a cylinder.

fiber will be due primarily to interception, impaction, or diffusion. Interception is the capture of a particle of negligible inertia because the streamline along which it moves passes the fiber within a distance equal to the particle's radius. Particles having significant inertia do not follow precisely the streamline with which they move initially, but diverge from it to a degree that depends on their size and density. The process is shown schematically in Fig. 4.2. The degree of divergence from its original streamline depends on the particle's stopping distance, which in turn depends on its mass and mobility, the upstream velocity U_0, and on the abruptness with which the streamlines change direction to pass around the fiber. For a given particle and a given fiber, Reynolds number also increases with U_0 and the streamlines change direction more abruptly, so that high values of Re greatly favor collection of particles by this process of impaction. Neither interception nor impaction is very effective when the particle is very small. In this case, diffusion due to Brownian motion becomes important, and the probability that a given particle will be captured by a fiber depends on the diffusion coefficient of the particle and the time in transit around the fiber. Each of these effects (which are discussed in more detail below) acts on all particles moving around a cylinder. When one

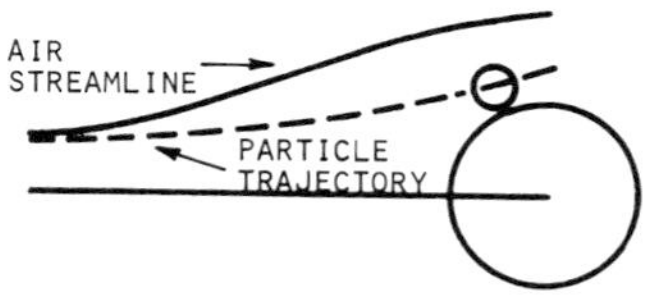

FIG. 4.2. Particle trajectory near a cylinder.

is discussed to the exclusion of the others, it means that the latter are negligible in the circumstances under consideration.

When filters are employed in air sampling, their effectiveness is spoken of in terms of "efficiency," since one is interested in the fraction of available particles that they retain. For air cleaning, however, the term "penetration" (which equals 1.0 – efficiency) is preferred because the amount passing through the filter is of primary interest. Order of magnitude changes in penetration may represent a change in efficiency that is insignificant in air sampling.

Before going into detail about the individual mechanisms, it will be necessary to define efficiency of collection. If the fiber is arranged perpendicular to the direction of flow, it will present to the flow a cross-sectional area equal to the fiber diameter times its length. Assume that all particles of a diameter D_p passing through the area defined by a width b and the total length of the fiber are captured by the fiber. If b is taken far enough upstream from the fiber for the flow lines to be essentially parallel, then the collection efficiency for the single, isolated fiber is defined as the ratio of the cross-sectional area of the original stream from which all particles are removed to the projected area of the fiber in the direction of the flow, i.e., it is b/D_f. The problem of estimating collection efficiency then becomes a matter of determining b under a given set of circumstances.

(b) *Interception.* A particle of diameter D_p will contact the fiber if its center comes within $D_p/2$ of the fiber. If the spreading of streamlines is ignored, the maximum efficiency possible due to interception is

$$E_R = b/D_f = (D_p/2 + D_f + D_p/2)/D_f = 1 + (D_p/D_f) = 1 + R,$$

where R is called the interception parameter. The efficiency due to interception can be calculated exactly for ideal flow. In this case, the equation of the streamlines about the fiber is

$$y/y_0 = (x^2 + y^2)/(x^2 + y^2 - 1),$$

where y_0 is the initial distance of the streamline from the axis of flow measured in units of the fiber radius. The streamline is closest to the cylinder when $x = 0$, at which point

$$y_0 = y - 1/y.$$

If $y \leqslant (D_f + D_p)/D_f$, then all particles approaching the cylinder initially at a value of $y_0 \leqslant 1 + R - 1/(1 + R)$ will be captured. The collection efficiency is, therefore,

$$E_R = b/D_f = y_0 = 1 + R - 1/(1 + R).$$

As R gets large, the efficiency approaches the maximum possible value of $1 + R$. There is no general expression for the interception effect as a function of

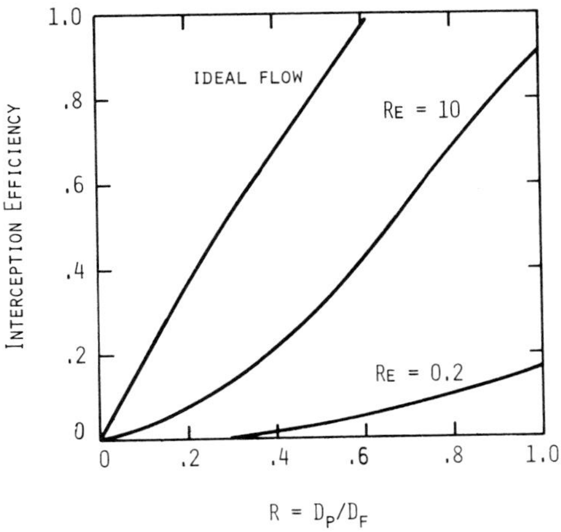

FIG. 4.3. Interception efficiency at different Reynolds numbers.

Reynolds number. It has been calculated by Davies and Peetz [2], using stepwise solutions of the flow patterns, for the three flow conditions shown in Fig. 4.1. Their results are shown in Fig. 4.3.

(c) *Impaction.* Impaction is particularly important for particles in the respirable size range. Collection efficiency usually has been calculated on the assumption that the trajectory of the particle's center must intersect the cylinder (i.e., $D_p/D_f = 0$). When the air flow begins to deflect, the particle's inertia produces a relative velocity between particle and air:

$$m \cdot d\overline{U}/dt = (\overline{V} - \overline{U})/Z, \tag{4.1}$$

where $\overline{U}$ is the vector velocity of the particle, $\overline{V}$ is that of the air, and Z is the particle's mobility. In a two-dimensional situation, such as this, Eq. (4.1) can be put in terms of the x and y components:

$$m\, dU_x/dt = (V_x - U_x)/Z,$$

and (4.2)

$$m\, dU_y/dt = (V_y - U_y)/Z.$$

Initially, $U_x = V_x = U_0$, the velocity of the undisturbed air stream approaching the cylinder, and $U_y = V_y = 0$. Since these equations must be solved by stepwise methods, it is wise to reduce the amount of computation needed by putting the variables into dimensionless form. This is done by measuring velocities in units of U_0, distances in units of $D_f/2$, and time in units of the time required to traverse a distance equal to the fiber radius, when moving at the velocity U_0:

$$\tilde{U}_x = U_x/U_0, \qquad \tilde{U}_y = U_y/U_0,$$
$$\tilde{V}_x = V_x/U_0, \qquad \tilde{V}_y = V_y/U_0, \tag{4.3}$$
$$\tilde{t} = t/(D_f/2U_0) = 2U_0 t/D_f.$$

Equations (4.2) then become

$$\text{Stk} \cdot (d\tilde{U}_x/d\tilde{t}) = \tilde{V}_x - \tilde{U}_x,$$

and (4.4)

$$\text{Stk} \cdot (d\tilde{U}_y/d\tilde{t}) = \tilde{V}_y - \tilde{U}_y.$$

The quantity Stk is called the Stokes number* (formerly, inertial or impaction parameter) of the particle and is equal to

$$\text{Stk} = 2mZU_0/D_f. \tag{4.4a}$$

For a particle moving initially at a velocity U_0, it is twice the ratio of the particle's stopping distance to the diameter of the cylinder. In this form, the solution of the equations for a given value of Re yields a plot of collection efficiency as a function of Stk that is valid for any combinations of U_0, D_f, D_p, and ρ_p for which the same Reynolds number applies. It must be pointed out that in some cases Stk is defined as mZU_0/D_f, so that it should be examined carefully when encountered in the literature.

Davies and Peetz [2] also made extensive calculations of the impaction effect. Their results for collection efficiency as a function of Stokes number for ideal flow and for Re = 0.2 and 10 are shown in Fig. 4.4(A). These curves were calculated under the assumption that $D_p/D_f = 0$. Davies and Peetz also took the finite size of the particle into consideration and obtained the results in Fig. 4.4(B) for various values of the interception parameter R at Re = 10.

(d) *Diffusion.* The diffusion mechanism, which is mainly responsible for the capture of very small particles, is characterized by a dimensionless parameter:

$$B = \Delta/D_f U_0, \tag{4.5}$$

where Δ is the diffusion coefficient of the particle. The corresponding collection efficiency is difficult to assess; experimental data for this effect are scarce, so

* When a dimensionless parameter has acquired sufficient prestige, it is dubbed a number, and given an appropriate name and symbol, the latter usually being the first two letters of the name. Stokes number is an exception, probably because of possible confusion with other abbreviations.

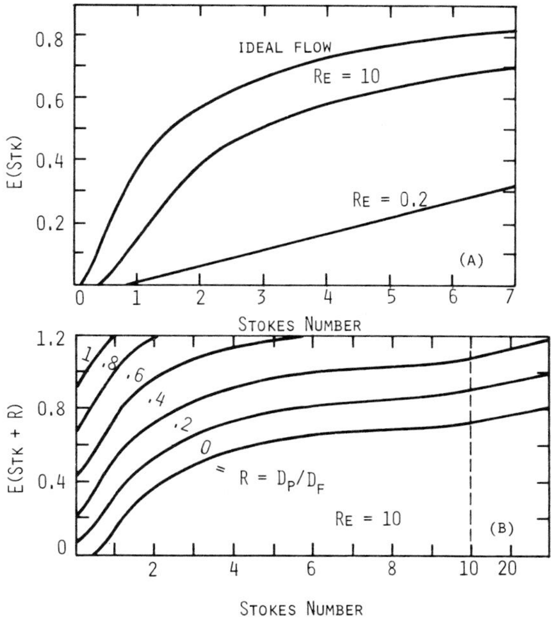

FIG. 4.4. Collection efficiencies of single cylinder as a function of Stokes number [2]. (A) For $D_p/D_f \simeq 0$; (B) for Re = 10. Courtesy of The Royal Society, London.

that theoretical results cannot be judged properly. For flow patterns similar to those of ideal flow, it appears that [3]

$$E_B \propto B^{1/2},$$

while for viscous flow,

$$E_B \propto B^{2/3}.$$

In the latter case, the proportionality factor is not a true constant, but varies slowly with Reynolds number. In both cases, different authors have deduced different numerical values for the proportionality factor [3]. Results involving diffusion are frequently related to the Peclet number, $\mathrm{Pe} = 1/B$.

Pasceri and Friedlander [7] were able to get a reasonably good correlation between experimental data and an expression based on the following relationships for single fiber efficiency by diffusion and by interception:

$$E_B \propto \mathrm{Re}^{1/6} \cdot B^{2/3} \qquad \text{and} \qquad E_R \propto \mathrm{Re}^{1/2} \cdot R^2.$$

Kirsh and Fuchs [36] found that the single fiber diffusional collection efficiency for arrays of cylinders having packing densities between 0.01 and 0.15 was

$$E_B = 2.7B^{2/3}.$$

(e) *Effect of Electrostatic Forces.* For a particle of charge q_p, approaching an uncharged fiber of diameter D_f and dielectric constant ξ, Kraemer and Johnstone [9] defined the following dimensionless parameter:

$$K_E = q_p^2(\xi - 1)/[3\pi\eta D_p D_f^2 U_0(\xi + 1)]. \tag{4.6}$$

This parameter depends on the ratio between the force acting on the particle due to its electrical image in the fiber and the drag force acting on it at a relative velocity U_0. Lundgren and Whitby [10] concluded that for viscous flow around an isolated cylinder of large dielectric constant, the efficiency of collection due to this force was

$$\begin{aligned} E_s &= 1.5K_E^{1/2} \\ &= 0.017 \cdot n/(D_f^2 D_p U_0)^{1/2}, \end{aligned} \tag{4.7}$$

where n is the number of charges on the particle, U_0 is in cm/sec, and the diameters of fiber and particle are in micrometers. For an aerosol at charge equilibrium, $n = 2.4 \cdot D_p^{1/2}$ and

$$E_s = 0.041/(D_f^2 U_0)^{1/2}. \tag{4.8}$$

The electrostatic effect introduces an uncontrolled variable into many studies of filtration efficiency. The DOP aerosols, produced by controlled condensation of a vapor, are formed essentially without charge, and it is likely that tests run with them are over before they can accumulate any appreciable charge. In some studies [11], the charge on latex particles as a result of their aerosolization by atomization from suspension has been dissipated in high concentrations of bipolar ions, leaving them close to the natural equilibrium charge. In others [12], no effort was made to discharge the particles, so that they undoubtedly carried high charges, thus leading to strong electrostatic forces.

(f) *Overall Collection Efficiencies.* Efforts have been made to express the overall collection efficiency in the form

$$E_T = f(R, \mathrm{Stk}, B, \mathrm{Re}),$$

but they are all "subject to certain reservations" [3]. In practice, one of the dimensionless parameters usually will be much larger than the others, and overall efficiency will be closely related to the collection efficiency for the corresponding process. For the range of particle sizes most likely to be of concern in hazard evaluation, experimental collection efficiencies correlate most readily with theoretical results for combined impaction and interception.

When the results for isolated cylinders are applied to arrays of cylinders, screens, or filters, the predicted efficiencies are sometimes grossly in error. This is due largely to the fact that at a given Reynolds number the flow pattern around an isolated cylinder differs significantly from that around a similar cylinder in an array of other cylinders. In effect, the presence of neighboring cylinders compresses the streamlines, limiting the flow of air to a much smaller region around the cylinder. Happel [4] has allowed for this effect in his cellular model for flow through an array of cylinders. Flow related to a given cylinder is considered to be confined to a symmetrical "cell" about the cylinder, the dimensions of the cell depending on the distance between cylinders. For fibrous filters, this distance is related to the packing density α which is given by

$$\alpha = \text{fiber volume/total volume} = 1 - p,$$

where p is the porosity of the filter. Happel's model permits calculation of the flow field around a cylinder, assuming fluid inertial forces are negligible, i.e., $\mathrm{Re} \simeq 0$. Harrop and Stenhouse [5] adapted the model to the problem of impaction efficiencies of single fibers in filters of different packing densities. Their results are shown in Fig. 4.5, which also includes the efficiency curves calculated by Davies and Peetz. The authors point out that the assumption

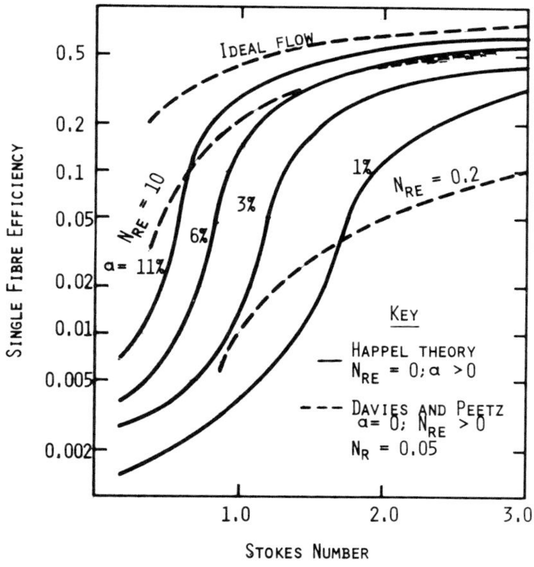

FIG. 4.5. Impaction collection efficiencies of single fibers in filters of different packing densities [5]. N_{Re} is the Reynolds number; α is the packing density; N_R is the interception parameter. Courtesy of Maxwell International Microforms Corp.

that Re = 0 tends to underestimate the collection efficiency, particularly at low Stokes numbers.

The results of the calculations of collection efficiency for a single fiber can be extended to a fibrous mat in the following way. Suppose that a number N of particles are incident on an element of the filter of thickness dH and cross-sectional area A. The total length of fiber contained in the filter element is

$$L = 4\alpha A \cdot dH/\pi D_f^2. \tag{4.9}$$

Air emerging from the previous element of filter is confined to the area $(1-\alpha)\cdot A$. All particles will be removed from an area equal to $E_s L D_f$, where E_s is the total efficiency with which particles are collected. Then the number of particles captured in the thickness dH is

$$-dN = \frac{NE_s LD_f}{(1-\alpha)A} = \frac{\alpha}{(1-\alpha)} \cdot \frac{4E_s}{\pi D_f} \cdot N \cdot dH = \gamma N \cdot dH, \tag{4.10}$$

and the number remaining in the air after it passes through the full filter is

$$N = N_0 \exp(-\gamma H), \tag{4.11}$$

where N_0 is the number of particles incident on the upstream side of the filter, H is the thickness of the filter, and γ is called its index of filtration. The efficiency of the filter as a whole, when collecting particles for which the single fibers have a collection efficiency E_s is

$$E = 1 - N/N_0 = 1 - \exp(-\gamma H). \tag{4.12}$$

If the fibers have a distribution of diameters, as is usually the case, the fiber diameter in the expression for γ should be replaced with the ratio $\overline{D_f^2}/\overline{D}_f$ of the mean square fiber diameter to the mean fiber diameter [6], or with an empirical effective diameter, D_{fe} [1].

The criterion for judging a filter has been defined [6] as the ratio

$$\frac{\gamma H}{\Delta P} = \frac{4\alpha E_s D_{fe}}{\pi(1-\alpha)\eta U_0 \cdot f(\alpha)}, \tag{4.13}$$

where ΔP is the pressure drop across the filter, and D_{fe} and $f(\alpha)$ are defined in Sect. 4.1.1g. Judging solely for filtration properties, the larger this ratio the better the filter. For values of α of practical significance, this ratio diminishes as α increases, if the other quantities are kept constant. For fixed values of D_{fe} and U_0, the particle size of the sampled aerosol affects the ratio exactly as it does E_s. The effects of changing D_{fe} and U_0 are not readily generalized, since both affect E_s.

Comparisons between theoretical efficiencies and experimental efficiencies have generally shown only qualitative agreement, if that [6]. That this is not entirely the fault of theory is apparent from the frequent disagreement between experimental results for the same filter [80]. Some examples of experimental data plotted with the corresponding theoretical efficiency curves are shown in Figs. 4.6(A), (B). The theoretical calculations are based on the equations by Stechkina *et al.* [8] for the overall collection efficiency of a fibrous filter.

Part of the observed differences between experimental results on a given filter are doubtless due to differences in the charge characteristics of the particles. This cannot be the case for DOP aerosols, however, which are almost devoid of charge. The agreement between experimental and theoretical efficiencies is not all that one might desire. However, the results serve to show that differences in filter characteristics, for filters presumed to be the same, lead to similar changes in theoretical and experimental efficiencies, indicating the need to report the results of efficiency tests in terms of α and $\Delta P/U_0$. Some of the available data on filter properties are given in Table 4.1.

(g) *Pressure Drop.* For air sampling purposes, the amount of energy required to move air through a filter is much less important than it is when filtering air to clean it. It is still a matter of some interest, however, particularly with respect to the manner in which it changes as deposits build up in the filter. There have been two theoretical approaches used in estimating pressure drop; one is based on the assumption that a filter is a porous mat for which the pressure drop–flow rate characteristics can be predicted from D'Arcy's law; the other attempts to predict pressure drop in terms of the summed viscous drag on the individual fibers. In either case, the relationship takes the form

$$\frac{\Delta P}{U_0} = \frac{\eta H \cdot f(\alpha)}{D_{fe}^2}, \tag{4.14}$$

at a linear flow rate U_0 into a filter of thickness H. On the basis of available experimental data, Davies [1] concluded that if $\alpha < 0.02$,

$$f(\alpha) = 64\alpha^{1.5}(1+56\alpha^3); \tag{4.15}$$

otherwise,

$$f(\alpha) = 70\,\alpha^{1.5}(1+52\alpha^3). \tag{4.16}$$

These equations permit calculation of D_{fe} from filter dimensions and pressure drop–flow rate data. Estimates of the effective fiber diameters for some of the filters used in sampling are shown in Table 4.1.

The pressure drop across a filter builds up with time as the filter becomes loaded with particles. With the exception of some recent work by Davies [13], the subject has received little theoretical attention, perhaps because it appears to be essentially a practical problem. Davies showed that the efficiency of a filter increases with time according to

$$\ln\{(1-E_t)/(1-E_0)\} = -[(\Delta P_t/\Delta P_0)^{1/2}-1]\cdot\gamma H. \tag{4.17}$$

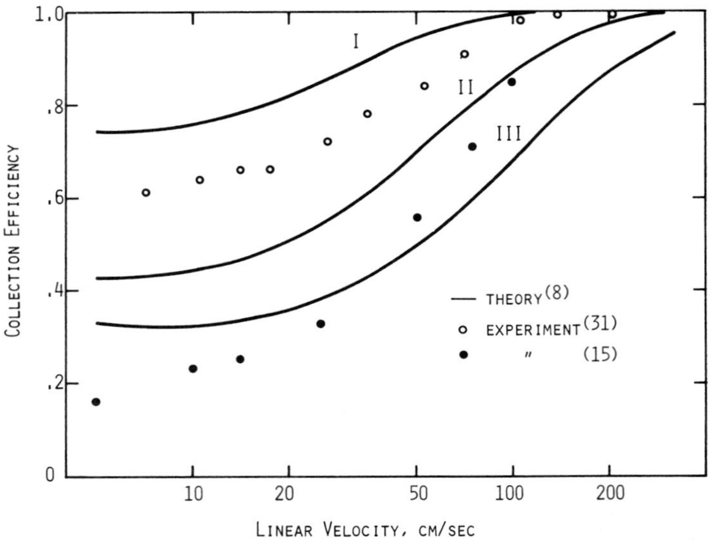

1.0
.8
.6
.4
.2
0
COLLECTION EFFICIENCY
I
II
III
THEORY(8)
EXPERIMENT(31)
" (15)
10
20
50
100
200
LINEAR VELOCITY, CM/SEC

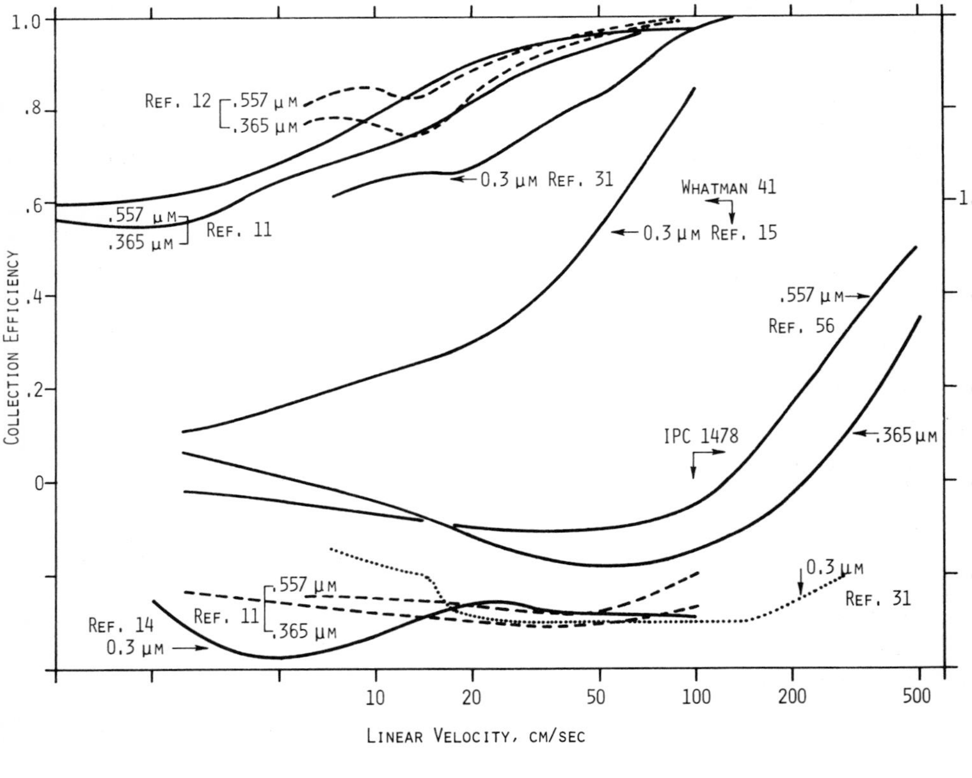

1.0
.8
.6
.4
.2
0
COLLECTION EFFICIENCY
REF. 12
.557 μM
.365 μM
.557 μM
.365 μM
REF. 11
0.3 μM REF. 31
WHATMAN 41
0.3 μM REF. 15
.557 μM
REF. 56
IPC 1478
.365 μM
.557 μM
REF. 11
.365 μM
REF. 14
0.3 μM
0.3 μM
REF. 31
10
20
50
100
200
500
LINEAR VELOCITY, CM/SEC

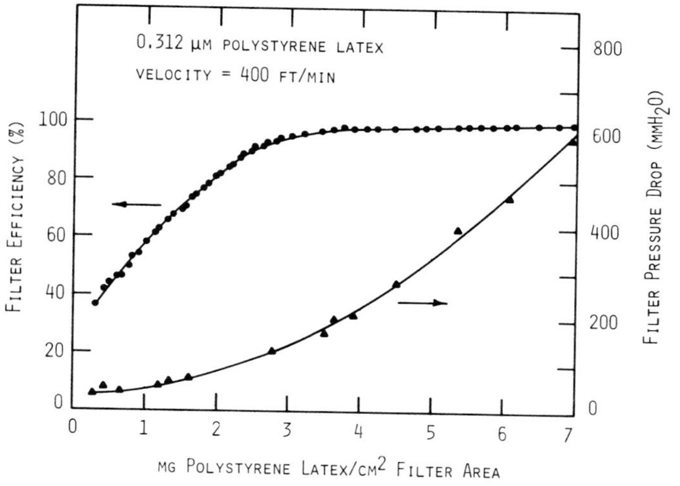

FIG. 4.7. Collection efficiency and pressure drop as a function of mass of particles incident on filter [14].

$E_0, \Delta P_0$ and $E_t, \Delta P_t$ are the filter efficiency and pressure drop at times zero and t, respectively, H is the filter thickness, and γ is the filter's index of efficiency discussed above. He also derived the rate at which the pressure drop changed with time. His equation, modified to relate the changing pressure to the accumulated weight, W, presented to the filter per unit area, is

$$d\,\Delta P/dW = k \cdot \Delta P. \tag{4.18}$$

The constant k depends on flow rate and the physical characteristics of both the filter and the particles being collected. These forms of Davies' equations contain the implicit assumption that flow rate is constant during the loading process. Integration of Eq. (4.18) yields

$$\ln(\Delta P_t/\Delta P_0) = k \cdot (W_t - W_0). \tag{4.19}$$

FIG. 4.6. Top: Collection efficiencies for Whatman 41 filter paper for 0.3 μm particles. Data for theoretical calculations are tabulated below:

Curve	H (mm)	α	$\Delta P/U$	D_f (μm)	D_p (μm)
I	0.25	0.245	0.93	2.8	0.3
II	0.18	0.350	0.38	6.4	0.3
III	0.10	0.245	0.38	2.8	0.3

Bottom: Experimental collection efficiencies for Whatman 41 and IPC 1478 filters.

TABLE 4.1

PHYSICAL CHARACTERISTICS OF SOME COMMON FILTERS

Filter	Type	Material	Thickness (mm)	Fiber or pore diameter (μm)	Packing density	$\Delta P/U_0$ $\left(\frac{\text{cm } H_2O}{\text{cm/sec}}\right)$	Effective[a] fiber diameter(μm)	Ref.
Whatman 40	Fiber	Cellulose	0.15	—	0.44	2.59 ± 0.08	3.5	[21, 15]
Whatman 41	Fiber	Cellulose	0.18	—	0.35	0.38 ± 0.04	6.4	[21, 15]
Whatman 41	Fiber	Cellulose	0.25	—	0.26	0.93 ± 0.01	3.0	[31]
Whatman 42	Fiber	Cellulose	0.18	—	0.38	8.38 ± 0.26	1.6	[21, 15]
Whatman 50	Fiber	Cellulose	0.10	—	0.69	7.35 ± 0.30	4.3	[21, 15]
Esparto	Fiber	Cellulose	1.37	—	0.13	0.39 ± 0.01	4.9	[31]
IPC–1478	Fiber	Cellulose	0.76	—	0.13	0.037 ± 0.003	11.8	[21, 31]
Fourstones Sample A	Fiber	Cellulose-asbestos	0.35	19	0.26	0.25	6.8	[20]
HV–70	Fiber	Cellulose-asbestos	0.23	0.1–35	0.23	1.09 ± 0.07	2.2	[15, 21]
HV–70	Fiber	Cellulose-asbestos	0.46	0.1–35	0.21	1.26 ± 0.05	2.6	[15, 21]
Microsorban	Fiber	Polystyrene	1.27	0.6–0.8	0.04	0.75	1.3	[21]
Whatman AGFA	Fiber	Glass	0.34	0.7	0.06	0.49	1.1	[20, 21]
Millipore AA	Membrane	Cellulose ester	0.15	0.80	0.19	2.2–4.6	—	[15, 21]
Millipore HA	Membrane	Cellulose ester	0.15	0.45	0.20	5.6–10.5	—	[15, 21]
Gelman AM–4	Membrane	Cellulose ester	0.20	0.80	0.11	3.65	—	[21]
Gelman AM–6	Membrane	Cellulose ester	0.20	0.45	0.17	—	—	[21]
Flotronics	Membrane	Silver	—	3.0	0.42	—	—	
Flotronics	Membrane	Silver	—	0.45	0.62	—	—	
Flotronics	Membrane	Silver	—	0.80	—	4.1	—	[24]

[a] Calculated using Eqs. (4.14) and (4.16).

These two equations provide information with which to estimate the average efficiency over the life of a filter.

Some data recently published by Stafford and Ettinger [14] provide an example of the effect of loading on filtration. Figure 4.7 shows the course of filter efficiency and pressure drop for IPC-1478 filter collecting 0.312 μm polystyrene latex particles. For each point, the authors tabulated the accumulated weight per unit area that had been presented to the filter. Their data plotted according to Davies' equations are shown in Fig. 4.8. Of the initial

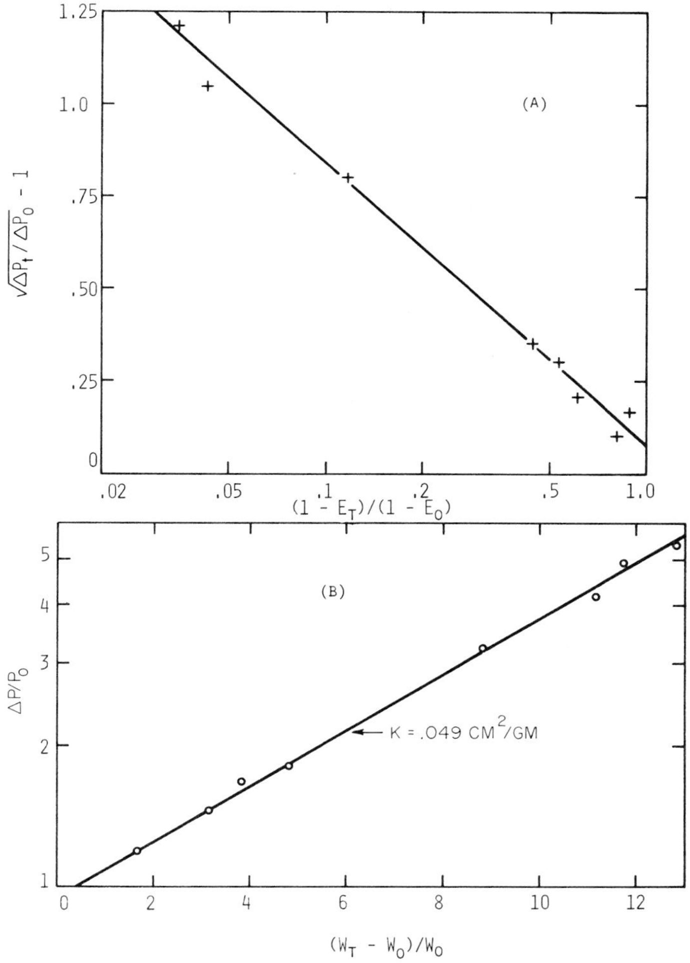

FIG. 4.8. Loading characteristics of IPC 1478 filter [14]. 0.312 μm particles, 203 cm/sec. Data plotted according to (A) Eq. (4.17); (B) Eq. (4.19).

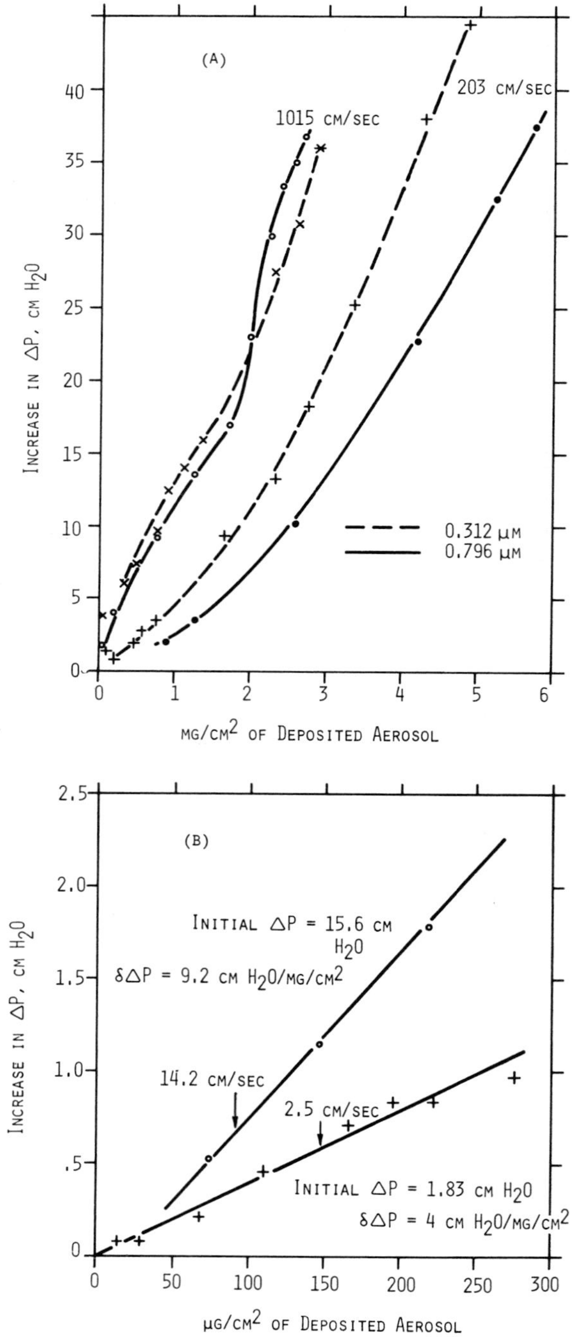

FIG. 4.9. Pressure increase with loading of filter. (A) IPC 1478 filter [14]; (B) CWS–6 filter [15].

incident weight, W_0, approximately 0.1 mg, corresponding to a particulate surface of collected material almost equal to the total fiber surface, had been retained on the filter. Thereafter, the filter properties of pressure drop and efficiency are characteristic of the collected particles rather than the fibrous mat. This interpretation is supported by the value of γ obtained from Fig. 4.8(A). For a filter thickness of 0.084 cm [14], γ equals 36.2, a value which would require impossibly high single fiber efficiencies for particles of 0.312 μm diameter. The value is reasonable, however, for the collection efficiency of spheres of this size. Further support for the interpretation comes from the observation by the same authors that the IPC filter, operating at approximately the same flow rate, shows no change in either collection efficiency or pressure drop while collecting 1.9 mg (equivalent to $6.7W_0$) of 0.3 μm diameter particles of DOP, a liquid which apparently wets the fibers.

The effects of flow rate and particle size of the collected substance are brought out in Fig. 4.9. These curves show the increase in pressure drop as a function of filter loading for IPC-1478 [14] and CWS-6 [15] filters. The former include data for polystyrene particles of 0.312 and 0.796 μm diameters; the latter are based on sampling of natural airborne aerosols.

4-1.2 Membrane Filters

Theoretical estimates of the collection efficiency of the membrane filter have been based on the assumption that the filter approximates a sheet perforated with straight, parallel, equidistantly spaced, identical circular pores (see Fig. 4.10) [58]. This model led to the following expression with which to calculate the total collection efficiency, E_T:

$$\begin{aligned} E_T &= E_I + E_D - E_I \cdot E_D, \\ E_I &= 2 \cdot E^* - E^{*2}, \\ E^* &= 2y(1-\sqrt{p})[1-y(1-e^{-1/y})], \\ y &= \mathrm{Stk} \cdot [\sqrt{p}/(1-\sqrt{p})]^{1/2}, \\ E_D &= 1 - \sum a_i \cdot \exp(-b_i N_D). \end{aligned} \tag{4.20}$$

In this case, the Stokes number [Eq. (4.4a)] is related to the pore diameter, p is the porosity of the filter, and $N_D = \pi H\Delta/Q_1$ is the diffusion parameter for particles of diffusivity Δ moving at a volumetric flow rate Q_1 through a

Fig. 4.10. Model of a membrane filter surface [58]. Circles represent pore openings. Courtesy of Staub-Reinhaltung der Luft.

pore of length H. For a total filter area A_f and pores of area A_1, the flow rate through a single pore, at a total filter flow rate Q_f is

$$Q_1 = Q_f A_1 / pA_f.$$

The summation in the equation defining E_D is given by the formula for penetration through a tube of circular cross section [Eq. (7.13)] by replacing μ in that formula with $2N_D$. At large values of Stk, $E_I \rightarrow (1-p) \simeq 0.1$–$0.3$ for membrane filters in common use.

Figure 4.11. shows theoretical efficiency curves for three sets of conditions

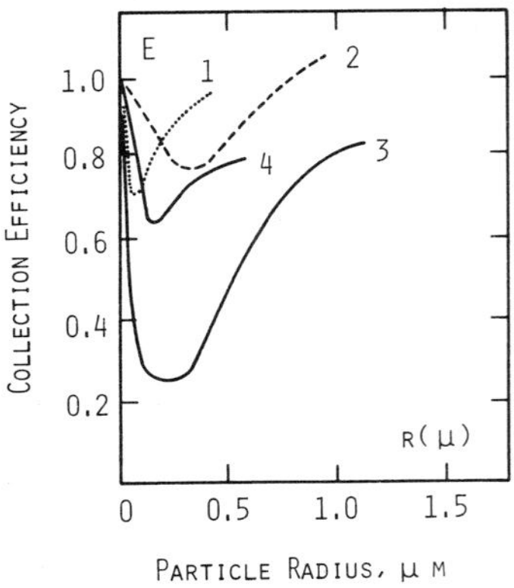

FIG. 4.11. Efficiency as a function of particle size for membrane filter having mean pore diameter of 1.8 μm [59]. Curve 3 ($L = 148$ μm, $q = 7$ cm/sec, $P = 0.78$, calculated curve). Curve 2 ($L = 300$ μm, $q = 7$ cm/sec, $P = 0.78$, calculated curve). Curve 1 ($L = 300$ μm, $q = 25$ cm/sec, $P = 0.78$, experimental curve). Curve 4 ($L = 300$ μm, $q = 25$ cm/sec, $P = 0.78$, calculated curve). Courtesy of Maxwell International Microforms Corporation.

and an experimental curve for one set. Agreement between theory and experiment was believed to be reasonable [59]. The calculated results, however, are difficult to reconcile with a maximum value for E_I of 0.22. The theoretical model is inadequate to deal with impaction, because it permits a fraction p of the total sample to pass directly through the capillaries without being subjected to any deflection due to their inertia. In reality, the pores follow a tortuous course through a filter and both impaction and interception must be important factors as the incoming air deflects to follow a pore.

Early experimental results on Millipore AA and HA membrane filters showed efficiencies as low as 0.44 for the former when used to sample 0.02 μm particles at a face velocity of 40 cm/sec [60]. Later studies [27, 61] have refuted this work, which is fortunate, since membrane filters of appropriate pore size are generally accepted as standards with which to compare the sampling efficiency of other filters.

The dielectric properties of membrane filters permit them to acquire a high electrostatic charge, which has been thought to contribute significantly to their collection efficiencies [22]. However, Megaw and Wiffen [61] concluded that efficiency did not depend on electrostatic forces for particles having diameters $\gtrsim 0.4$ μm. Binek and Przyborowski [62] reported that uncharged membrane filters permitted significantly greater penetration by particles having diameters in the range from 0.04 to 0.7 μm. However, at an input concentration of 4×10^4 particles/cm^3, penetration was of the order of 10 particles/cm^3, so that the significant changes in penetration had negligible effect on efficiency. As far as collection efficiency is concerned, the electrostatic characteristics of membrane filters apparently represent a redundant advantage.

The pressure drop across a membrane filter can be estimated using Eq. (4.14) if the right-hand term is divided by a slip factor, $1 + 0.23/D_{fe}$, where D_{fe} is the pore diameter in micrometers [63]. For Millipore AA and HA filters, the calculated values of $\Delta P/U$ are respectively 2.7 and 8.1 cm H_2O/cm/sec, compared to average experimental values of 2.2–4.6 and 5.6–10.5.

4-1.3 Nuclepore Filters

The passage of a fission fragment through a thin plastic sheet leaves behind a trail of damaged plastic that is more easily attacked by sodium hydroxide solution than is the intact plastic. The number of trails produced in a given sheet is equal to the number of fragments that are allowed to traverse it. The size of the hole etched around the trail depends on the concentration of reagent in the solution to which the film is exposed and the duration of the exposure. By controlling these factors, it is possible to produce plastic sieves having cylindrical pores of very uniform diameter distributed randomly over the sieve surface [79]. The open area of the sieve must be kept below about 2% of its total area if overlap of pores is to be avoided. For the most part, the axis of the pore is nearly normal to the sieve surface. Data on some of the sieves that are available commercially are given in Table 4.2.

TABLE 4.2

Characteristics of Nuclepore Filters[a]

Pore diameter (μm)	Pores (cm^{-2})	Thickness (μm)	Porosity
0.2	$(2.55–3.45) \cdot 10^8$	12	0.094
0.4	$(0.85–1.15) \cdot 10^8$	12	0.126
0.6	$(2.55–3.45) \cdot 10^7$	12	0.085
0.8	$(2.55–3.45) \cdot 10^7$	12	0.151
1.0	$(1.70–2.30) \cdot 10^7$	12	0.158
3.0	$(1.70–2.30) \cdot 10^6$	12	0.141
5.0	$(3.40–4.60) \cdot 10^5$	11	0.079
8.0	$(0.85–1.15) \cdot 10^5$	11	0.050

[a] Data provided by S. C. Furman, Nuclear Energy Division, General Electric Company.

Equations (4.20) predict the performance of nuclepore filters more accurately than they do the performance of membrane filters because the structure of the former is more similar to that shown in Fig. 4.10. The curves of efficiency as a function of particle size are similar to those of Fig. 4.11, so that the nuclepore filter has some useful size-selective characteristics.

4-1.4 Important Secondary Filter Characteristics

The ability of a filter to collect particles efficiently and with a minimum energy requirement is not the only criterion of a filter's usefulness. After sampling has ceased or, in some cases, while it is still in progress, it is necessary to analyze the collected material for the presence of the toxic agent of interest. The available methods include gravimetric, chemical, and spectographic analysis, radioassay, optical analysis of the deposit by absorption [16] or reflectance [17] methods, measurement of x-ray diffraction [18], neutron activation analysis, and measurement of beta ray absorption by the deposit [19]. The choice of analytical method depends on the substance to be assayed and the sensitivity required to assay it in the amount expected to be collected. The choice of filter must include a consideration of the extent to which it facilitates the analytical method.

Gravimetric analysis requires a filter that shows no weight change other than that due to the collected particles. The most serious source of interference is water vapor. At 100% relative humidity, cellulose fiber papers show an increase of about 17% [20] over their dry weight, compared to 0.1% for glass fiber papers. Membrane filters, either of vinyl esters or metal, show no effect of humidity and those of cellulose esters show very little effect. For radioassay or chemical analysis of inorganic substances requiring separation of the deposit from the filter, the ash content or the resistance of the filter to strong acids or bases is important. Data concerning ash content have been tabulated by Smith and Surprenant [15] and by Lippmann [21]. Strong reagents can be used to dissolve samples from cellulose ester membrane filters without disintegration of the filter [22]. Separation of organic samples from filters by means of benzene extraction requires that the filter contribute only a negligible quantity to the extract. Pate and Tabor [23] have summarized the characteristics of a variety of glass fiber filters in this regard. Silver metal membrane filters have been found particularly useful for collection of coal tar pitch volatiles [24, 25] because the change in filter weight due to benzene extraction was less than 0.04 mg. The metal filters were very much better for this purpose than ordinary membrane filters or filters containing glass or cellulose fibers.

The advantages exhibited by cellulose ester membrane filters in x-ray diffraction analysis [18] are even more apparent with metal membrane filters.

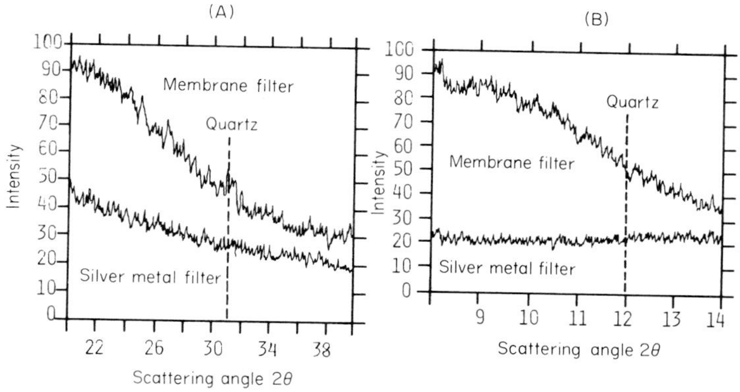

FIG. 4.12. Patterns of x-ray scattering by regular and silver membrane filter blanks [26]. (A) Cobalt x rays, 1.79 Å, 27 kV, 20 mA; (B) molybdenum x rays, 0.71 Å, 50 kV, 20 mA. The ordinate is intensity. Courtesy of Staub-Reinhaltung der Luft.

Leroux and Powers [26] have demonstrated the superiority of the latter for the analysis of quartz samples. Figure 4.12 shows the intensity of diffracted x rays from unused metal membrane and ordinary membrane filters. The improvement seen with the metal membrane filter when cobalt radiation is used is enhanced by the use of molybdenum radiation. The method is adequate for samples containing as little as 10 $\mu g/cm^2$ of quartz.

The depth to which particles penetrate before being captured by a filter is of importance in the radioassay of alpha-emitting nuclides. Madelaine and Parnianpour [27] studied this effect for room aerosols tagged with the decay products of thoron and for aerosols produced in filtered air by the addition of thoron, using a modification of Sisefsky's [28] technique for successively peeling off thin layers of the filter and analyzing the residual activity of the remaining filter. They studied two fibrous filters and two membrane filters. Lössner [29] measured the residual energy of alpha particles leaving the filter as a means of estimating the depth of filter to which 0.55 μm diameter SiO_2 particles tagged with thoron decay products penetrated. Some of his results and those of Madelaine and Parnianpour are shown in Fig. 4.13. The characteristics of the filters involved are shown in Table 4.3. The AF-150 is seen to be very similar to the Millipore AA, but the penetration patterns of the two differ significantly. All work reported by Madelaine and Parnianpour was done at a face velocity of 11 cm/sec, while the measurements on AF-150 were made at 40 cm/sec. Lössner's results at other velocities indicate that the slope of the curve for AF-150 would have been even smaller at 11 cm/sec. The effect of particle size appears to be much less marked for fibrous filters (Fig. 4.13B). Filters of comparable gross efficiency (A and C; B and D) are

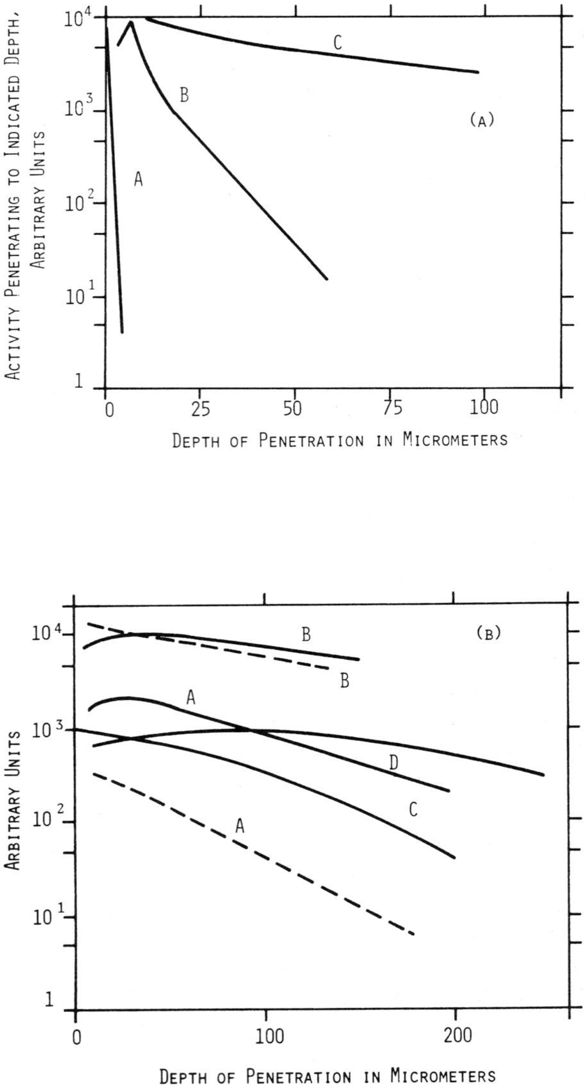

FIG. 4.13. Depth of penetration of particles into various filter media. (A) Curve A = Millipore Type VC, Curve B = Millipore Type AA [27], Curve C = AF-150 [29].
(B) Curve A = pink filter, Curve B = blue filter, S lid Curve: Room aerosol 43% < 0.05 μm, 100% < 0.5 μm, Dashed Curve: Thoron in filtered air, < 0.003 μm [27]. Curve C = AAF Glass filter, Curve D = Air monitoring filter, 0.55 μm SiO_2 [29].

TABLE 4.3

CHARACTERISTICS OF FILTERS USED IN PENETRATION STUDIES

Filter	Type	Pore or fiber diameter (μm)	Thickness (mm)	Area density (mg/cm^2)	Porosity $1-\alpha$	Fraction of alpha activity absorbed	Activity	Ref
AF–150	Membrane	0.7–1.0	0.150	5.0	0.80	—	Th C–C′	[29]
AAF	Glass fiber	$\simeq 1.0$	0.250	5.6	0.90	—		[29]
Air monitoring	Cellulose fiber	15–25	0.250	11.0	0.70	—		[29]
S–P blue	Cellulose fiber	7	0.175	—	—	—	Th C–C′	[27]
S–P pink	Cellulose-asbestos fibers	0.7	0.400	16.2	0.74	0.33	Th C–C′	[27]
Millipore VC	Membrane	0.1 ± 0.08	0.130	—	—	$\simeq 0$	Th C–C′	[27]
Millipore AA	Membrane	0.8 ± 0.05	0.150	4.7	0.81	$\simeq 0$	Th C–C′	[27]
Whatman GF/A	Glass fiber	0.7	0.450	5.0	0.95	$\simeq 0$	Pu–239	[20]
Whatman 41	Cellulose fiber	—	0.018	9.1	0.65	0.40	Th C–C′ Ra A–C′	[30]
Whatman Ap/A	Polystyrene fiber	—	0.750	—	—	$\simeq 0$	Th C–C′ Ra A–C′	[30]
Microsorban	Polystyrene fiber	0.6–0.8	1.88	—	—	0.43	Th C–C′ Ra A–C′	[30]
Gelman E	Glass fiber	—	0.625	—	—	0.09	Th C–C′ Ra A–C′	[30]
HV–70	Cellulose-asbestos fiber	0.1–35	0.225		0.77	0.18	Th C–C′ Ra A–C′	[30]
HV–70	Cellulose-asbestos fiber	0.1–35	0.500		0.79	0.28	Th C–C′ Ra A–C′	[30]

more similar in their penetration patterns, despite the rather considerable differences in particle size.

The more deeply a particle penetrates into a filter, the smaller the residual energy, at the surface, of any alpha radiation it emits. The diminished energy lessens the probability that the alpha particle will be detected, effectively lowering the sensitivity of the method and making it necessary to use correction factors. These factors introduce another source of experimental error. Detection losses due to absorption of alpha particle energy by the filter have been determined on intact filters by Lindeken [30], whose results are included in Table 4.3.

4-2 Electrostatic Precipitation

The use of electrostatic forces to remove particles from air was apparently first demonstrated by Hohlfeld in 1824, when he applied a high potential to a wire suspended in a bottle filled with smoke and rapidly precipitated the smoke particles on the bottle. The phenomenon attracted little attention and it was not until 1884 that its practical possibilities were recognized by Lodge [32], who attempted to adapt it to the cleaning of flue gases. Lodge's work led to experiments by Cottrell in the U.S., resulting in the development of the so-called Cottrell precipitator, now widely used as a means of cleaning the effluents from various manufacturing processes [33]. The process was not applied to air sampling until 1919 when Tolman *et al.* [34] built a small glass precipitator and used it to collect smokes. The process was first applied to industrial hygiene sampling by Bill [35]. Subsequently, a number of precipitators were described [37], the most successful being that designed by Barnes and Penney [38], which was the basis for models now available commercially.

Electrostatic precipitation is a two-stage process. In the first stage, the particles are charged in a unipolar ion field and in the second stage, a strong

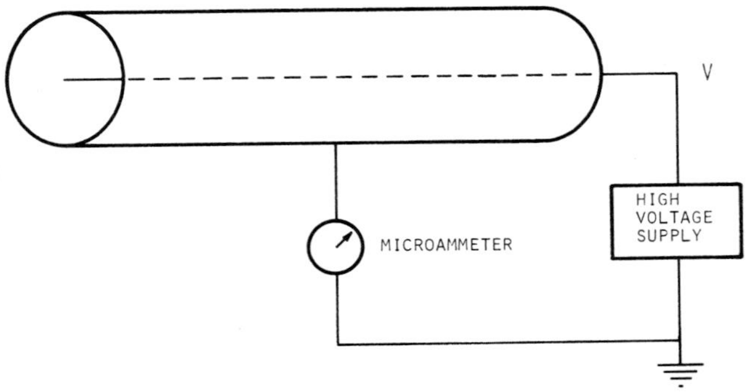

FIG. 4.14. Schematic of concentric cylinder electrostatic precipitator.

electric field precipitates the charged particles on a suitable collecting surface. In practice, the second stage has usually begun before the first is completed. The charging process requires the continuous production of large numbers of ions, for which a corona discharge is normally used. In the typical electrostatic precipitator, the electrodes form a concentric cylinder condenser, with a central, high-voltage electrode of very small diameter and a grounded collecting electrode in the form of a cylindrical shell of much greater diameter. A schematic diagram of such an instrument, arranged to permit measurement of precipitator current as a function of voltage, is shown in Fig. 4.14.

4-2.1 Formation of a Corona

If R_i and R_o are, respectively, the radii of the inner and outer cylinders of the precipitator diagrammed in Fig. 4.14, the electrical field strength between the cylinders at a distance r from the axis of the central electrode is

$$\mathscr{E}(r) = V/[r \ln(R_o/R_i)], \tag{4.21}$$

provided there is no appreciable flow of current. V is the dc potential difference between the cylinders. The maximum field strength occurs at the surface of the inner electrode, where $r = R_i$. In the electrostatic precipitator, R_i is small and $R_o \gg R_i$, so that the field strength at the surface of the center electrode is much greater than throughout most of the precipitator.

Typical current-voltage characteristics of the arrangement of Fig. 4.14 are shown in Fig. 4.15. At first, only an insignificant current due to ions occurring naturally in air is seen. As the voltage is increased, these ions acquire sufficient energy to create other ions by collision with air molecules. When the rate of ionization exceeds the rate of recombination, the formation of ion "avalanches" begins, the collision ionization is augmented by photoionization of gas molecules and at a certain critical voltage V_c, a self-sustaining current develops. Thereafter, the current increases rapidly with increasing voltage until a potential is reached at which arcing between the electrodes occurs.

The field strength necessary to produce the avalanches occurs only near the surface of the center electrode in a small region defined by a violet glow, termed the "corona discharge." Ions of both polarities occur within the

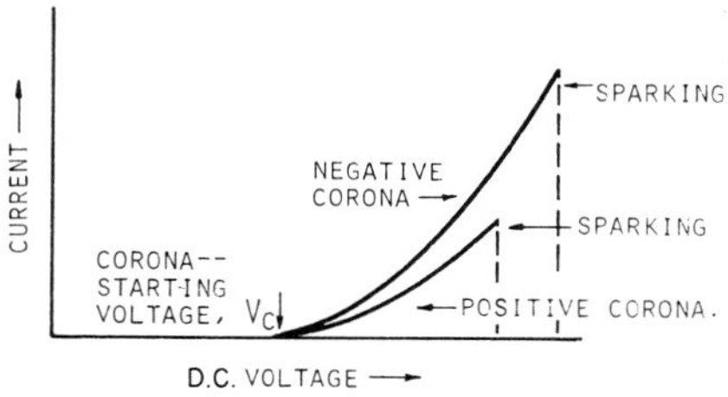

Fig. 4.15. Current-voltage characteristics for arrangement of Fig. 4.14.

corona, but beyond it are found only ions of the same polarity as the center electrode, suggesting that the field strength at the outer edge of the corona is the minimum at which a discharge can exist in the gas. When the corona is initiated, the critical field strength at the surface of the electrode is

$$\mathscr{E}_c = \mathscr{E}(R_i, V_c) = V_c/[R_i \ln(R_o/R_i)]. \tag{4.22}$$

It has been found experimentally that the critical field strength is also given by [39]

$$\mathscr{E}_c = a + b/\sqrt{R_i}, \tag{4.23}$$

where a and b are constants depending on the gas. The constant a is the minimum field strength at which the corona can exist and b is related to the density of the air. For positive and negative coronas, a equals 33.7 and 31.0 kV/cm, respectively, and the corresponding values of b are $8.13\sqrt{\beta}$ and $9.54\sqrt{\beta}$ [39] where β is the density of air relative to its density at 760 mm Hg and 25°C. Combining Eqs. (4.22) and (4.23), the field strength a is found to be

$$a = \frac{V_c}{\ln(R_o/R_i)} \cdot \frac{1}{(R_i + b\sqrt{R_i/a})} = \frac{\mathscr{E}_c R_i}{R_i + b\sqrt{R_i/a}}, \tag{4.24}$$

Therefore, $R_i + b\sqrt{R_i/a}$ is the value of r at which the corona ceases and $b\sqrt{R_i/a}$ represents the corona's thickness.

The space charge due to the presence of unipolar ions in the region between the corona and the outer electrode distorts the field predicted by Eq. (4.21). If there are no particles present in that region, the field distribution is given by [39]

$$\mathscr{E} = \left\{\left(\mathscr{E}_c \frac{R_i}{r}\right)^2 + \frac{2I}{\mu}\left(1 - \frac{R_i^2}{r^2}\right)\right\}^{1/2}, \tag{4.25}$$

where I is the current per unit length of electrode and μ is the electrical mobility (velocity per unit field strength) of the ions. The charge density in the space between corona and outer electrode is

$$C \cdot \varepsilon = (I/2\pi\mu) \cdot (1/r\mathscr{E}). \tag{4.26}$$

C is the concentration of ions at r, and ε is the electronic charge. Charged particles, which have a mobility very much smaller than that of ions, can have a marked effect on the field distribution in precipitators used for cleaning purposes [40]. For particles of dielectric constant ξ having a surface area concentration of S cm^2/cm^3 of air, the current I at a given value of ξ will be decreased by the factor

$$1 + [2\xi/(\xi + 2)] \cdot S \cdot R_o. \tag{4.27}$$

The factor will not differ significantly from unity for most sampling applications.

Precipitators are usually operated with the center electrode at a negative voltage, although both positive [41] and alternating [42] voltages have been used. Negative voltages are preferred because the negative corona begins at a lower value of $\mathscr{E}_c$, produces larger currents, and can sustain higher voltages before arcing over. The two coronas differ in appearance, the positive corona presenting a uniform glow along the length of the electrode, whereas the negative corona, even at quite low currents, exhibits numerous bright points along the electrode.

The preceding results have been used to calculate some of the characteristics of a precipitator [43] similar in design to that described by Barnes and Penney [38]. The pertinent data are:

$$R_i = 0.03 \quad \text{cm}, \qquad R_o = 1.85 \quad \text{cm}, \qquad I = 39.5 \quad \text{A/cm}, \qquad V = 14 \quad \text{kV}.$$

Electric fields and ion concentrations are shown in Fig. 4.16.

4-2.2 Charging of Particles

The ion current through an element of area dS at a distance r from the center of a particle of diameter D is

$$j = (\mu \mathscr{E}_r N_r - \Delta\, dN/dr)\, dS, \tag{4.28}$$

where μ is the electrical mobility of the ions, $\mathscr{E}_r$ is the electrical field strength, N is the ion density, and Δ is the diffusion coefficient of the ions. $\mathscr{E}_r$ is the vector sum of the field due to whatever charge may be on the particle, the field due to the electrical image of the ion induced on the particle, and any externally applied field. This equation cannot be solved as it stands, partly because it is not generally possible to set up expressions for N and dN/dr. Although both terms are significant in practice, theoretical studies usually consider two limiting situations. The first, known as field charging, occurs when $\mu \mathscr{E}_r N_r$ is large compared with $\Delta(dN/dr)$; the second, known as diffusion charging, occurs when the externally applied field is essentially zero.

Field Charging. Rohmann [44] was apparently first to derive an expression for the charging of a particle moving through a cloud of unipolar ions in an externally applied, homogeneous electric field, $\mathscr{E}$. He assumed that movement of the ions was due solely to electrostatic forces and that electrical image forces were negligible. The equation he derived for the number of electron charges on a spherical particle of diameter D was

$$n_E = \frac{3\xi}{\xi + 2} \cdot \frac{\mathscr{E} D^2}{4\varepsilon} \cdot \frac{\pi i t}{\mathscr{E} + \pi i t}, \tag{4.29}$$

where ξ is the dielectric constant of the particle, ε is the electron charge, i is the

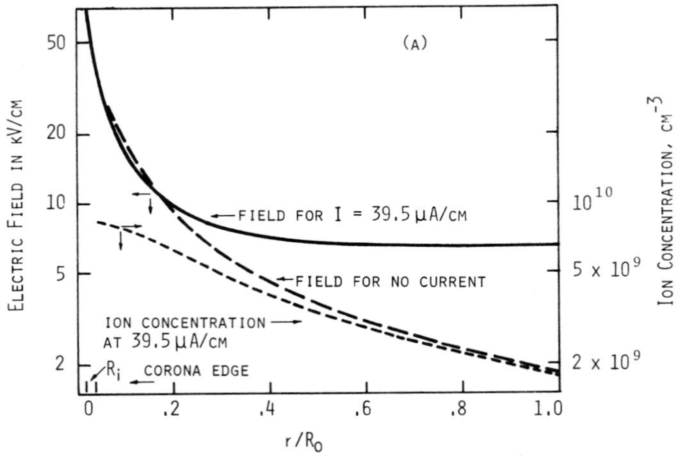

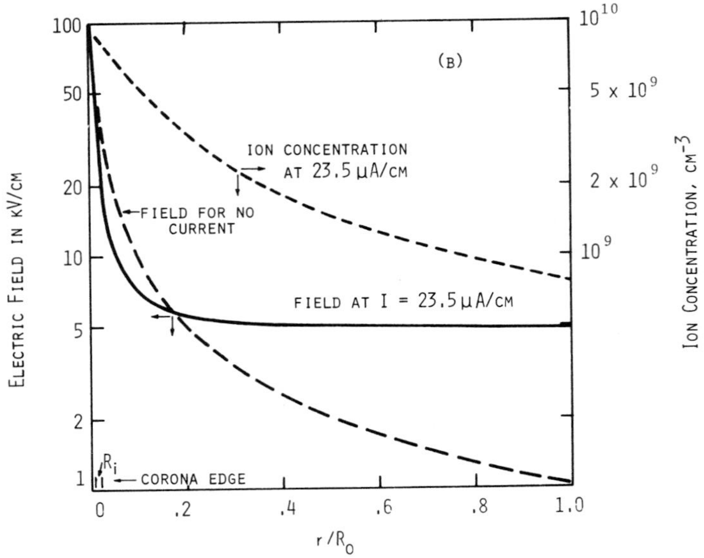

FIG. 4.16. Electric fields and ion concentrations in electrostatic precipitator. (A) Low flow rate [43]; (B) high flow rate [57].

current per unit area in the charging region, and t is the time of charging. Rohmann's limited experimental data were in agreement with this theory.

Pauthenier and Moreau-Hanot [45], using a different experimental technique, made a more extensive study of the charging process and verified the formula for steel balls having radii greater than 1 mm and for metal spheres having radii in the range 10–100 μm. They rederived the charging equation, making the same assumptions that Rohmann had made but using a somewhat different mathematical method, and arrived at the same result. They pointed out that the equation was accurate only for particles having radii greater than a few micrometers; at 10 μm, the error due to neglecting image forces was 1%, at 3 μm, it was 10%, and it increased as the particle size decreased, causing the charge to be underestimated.

Fuchs *et al.* [46] also confirmed the field charging formula, using particles having diameters between 1 and 4 μm. On the other hand, Dötsch *et al.* [47] found that particles of 1.3 μm diameter acquired charges exceeding n_E by a factor of 1.6 when charged at comparable voltage gradients and ion density × time values. At lower voltage gradients the factor was greater than 3.0. For particles of the same size and material and similar voltage gradients, but at higher values of Nt, Hewitt [48] found charges exceeded n_E by a factor of 2.5.

Diffusion Charging. Even when there is no externally applied electric field, a particle may receive a charge as the result of collisions with ions in random motion due to their thermal energy. Arendt and Kallman [49] developed a differential equation for diffusion charging that was valid when the particle's charge was large enough to keep the charging rate very small. The equation, which could be integrated only by approximate methods, predicted a quasi-limiting charge that was approximately proportional to the diameter of the particle. This was confirmed experimentally by Arendt and Kallman, by Schweitzer [50], and by Sachsse [51], for particles in the range 0.4–8 μm diameter.

By making the simplifying assumption that the ion density near the particle surface was given by

$$N = N_0 \cdot \exp(-2n\varepsilon^2/DkT), \tag{4.30}$$

where k is Boltzmann's constant, T is the absolute temperature, and N_0 is the ion density at a great distance from the particle, White [52] was able to derive the following expression for the charging of a particle due to the diffusion process:

$$n_D = (DkT/2\varepsilon^2)\ln[1+(\pi D\bar{v}N_0\,\varepsilon^2 t/2kT)], \tag{4.31}$$

where $\bar{v}$ is the root-mean-square velocity of the ions. Like the equation of Arendt and Kallman, this equation is valid only after the charging rate has become very slow, since the relationship assumed for N is only true for a condition in which there is no net movement of ions.

Charging of Small Particles in an Electric Field. In the electrostatic precipitator, charging is a complex combination of both diffusion and ion bombardment. The total charge n_T varies with particle size approximately as follows:

$n_T \simeq n_E$, for $D > 6\ \mu m$, with little error;

$n_T > n_E$, for $D < 6\ \mu m$, with the difference increasing rapidly below $2\ \mu m$;

$n_T \simeq n_D$, for $D \gtrsim 0.4\ \mu m$ and *long* charging times.

Since charging time is quite short in precipitators, it is doubtful if the last condition is ever realized. A number of investigators [48, 53, 54, 55] have found that small particles, which might be expected to charge according to Eq. (4.31), actually take a much higher charge than predicted by either n_D or n_E when charged in an electric field. In general, the relationship between charge and size can be expressed as

$$n = aD^b.$$

An example of the relationship is shown in Fig. 4.17, for submicron sodium chloride particles [53]. Both a and b appear to depend on both voltage gradients and Nt values. The results of other investigators can be expressed in a similar manner, as shown in Fig. 4.18. Particles charged in a field of negative ions apparently acquire a larger charge than they do under similar conditions in a field of positive ions [54].

4-2.3 Electrical Mobility of Particles

The force acting on a charged particle in an electric field was given above as $F_E = \mathscr{E}q$. Since $q = n\varepsilon$, the terminal velocity U_E due to that force will occur when

$$\mathscr{E}n\varepsilon = U_E/Z,$$

where Z is the mechanical mobility of the particle. The particle's electrical mobility is defined as it is for an ion, that is, as the velocity per unit field strength:

$$\mu_p = U_E/\mathscr{E} = n\varepsilon Z \propto D^{b-1} \cdot K_s \tag{4.32}$$

where K_s is the particle's slip factor. When particles are large enough to charge according to Eq. (4.29), $b = 2$, and the mobility increases with increasing size. At smaller sizes, however, $1 < b < 2$, and the mobility goes through a

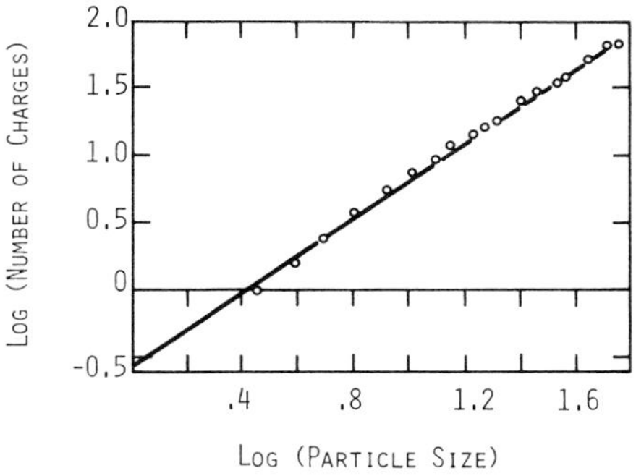

FIG. 4.17. Charge as a function of size for sodium chloride particles [53]. $E = 2.36$ kV/cm, $Nt = 3.4 \times 10^5$ sec/cm^3, size $\propto$ side of cube.

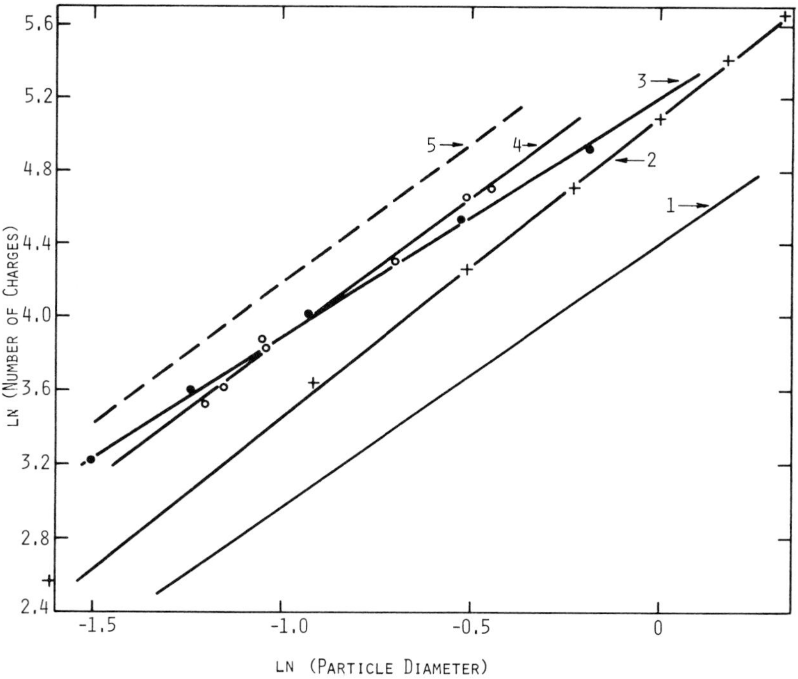

FIG. 4.18. Charge as a function of diameter (in μm) for various charging conditions.

Curve	Equation	E (kV/cm)	Nt (sec/cm^3)	Ref.
1	$n_- = 81D^{1.43}$	2.36	3.4×10^5	[53]
2 (+)	$n_+ = 162D^{1.64}$	3.60	10^7	[48]
3 (●)	$n_+ = 179D^{1.3}$	2.73	1.8×10^6	[55]
4 (○)	$n_+ = 218D^{1.52}$	2.3	6.4×10^7	[54]
5	$n_- = 297D^{1.52}$ (est.)	2.3	6.4×10^7	[54]

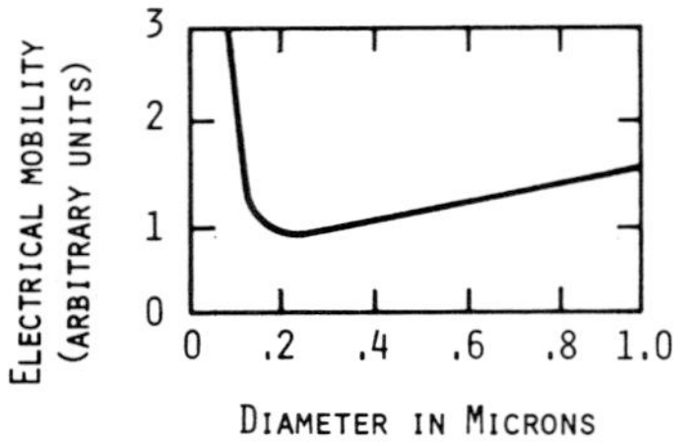

FIG. 4.19. Electrical mobility as a function of size for charge proportional to $D^{1.4}$ [53].

minimum value as the diameter decreases. This is demonstrated in Fig. 4.19 for $b = 1.4$. Because of this relationship, very small particles that acquire a charge and very large particles precipitate most rapidly in the electrostatic precipitator, while particles in the neighborhood of 0.2 μm are the most difficult to precipitate. Very small particles may acquire, on the average, only one or two charges, and a certain fraction passes through the precipitator uncharged. In the absence of an electric field, particles charge approximately according to Eq. (4.31), so that $b = 1$ and $\mu_p \propto K_s$.

4-2.4 COLLECTION EFFICIENCY OF ELECTROSTATIC PRECIPITATORS

The efficiency with which particles are collected depends on the relative values of the transit time of air passing through the precipitation region and the time required for a charged particle to reach the outer electrode. If flow through the precipitator is laminar, then collection efficiency E, for particles of a given size, assuming they all acquire the same electrical mobility, is

$$E = (R_o^2 - r^2)/(R_o^2 - R_i^2) \simeq 1 - (r^2/R_o^2),$$

where r is defined by

$$(R_o - r)/\mu_p \bar{\mathscr{E}} = L/U$$

for $r \geqslant R_i$. The term on the left is the time it takes a particle of mobility μ_p to move from r to R_o in an average electric field $\bar{\mathscr{E}}$. The term on the right is the transit time for air moving through a precipitator of length L at an average linear velocity U. In general, the minimum value of μ_p will be greater than 0.001 cm^2/volt-sec, and collection of particles will be essentially complete if

$$L/U > 10^3 \cdot R_o/\bar{\mathscr{E}}.$$

The precipitator used as an example for Fig. 4.16(A) had transit times L/U greater than 0.5 sec for volumetric flow rates less than 23.4 liters/min, indicating complete collection of particles should occur. Experimentally, collection efficiencies for aerosols having mass median diameters between 0.2 and 0.7 μm were found to be greater than 99.9%, with less than 0.1% of the aerosol deposited on surfaces other than that of the collector tube.

If the air flow is turbulent, the efficiency of collection is given by [64]

$$E = 1 - \exp(-2\mu_p \mathscr{E} L / R_o U). \tag{4.33}$$

In this case, the particle's electrical velocity $\mu_p \mathscr{E}$ is that close to the collecting electrode. This equation is primarily of interest when electrostatic precipitation is used to clean air. However, some precipitators, designed for special sampling purposes, operate under conditions of turbulent air flow [57]. One of these is the large-volume air sampler used as an example in Fig. 4.16(B). It was found to have the following collection characteristics:

Air flow rate (liters/min)	Reynolds number	Transit time (sec)	Average collection efficiency (%)	Estimated μ_p (cm^2/volt-sec)
700	14,120	0.056	39	0.0031
500	10,120	0.078	51	0.0033
250	5060	0.156	91	0.0055

The electrical mobilities were calculated using Eq. (4.33). They are compatible with the particle sizes (0.5–1.5 μm) and Nt values involved.

The effective collection efficiency of the electrostatic precipitator can be reduced by re-entrainment of deposited particles. This is more likely to be a problem when large deposits build up or when the particles are loosely aggregated. The formation of ozone in the corona region may cause undesirable chemical reactions.

4-3 Optical Methods

When a beam of collimated light passes through a medium in which particles are suspended in a uniform distribution, its intensity is diminished in a distance dL by an amount

$$dI = -I \cdot n_T \cdot \overline{AE} \cdot dL, \tag{4.34}$$

where I is the intensity of the beam entering dL, n_T is the number of particles per unit volume, and $\overline{AE}$, the average product of particle cross section A and extinction coefficient E, is given by

$$\overline{AE} = (1/n_T) \sum n(A) \cdot A \cdot E(A).$$

Integration of (4.34) leads to the Bouguer (or Bouguer–Beer, or Lambert–Beer) law:

$$I = I_0 \cdot \exp(-n_T \cdot \overline{AE} \cdot L), \tag{4.35}$$

where I_0 is the intensity of the light upon first entering the medium. If I and I_0 are measured, this equation makes it possible to calculate the total scattering area of the particles per unit volume of air. If the particles are of irregular shape and of sufficient size to be in the region in which $E \simeq 2$ (see Fig. 2.18), then

$$n_T \cdot \overline{AE} \simeq 2n_T \cdot \bar{A},$$

where $\bar{A}$ is the average cross-sectional area of the particles when viewed in the direction of the beam. Since much of the light scattered by particles in this size range is in the forward lobe, considerable care must be exercised to be sure that only transmitted light is included in the measurement of I. Hodkinson [65] recommends an arrangement similar to that shown in Fig. 4.20 except that a collimator lens is used at C, thus providing a broad beam of parallel light across the dust particles, which fill the region between C and B′. The scattered light in the forward lobe can be eliminated adequately by limiting the diameter of the apertures to 0.122 λ/D, for light of wavelength λ and particles of maximum diameter D.

When the suspending medium is air, very large values of L are needed for most practical applications because of the relatively low concentrations of airborne particles normally encountered. The same information can be acquired by measuring some known fraction of the light actually scattered. Instruments designed for this purpose are known as photometers. Diagrams of typical instruments are shown in Fig. 4.20. The significance of their measurements depends on the position and size of the solid angle from which scattered light is gathered. If the solid angle lies in the primary forward lobe, where a large fraction of the scattered light is due to Fraunhofer diffraction, the effect of refractive index is least pronounced, and particles of a given cross-sectional area give a similar response, even if they are opaque or strongly absorbing. Outside that region, the effects of index of refraction and opacity become more pronounced. They are most pronounced at 90°, which is the angle at which

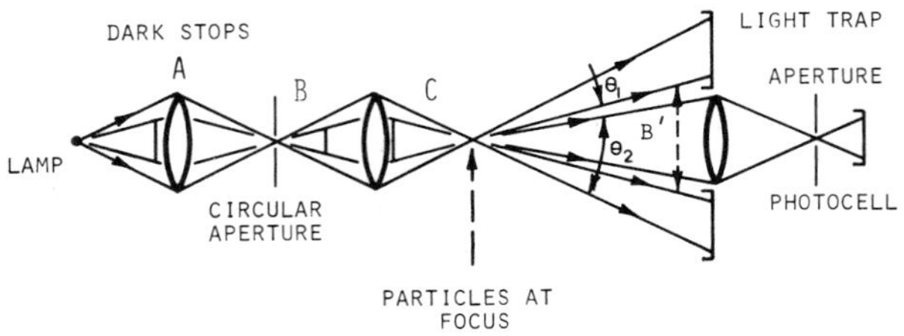

FIG. 4.20. Optical arrangements for aerosol photometers [65]. Dark stops: At C only, aerosol photometer of Sinclair (1953). At B only, particle counter of Gucker and Rose (1954). At A, B, and C, aerosol photometer of Clarenburg and Princen (1963). Courtesy of Academic Press, London.

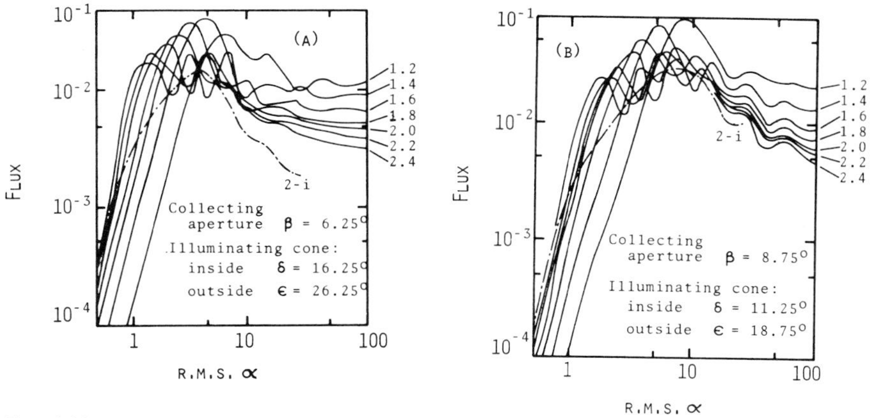

FIG. 4.21. Response curves for photometers of Fig. 4.20 [65]. (A) Dark stops at C only; (B) dark stops at B only or at A, B, and C (curve parameters are indexes of refraction). Courtesy of Academic Press, London.

scattered light is traditionally measured with Tyndall meters. The typical responses of photometers having different angles of collection are shown in Fig. 4.21.

4-4 Piezoelectric Microbalance Methods

When a piezoelectric crystal is subjected to an electric field, a mechanical stress proportional to the strength of the electric field is produced. If a thin quartz disk is made part of an oscillator circuit, it will vibrate at a characteristic frequency that depends on the orientation of the disk with respect to the axes of symmetry of the crystal from which it was cut, on its thickness and mass, and on the temperature at which it operates. The orientation referred to as "AT cut" minimizes the effect of temperature on the characteristic frequency.

For a given disk, the characteristic frequency f is inversely proportional to its mass M. If particles of mass ΔM are deposited on the disk surface, where they vibrate in phase with the crystal, the characteristic frequency will be altered by an incremental amount:

$$\Delta f = -Cf^2 \Delta M,$$

where C is a constant related to the type and thickness of the crystal and to its uncontaminated mass. For AT crystals, $C \simeq 2.27 \times 10^{-6}/A$ sec/g, where A is the area of the crystal face in square centimeters. When $\Delta M = 0$, the characteristic frequency of an AT crystal of thickness τ cm is

$$f = 1.66 \times 10^5/\tau \qquad \text{Hz (1 Hz = 1 cycle/sec)},$$

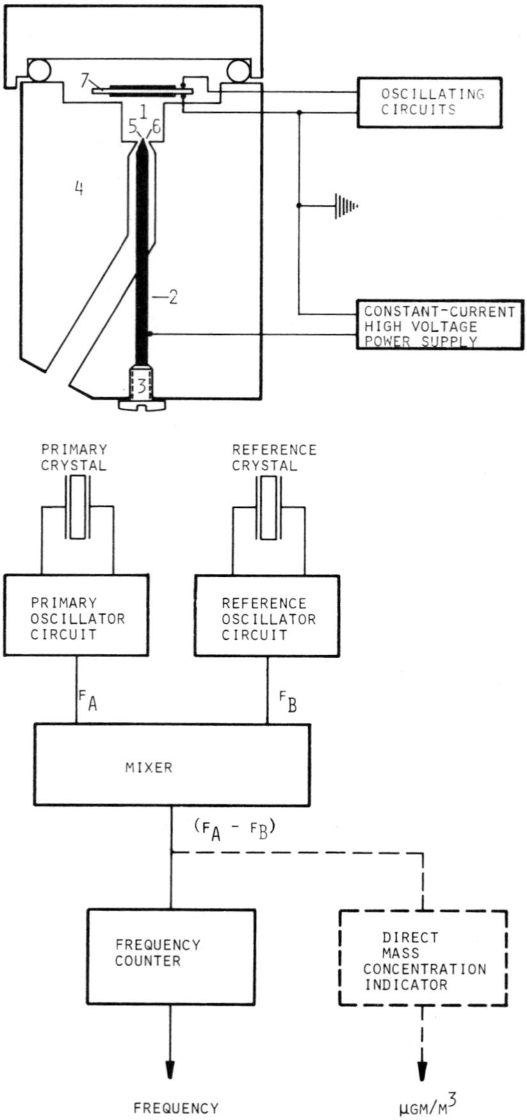

FIG. 4.22 Top: Side view of electrostatic precipitator and piezoelectric microbalance. (1) precipitator chamber, 0.25 in. diameter by 0.25 in. high; (2) corona needle; (3) needle adjusting screw; (4) Teflon precipitating block; (5) high corona region; (6) entrance nozzle to precipitating chamber; (7) piezoelectric crystal sensor [77]. Courtesy of American Industrial Hygiene Association. Bottom: Piezoelectric microbalance electronic block diagram. Courtesy American Industrial Hygiene Association.

and the frequency shift per unit mass of contamination deposited on the crystal surface is

$$\Delta f/\Delta M = -6.25 \times 10^4/A\tau^2 \qquad \mathrm{Hz/g}.$$

For crystals having $A = 1\ \mathrm{cm}^2$, $\tau = 0.02$ cm, 10^{-8} g of contamination will cause a frequency shift of ~ 1.6 Hz, which can be determined accurately, permitting measurement of very small masses.

This method of measurement has been applied to the determination of mass concentrations of particulate material deposited on a quartz crystal by electrostatic precipitation [77] or by impaction [78]. Figure 4.22 (top) is a diagram of a point-to-plane precipitator (see Chapter 5) in which the plane is one of the surfaces of a thin quartz disk. Air enters through the annular space around the point, passes through the corona region, and exits around the periphery of the crystal. Airborne particles are charged in the corona region and deposited on a thin metal coating on the crystal surface that also serves as one of the crystal electrodes. Collection efficiencies close to unity are possible at flow rates up to 5 liters/min.

Figure 4.22 (bottom) is a block diagram of the instrument's electronic circuitry. The signal from the detector crystal is mixed with that from a reference crystal and the beat frequency monitored for changes in frequency. This arrangement minimizes any effect of temperature changes on frequency and yields a resolution of ± 1 Hz, equivalent to $\sim \pm 10^{-8}$ g. Mass concentrations of typical atmospheric aerosols can be measured with 5% accuracy in about 10 sec [77]. The instrument that collects particles by impaction can detect the deposition of single particles having masses $\lesssim 10^{-11}$ g.

4-5 Measurement of Number Concentration

Early studies of hazards in the dusty trades and in coal mines (see Chapter 8) turned up a correlation between number concentration of airborne particles in a characteristic size interval and the occurrence of lung diseases. This led to the adoption of standard techniques for measuring number concentrations. In the U.S., the samplers chosen for this purpose were the Greenburg–Smith (standard) impinger and the midget impinger, and the characteristic size interval was described as 0.5–10 μm [66]. In Great Britain, the thermal precipitator became the standard sampling instrument for a size range of 0.5–5 μm. The following discussion is devoted mainly to the impingers; Chapter 5 includes a discussion of the thermal precipitator.

Figure 4.23 includes diagrams of modified versions of the Greenburg–Smith [67] and midget [68] impingers. Significant dimensions and operating characteristics of the two instruments are summarized in Table 4.4. The theoretical

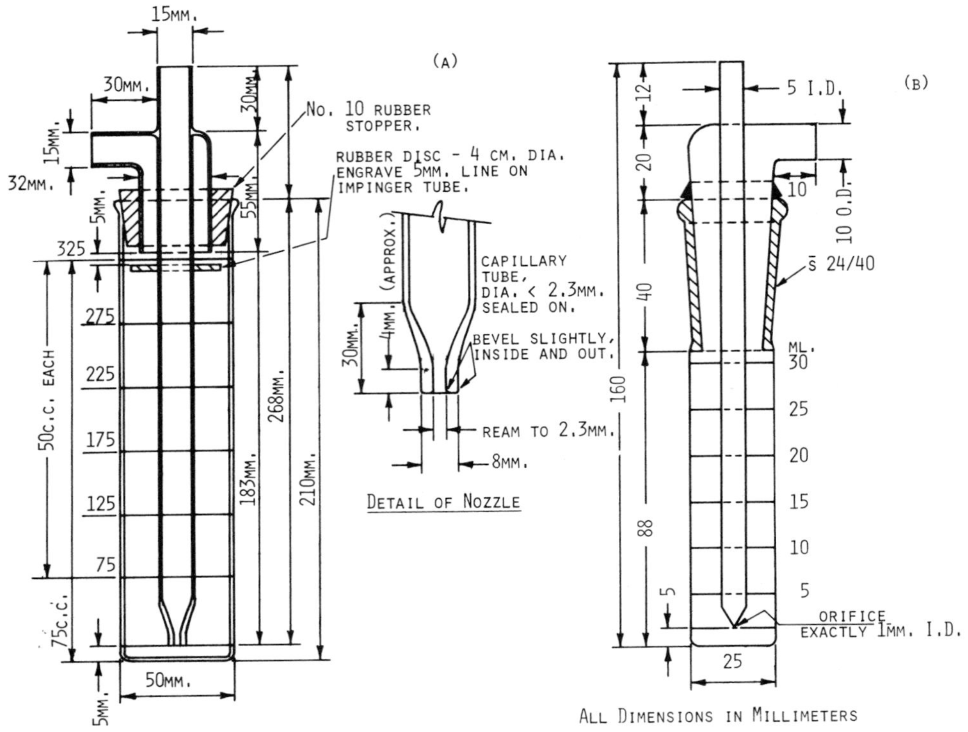

FIG. 4.23. Diagrams of impingers. (A) Modified Greenburg–Smith (standard) impinger [69], courtesy of American Industrial Hygiene Association; (B) all-glass midget impinger [70], courtesy of American Chemical Society.

TABLE 4.4

DETAILS OF IMPINGERS

	Standard impinger	Midget impinger
Orifice diameter	0.23 cm	0.1 cm
Orifice-flask separation	0.5 cm	0.5 cm
Flask diameter	5.0 cm	2.5 cm
Standard flow rate	1.0 cfm (28.3 liters/min)	0.1 cfm (2.83 liters/min)
Velocity at orifice[a]	120 m/sec	60.4 m/sec
ΔP at standard flow rate:		
Theoretical	6.4 cm Hg	1.6 cm Hg
Measured	7.2 cm Hg	2.2 cm Hg

[a] Assuming adiabatic expansion of air.

values of ΔP refer to velocities in the orifice; the measured values include the effects of other resistances in the system and of any contraction of the air stream after it leaves the orifice.

Despite the wide use of impingers, there is little information available concerning their collection efficiency as a function of particle size. However, their sampling characteristics should be similar to those of other round-jet impaction systems that have the same ratio between the jet-collector separation distance S and the orifice diameter W. Figure 4.24 is an experimental curve showing efficiency as a function of aerodynamic diameter for $S/W = 5$ [71], the appropriate ratio for the midget impinger. The corresponding curve for the standard impinger ($S/W = 2$) has been omitted because it differs only slightly from that for the midget impinger. Davies *et al.* [72] sampled a quartz aerosol using both a midget impinger and a thermal precipitator. They determined the size distribution and relative numbers of particles deposited in each sampler and deduced collection efficiencies for several sizes Their data, corrected for density and shape, have been included in Fig. 4.24. Jacobson *et al.* [73] determined the gross collection efficiency of the midget impinger relative to membrane filters for coal particles $\geqslant 1$ μm (equivalent volume diameter) and, on the same samples, for particles $\geqslant 0.68$ μm. From these data, the average efficiencies with which the impinger collects particles in the range 0.68–1.0 μm have been calculated. They are plotted in the same figure against the midpoint of that range, expressed as the aerodynamic diameter. Davies' data indicate a significant amount of rebound for diameters $\lesssim 1$ μm.

Impingers are used in aerosol sampling almost exclusively for the determination of number concentrations, so it is important to ensure that spurious

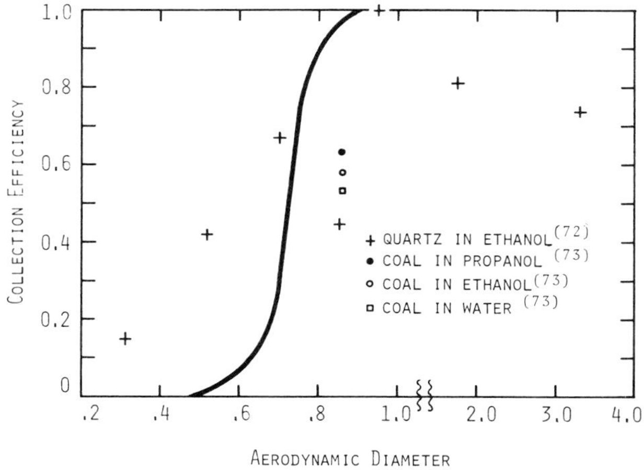

FIG. 4.24. Estimated collection efficiency curve for midget impinger (note change in abscissa scale).

counts are not introduced as a result of the collection process. There is conflicting information about this point for both the standard and midget impingers. Davies *et al.* [72] found that the midget impinger did not fracture quartz particles, but caused so much disaggregation of coal particles that they were unable to compare its sampling performance with that of the thermal precipitator. Jacobson *et al.* [73], on the other hand, compared the air concentration of coal particles, measured using the midget impinger, with that obtained from membrane filter samples, and found no evidence of either disaggregation of agglomerates or fracturing of particles. Similarly, studies of the fracture of particles in the standard impinger have yielded equivocal results. Anderson [74], like several earlier investigators, concluded that impingers fracture the larger particles causing dust counts to be erroneous "as they always give a higher count than actually exists in the air sampled." Silverman and Franklin [75], however, claimed that fracture did not occur until jet speeds were much higher than that occurring in the impinger at its normal operating flow rate. The meaning of their results is open to question, however, since they not only record jet speeds well in excess of the speed of sound, but show flow rates about twice the maximum that can be attained by sucking air through an orifice 0.23 cm in diameter.

Several comparisons have been made between number concentrations based on midget impinger samples and on membrane filter or thermal precipitator samples. Some of the results are summarized in Table 4.5. The relationships

TABLE 4.5

MIDGET IMPINGER EFFICIENCY FOR MEASUREMENT OF NUMBER CONCENTRATION

Dust	Reference sampler	Collection liquid	Size range (μm)	Impinger concentration, C_{MI}	Ref
Quartz	Thermal precipitator	Ethanol	> 0.2	$0.63\ C_{TP}$	[72]
Coal	Membrane filter	Ethanol	> 1.0	$0.96\ C_{MF}$	[73]
Coal	Membrane filter	Propanol	> 1.0	$0.91\ C_{MF}$	[73]
Coal	Membrane filter	Water	> 1.0	$0.77\ C_{MF}$	[73]
Polystyrene	Membrane filter	Ethanol	> 1.1	$1.05\ C_{MF}^{1.16}$	[76]

for quartz and coal, when the collection liquid is ethanol, are in reasonable agreement, considering the size ranges involved. The fact that the midget impinger yields a higher count of polystyrene particles than does the membrane filter was attributed, in part, at least, to overlap of particles on the filter [76].

References

1. C. N.Davies, The Separation of Airborne Dust and Particles, *Proc. Inst. Mech. Eng. B*, *1B:* 185–198 (1952).
2. C. N. Davies and C. V. Peetz, Impingement of Particles on a Transverse Cylinder, *Proc. Roy. Soc.*, *A234:* 269–295 (1956).
3. J. Pich, Theory of Aerosol Filtration, in C.N. Davies (Ed.), *Aerosol Science*, Academic Press, London, 1966.
4. J. Happel, Viscous Flow Relative to Arrays of Cylinders, *AIChE J.*, *5:* 174–177 (1959).
5. J. A. Harrop and J. I. T. Stenhouse, The Theoretical Prediction of Inertial Impaction Efficiencies in Fibrous Filters, *Chem. Eng. Sci.*, *24:* 1475–1481 (1969).
6. C. Y. Chen, Filtration of Aerosols by Fibrous Media, *Chem. Rev.*, *55:* 595–623 (1955).
7. R. E. Pasceri and S. K. Friedlander, The Efficiency of Fibrous Aerosol Filters: Deposition by Diffusion of Particles of Finite Diameter, *Can. J. Chem. Eng.*, *38:* 212–213 (1960).
8. I. B. Stechkina, A. A. Kirsh, and N. A. Fuchs, Investigations of Fibrous Filters for Aerosols. Calculations of Aerosol Deposition in Model Filters in the Region of Maximum Particle Breakthrough, *Koll. Zh.*, *31:* 121–126 (1969).
9. H. F. Kraemer and H. F. Johnstone, Collection of Aerosol Particles in Presence of Electrostatic Fields, *Ind. Eng. Chem.*, *47:* 2426–2434 (1955).
10. D. A. Lundgren and K. T. Whitby, Effect of Particle Electrostatic Charge on Filtration by Fibrous Filters, *Ind. Eng. Chem. Proc. Res. Develop.*, *4:* 345–349 (1965).
11. D. Rimberg, Penetration of IPC-1478, Whatman 41, and Type 5G Filter Paper as a Function of Particle Size and Velocity, *Amer. Ind. Hyg. Ass. J.*, *25:* 394–401 (1969).
12. C. L. Lindeken, R. L. Morgin, and K. F. Petrock, Collection Efficiency of Whatman 41 Filter Paper for Submicron Aerosols, *Health Phys.*, *9:* 305–308 (1963).
13. C. N. Davies, The Clogging of Fibrous Aerosol Filters, *Aerosol Sci.*, *1:* 35–40 (1970).
14. R. G. Stafford and H. J. Ettinger, *Efficiency of IPC-1478 Filter Paper Against Polystyrene Latex and Dioctyl Phthalate Aerosols*, USAEC Rep. LA-4356, 1970.
15. W. J. Smith and N. F. Surprenant, Properties of Various Filtering Media for Atmospheric Dust Sampling, *Proc. ASTM*, *53:* 1122–1135 (1953).
16. N. Stanley, The Penetration of Filter Paper by Coal-Dust Particles in the Respirable Size Range, *Ann. Occup. Hyg.*, *4:* 295–299 (1962).
17. M. Katz and H. P. Sanderson, Filtration Methods for Evaluation of Aerosol Contaminants, in *ASTM Spec. Publ. No. 250*, 1958.
18. D. Lennon and J. Leroux, Applications of X-Ray Diffraction Analysis in the Environmental Field, *A.M.A. Arch. Ind. Health Occup. Med.*, *8:* 359–370 (1953).
19. A. J. Williams, The Use of a Radioactive Beta Source to Monitor High Airborne-Dust Concentrations, *J. Sci. Instrum.*, *44:* 562–563 (1967).
20. D. C. Stevens and R. F. Hounam, A Fine Glass Fibre Paper for Air Sampling, *Ann. Occup. Hyg.*, *3:* 58–63 (1961).
21. M. Lippmann, Filter Holders and Filter Media, in *Air Sampling Instruments*, A.C.G.I.H., Cincinnati, Ohio, 1966.
22. H. J. Paulus, N. A. Talvities, D. A. Fraser and R. G. Keenan, Use of Membrane Filters in Air Sampling, *Amer. Ind. Hyg. Ass. Quart.*, *18:* 267–273 (1957).
23. J. B. Pate and E. D. Tabor, Analytical Aspects of the Use of Glass Fiber Filters for the Collection and Analysis of Atmospheric Particulate Matter, *Amer. Ind. Hyg. Ass. J.*, *23:* 145–150 (1962).
24. R. T. Richards, D. T. Donovan, and J. R. Hall, A Preliminary Report on the Use of Silver Metal Membrane Filters in Sampling for Coal Tar Pitch Volatiles, *Amer. Ind. Hyg. Ass. J.*, *28:* 590–594 (1967).

25. V. Masek, The Use of Metal Membrane Filters in Sampling for Coal Tar Pitch Volatiles in Coke Oven Plants, *Amer. Ind. Hyg. Ass. J.*, *31:* 641–644 (1971).
26. J. Leroux and C. A. Powers, Quantitative röntgenographische Analyse von Quarz in Staubproben auf Silbermembranfiltern, *Staub*, *29:* 197–200 (1969).
27. G. Madelaine and H. Parnianpour, Contribution a l'Etude de la Penetration des Aerosols dans les Filtres, *Ann. Occup. Hyg.*, *10:* 31–38 (1967).
28. J. Sisefsky, A Method for Determination of Particle Penetration Depths in a Filter, *Nature* (*London*), *182:* 1437 (1958).
29. V. Lössner, Die Bestimmungen der Eindringtiefe von Aerosolen in Filtern, *Staub*, *24:* 217–221 (1964).
30. C. L. Lindeken, Use of Natural Airborne Radioactivity to Evaluate Filters for Alpha Air Sampling, *Amer. Ind. Hyg. Ass. J.*, *22:* 232–237 (1961).
31. L. B. Lockhart, R. L. Patterson and W. L. Anderson, *Characteristics of Air Filter Media Used for Monitoring Airborne Radioactivity*, NRL Rep. 6054, 1964.
32. O. J. Lodge, On Lord Rayleigh's Dark Plane, *Nature* (*London*), *28:* 297–299 (1883).
33. F. G. Cottrell, The Electrical Precipitation of Suspended Particles, *J. Ind. Eng. Chem.*, *3:* 242–250 (1911).
34. R. C. Tolman, L. H. Reyerson, A. P. Brooks and H. D. Smyth, An Electrical Precipitator for Analyzing Smokes, *J. Amer. Chem. Soc.*, *41:* 587–589 (1919).
35. J. P. Bill, An Electrostatic Method of Dust Collection as Applied to the Sanitary Analysis of Air, *J. Ind. Hyg.*, *1:* 323–342 (1919).
36. A. A. Kirsh and N. A. Fuchs, Investigations of Fibrous Aerosol Filters. Diffusional Deposition of Aerosols in Fibrous Filters, *Kolloid Zh.*, *30:* 836–841 (1968).
37. M. Lippmann, Electrostatic Precipitators, in *Air Sampling Instruments*, A.C.G.I.H., Cincinnati, Ohio, 1966.
38. E. C. Barnes and G. W. Penney, An Electrostatic Dust Weight Sampler, *J. Ind. Hyg. Toxicol*, *20:* 259–265 (1938).
39 R. Seeliger, Die physikalischen Grundlagen der elektrischen Gasreinigung, *Z. Tech. Phys.*, *7:* 49–71 (1926).
40. A. Winkel and A. Schutz, Electrical Separation of Finely Divided Iron Oxide Dust at High Temperatures with Particular Regard to the Specific Electrical Resistance of the Dust. *Staub*, *22:* 343–390 (1962).
41. D. G. Beadle, P. H. Kitto and P. J. Blignaut, Portable Electrostatic Dust Sampler with Electronic Air Flow, *AMA Arch. Ind. Hyg. Occup. Med.*, *10:* 381–389 (1954).
42. P. Drinker, Alternating Current Precipitator for Sanitary Air Analysis, *J. Ind. Hyg. Toxicol.*, *14:* 364–370 (1932).
43. K. E. Lauterbach, T. T. Mercer, A. D. Hayes and P. E. Morrow, Efficiency Studies of the Electrostatic Precipitator, *AMA Arch. Ind. Hyg. Occup. Med.*, *9:* 69–75 (1954).
44. H. Rohmann, Methode zur Messung der Grösse von Schwebeteilchen, *Z. Phys.*, *17:* 253–265 (1923).
45. M. M. Pauthenier and M. Moreau-Hanot, La Charge des Particules Sphérique dans un Champ Ionisé, *J. Phys. Radium*, *Ser.*, *7*, *3:* 590–613 (1932).
46. N. Fuchs, I. Petrajanoff, and B. Rotzeig, On the Rate of Charging of Droplets by an Ionic Current, *Trans. Faraday Soc.*, *32:* 1131–1138 (1936).
47. E. Dötsch, H. A. Friedrich, O. Knacke, and J. Krahe, Zur Kinetik der elektrischen Aufladung eines Aerosols, *Staub*, *29:* 282–286 (1969).
48. G. W. Hewitt, The Charging of Small Particles for Electrostatic Precipitation, *A.I.E.E. Trans.*, *76:* 300–306 (1957).
49. P. Arendt and H. Kallman, Über den Mechanismus der Aufladung von Nebelteilchen, *Z. Phys.*, *35:* 421–441 (1926).

50. H. Schweitzer, Über die Aufladung kleiner Schwebeteilchen in der Korono-Entladung, *Ann. Phys., 4:* 33–48 (1930).
51. H. Sachsse, Über die elektrischen Eigenschaften von Staub und Nebel, *Ann. Phys., 14:* 396–412 (1932).
52. H. J. White, Particle Charging in Electrostatic Precipitation, *A.I.E.E. Trans., 70:* 1–6 (1953).
53. T. T. Mercer, *Charging and Precipitation Characteristics of Submicron Particles in the Rohmann Electrostatic Precipitator*, USAEC Rep., No. UR–475, 1956.
54. G. W. Penney and R. C. Lynch, Measurements of Charge Imparted to Fine Particles by a Corona Discharge, *A.I.E.E. Trans., 76:* 294–299 (1957).
55. B. Y. H. Liu and K. T. Whitby, Particle Charging at Low Pressures, *Proc. AEC Conf. Radioactive Fallout Nucl. Weapons Tests*, 260–280, 1964.
56. S. C. Stern, H. W. Zeller and A. I. Schekman, The Aerosol Efficiency and Pressure Drop of a Fibrous Filter at Reduced Pressures, *J. Colloid Sci., 15:* 546–562 (1960).
57. S. C. Stern, D. R. Steele and O. E. A. Bolduan, A Large-Volume Electrostatic Air Sampler, *A. M. A. Arch. Ind. Health Occup. Med., 18:* 30–33 (1958).
58. K. Spurny and J. Pich, Zur Frage der Filtrations-Mechanismen bei Membranfiltern, *Staub, 24:* 250–256 (1964).
59. K. Spurny and J. Pich, The Separation of Aerosol Particles by Means of Membrane Filters by Diffusion and Inertial Impaction, *Int. J. Air Water Pollution, 8:* 193–196 (1964).
60. J. J. Fitzgerald and C. G. Detwiler, Optimum Particle Size for Penetration through the Millipore Filter, *AMA Arch. Ind. Health, 15:* 3–8 (1957).
61. W. J. Megaw and R. D. Wiffen, The Efficiency of Membrane Filters, *Int. J. Air Water* Pollution, *7:* 501–509 (1963).
62. B. Binek and S. Przyborowski, Der Einfluss der elektrostatischen Ladung von Membranfiltern auf ihre Abscheidewirkung, *Staub, 25: 533–535* (1965).
63. P. J. Rigden, The Specific Surface of Powders, A Modification of the Theory of the Air Permeability Method, *J. Soc. Chem. Ind., 66:* 130–136 (1947).
64. W. Deutsch, Bewegung und Ladung der Eletrizitätsträgen im Zylinderkondensator, *Ann. Phys. 68:* 335–344 (1922).
65. J. R. Hodkinson, The Optical Measurement of Aerosols, in C. N. Davies (Ed.), *Aerosol Science*, Academic Press, New York, London, 1966.
66. S. A. Roach, E. J. Baier, H. E. Ayer and R. L. Harris, Testing Compliance with Threshold Limit Values for Respirable Dusts, *Amer. Ind. Hyg. Ass. J., 28:* 543–553 (1967).
67. L. Greenburg and G. W. Smith, *A New Instrument for Sampling Aerial Dust*, R. I. 2392, U. S. Dept. of Interior, Bureau of Mines, 1922.
68. J. B. Littlefield and H. H. Schrenk, *Bureau of Mines Midget Impinger for Dust Sampling*. R. I. 3360, U. S. Dept. of Interior, Bureau of Mines, 1937.
69. T. Hatch, H. Warren and P. Drinker, Modified Form of the Greenburg-Smith Impinger for Field Use, with a Study of Its Operating Characteristics, *J. Ind. Hyg., 14:* 301–311 (1932).
70. Division of Industrial Hygiene, N.I.H., U. S. Public Health Service All-Glass Midget Impinger Unit, *Ind. Eng. Chem.* (*Anal. Ed.*), *16:* 346 (1944).
71. T. T. Mercer and R. G. Stafford, Impaction from Round Jets, *Ann. Occup. Hyg., 12:* 41–48 (1969).
72. C. N. Davies, M. Aylward and D. Leacey, Impingement of Dust from Air Jets, *A.M.A. Arch. Ind. Hyg. Occup. Med., 4:* 354–397 (1951).

73. M. Jacobson, S. L. Terry and D. A. Ambrosia, Evaluation of Some Parameters Affecting the Collection and Analysis of Midget Impinger Samples, *Amer. Ind. Hyg. Ass. J., 31:* 442–445 (1970).
74. E. L. Anderson, Effect of Certain Impingement Dust Sampling Instruments on the Dust Particles, *J. Ind. Hyg. Toxicol., 21:* 39–47 (1939).
75. L. Silverman and W. Franklin, Shattering of Particles by the Impinger, *J. Ind. Hyg. Toxicol., 24:* 80–82 (1942).
76. F. M. Renshaw, J. M. Bachman and J. O. Pierce, The Use of Midget Impingers and Membrane Filters for Determining Particle Counts, *Amer. Ind. Hyg. Ass. J., 30:* 113–116 (1969).
77. J. G. Olin, G. J. Sem and D. C. Christenson, Piezoelectric-Electrostatic Aerosol Mass Concentration Monitor, *Amer. Ind. Hyg. Ass. J., 32:* 209–220 (1971).
78. R. L. Chuan, An Instrument for the Direct Measurement of Particulate Mass, *J. Aerosol Sci., 1:* 111–114 (1970).
79. R. L. Fleischer, P. B. Price, and E. M. Symes, Novel Filter for Biological Materials, *Science, 143:* 249–250 (1964).
80. R. G. Stafford and H. J. Ettinger, Comparison of Filter Media against Liquid and Solid Aerosols, *Amer. Ind. Hyg. Ass. J., 32:* 319–326 (1971).

5

Sampling for Geometric Size Measurement

The direct observation of particles for size measurement requires that they be deposited on a surface suitable for examination in a microscope. It is of some advantage if the sampling method achieves this directly, avoiding the problem of transferring particles from one surface to another. Thermal and electrostatic precipitators are especially valuable in this application. For optical microscopy, membrane filters also permit direct examination of the deposited particles in many cases. Techniques are also available for transferring a sample from a membrane filter to a substrate that can be used in the electron microscope.

When making a determination of particle size distribution, the number of particles that is actually measured represents a very small fraction of those that are collected. If short-term samples are required, or if it is undesirable to sample large volumes of air, then it is advantageous to use an instrument that deposits the particles in a small area, permitting an adequate sample to be collected from a small volume of air in a short time. To characterize the capability of a given instrument in this respect, Baum [1] defined a figure of merit FM as

$$\mathrm{FM} = \frac{F}{A} \cdot E \qquad (5.1)$$

where F is the volumetric flow rate, A is the deposition area, and E is the efficiency with which particles are collected. This has units of centimeters per second and is analogous to deposition velocity. The larger the figure of merit, the less time that is required to collect a sample of sufficient density to permit a determination of size distribution.

The figure of merit alone is not an adequate criterion with which to judge methods for sampling to establish geometric size distributions. Since the distribution of particles over the sampling area often varies with particle size, the ratio of the total sampling area to the area included in the microscope analysis may be of concern also. The standard electrostatic precipitator used in sampling for mass concentration has a figure of merit of about 10, but its application for measurement of size distributions is hampered by the need to examine a number of different locations along the deposit before a reasonably representative sample of the particle sizes can be obtained [2]. It is much easier to cope with the segregation of particles in a deposit no larger than an electron microscope grid than it is in a deposit covering a length of several centimeters.

When sampling to obtain a size distribution averaged over a long period, it may be a disadvantage to use an instrument having a high figure of merit, because serious overlap of particles is possible. Attempts to circumvent this problem by flow rate adjustments may introduce other, more serious, problems related to the collection efficiency of the sampler as a function of particle size. These factors can be given proper consideration only if the operating characteristics of the instruments and methods are understood.

5-1 Thermal Precipitation

In 1870 Tyndall [3] reported the observation of a narrow dark region above a heated body in a dusty atmosphere. Several years later, Aitken [4] and Lodge and Clark [5], observed that the dark space completely surrounded the heated body. Lodge and Clark found that the dark space increased with decreasing pressure, if the temperature was kept constant. Aitken made a systematic study of the phenomenon, eliminating the various effects that had been proposed to explain it, and concluded that it was due to a thermal repulsion resulting from an unequal molecular bombardment of the particle on its hot and cold sides. As a matter of practical significance, he pointed out that the radiant heat from fireplaces kept furniture at a higher temperature than the surrounding air, causing a repulsion of particles from furniture surfaces so that rooms heated with fireplaces were cleaner than those heated with stoves or hot air.

The dust-free space around rods and plates was further studied by Watson

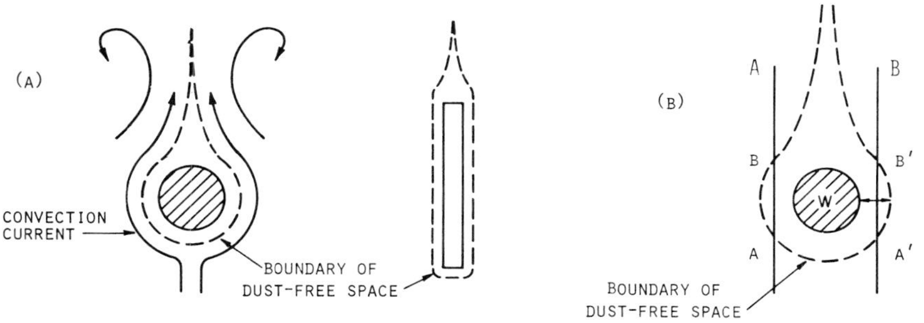

FIG. 5.1. Dust-free spaces around hot bodies [6]. Courtesy of The Faraday Society.

[6], who measured its thickness as a function of the temperature difference between the hot body and the surfaces enclosing it (see Fig. 5.1A). He found that the thickness of the dark space A increased at a rate that became essentially linear for temperature differences greater than about 70°C:

$$A = k \cdot \Delta T^{\alpha} \quad \text{cm.}$$

ΔT is in degrees centigrade. The values of k and α for the different hot bodies are tabulated below.

Hot body	k	α
Copper plate, 6 cm × 3 cm	0.0061	0.51
Copper plate, 3.1 cm × 3 cm	0.0054	0.55
Copper rod, 0.9 cm diameter	0.0035	0.49
Copper rod, 0.46 cm diameter	0.0027	0.55
Nichrome wire, 0.0254 cm diameter	0.0015	0.52

Watson studied clouds of magnesium oxide, flint dust, carbon black, tobacco smoke, and sulphur smoke, for which the particles sizes ranged from 0.1 to 2.0 μm, and found that neither the substance nor the size affected the value of A. He described the thermal precipitator as essentially a hot body having collecting surfaces placed close enough to it to intercept the dust-free space (Fig. 5.1B); any particles contained in air passing between the collecting surfaces will be deposited on them.

Watson's results were really applicable only to the particular experimental arrangement he used to obtain them. The thickness of the dark space calculated according to his empirical relationship cannot be relied on for design purposes because it does not include in its consideration the temperature gradient or

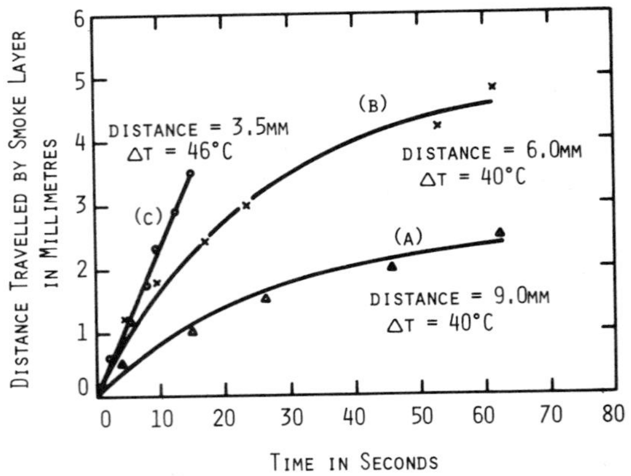

FIG. 5.2. Effect of plate separation on the movement of smoke particles in a temperature gradient [7].

the rate of flow of air through the precipitation region. Paranjpe [7] studied the convection currents between two parallel plates at different temperatures and found that the thickness of the dark space was greatly increased when the distance between the plates was decreased. For a temperature difference of 55°C, all convective action was destroyed when the plates were separated by 0.34 cm or less. At such separations, smoke particles moved from the hot plate to the cold plate at a constant velocity (Fig. 5.2), while at larger separations the velocity gradually lessened as the particle approached the cold plate. He also found that at a separation of 0.34 cm a constant velocity was observed at other temperature gradients. The relationship between velocity and temperature gradient ∇T was

$$U = 1.756 \times 10^{-4} \cdot \nabla T = 1.32 \times 10^{-4} \cdot K_s \cdot \nabla T \qquad \text{cm/sec.}$$

(The particles with which he was working had diameters of 0.5 μm and a slip factor K_s of 1.33). His results emphasize the need to base precipitator design on considerations of the transit time through the precipitation zone and the maximum time required for particles to be deposited.

5-1.1 Thermal Forces on Airborne Particles

According to Waldman [8], the thermal force acting on a small particle (radius less than the mean free path of the ambient gas molecules) in a temperature gradient is

$$F_{\text{th}}(D < 2\lambda) = -(P \cdot \lambda / T) \cdot D^2 \cdot \nabla T, \tag{5.2}$$

where P, λ, and T are the pressure, molecular mean free path, and temperature of the gas, and D is the particle diameter. The product $P \cdot \lambda$ depends only on temperature for a given gas. The negative sign is required because F is in the direction of increasing x when ∇T $(=dT/dx)$ is negative. Cawood [9] previously had derived an expression that differed from this only by the factor $8/\pi$. Rosenblatt and LaMer [10], in reporting the results of an experimental study of the force acting on tricresyl phosphate particles ($D > 2\lambda$) in a temperature gradient, pointed out that it was quantitatively the same as the force acting on a vane radiometer, a force that arises from the so-called "thermal creep" of gas molecules that occurs over the surface of an unequally heated body. The molecules move in the direction of increasing temperature, causing an increased pressure on the warmer side of the body. For a tangential temperature gradient $\partial T/\partial S$, the creep velocity is [11]

$$u = -\frac{3\eta}{4\rho T} \cdot \frac{\partial T}{\partial S} = -\frac{1.5}{\pi} \cdot \frac{P \cdot \lambda^2}{\eta T} \cdot \frac{\partial T}{\partial S},$$

where η is the gas viscosity. Epstein [12] calculated the radiometric force, including the thermal creep in his boundary conditions, and obtained the relationship

$$F_{\text{th}}(D > 2\lambda) = -9.0 \cdot \frac{P\lambda^2 D}{T} \cdot \frac{H_{\text{A}}}{2H_{\text{A}} + H_{\text{P}}} \cdot \nabla T, \tag{5.3}$$

where H_{A} and H_{P} are the thermal conductivities of air and particle, respectively. Rosenblatt and LaMer verified the effects of temperature gradient, pressure, and particle size on the thermal force, but the forces they observed were generally higher than predicted by Epstein's equation. Saxton and Ranz [13] measured the thermal forces on particles of a paraffin oil and of castor oil at atmospheric pressure and obtained results that were in good agreement with theory, leading to the conclusion that the theory was adequate for particles having diameters larger than twice the mean free path of the gas molecules.

As the use of thermal precipitators increased, it became apparent that they were highly efficient for a wide variety of particulate materials. This situation posed a theoretical anomaly that was not appreciated until Schadt and Cadle [14] made a comparative study of the collection efficiencies of several sampling devices. They found that particles of substances having large thermal conductivities were subjected to thermal forces 30 to 50 times greater than predicted by Epstein's equation. Their results prompted additional theoretical work, particularly by Brock [15, 16], who observed that the viscous slip of gas molecules at the particle surface had been omitted by Epstein in setting up his boundary conditions. He derived the following expression for the

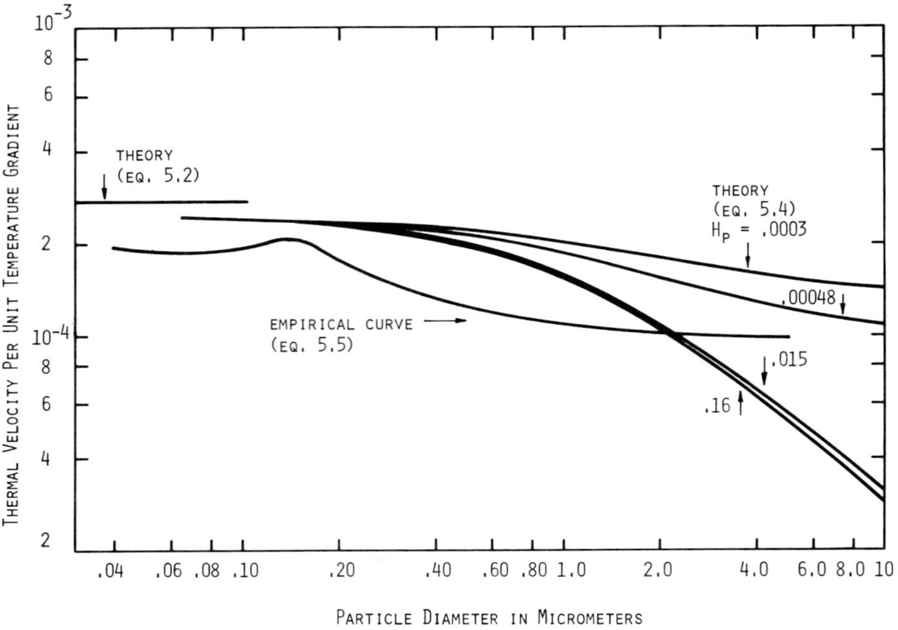

FIG. 5.3. Thermal velocities as a function of particle diameter. Calculated from the indicated equations for force per unit temperature gradient. (Velocity is in centimeters per second for thermal gradient in degrees centigrade per centimeter. Curves refer to air at STP.)

thermal force:

$$F_{\mathrm{th}}(D > 2\lambda) = -9.0 \cdot \frac{P\lambda^2 D}{T} \frac{(H_A/H_P) + C_t(2\lambda/D)}{[1+3C_m(2\lambda/D)][1+2(H_A/H_P)] + [2C_t(2\lambda/D)]} \cdot \nabla T \tag{5.4}$$

The constants C_t and C_m are related respectively to the thermal accommodation coefficient and the reflection coefficient of molecules striking the particle surface. The corresponding thermal velocity, $U_{\mathrm{th}} = F_{\mathrm{th}} \cdot Z$, has been calculated for several values of H_P, taking $C_t = 2.5$ and $C_m = 1.0$. The results are shown in Fig. 5.3. They have been plotted as U_{th}/G against the diameter in micrometers where

$$G = \nabla T(300/T).$$

The available data concerning thermal forces acting on particles in air have been plotted in Fig. 5.4. The ordinate is the ratio between the thermal force per unit temperature gradient, normalized to 300°K, and the square of the particle diameter; the abscissa is the ratio of the mean free path of air molecules to the particle radius. There is no clear effect of thermal conductivity. The single points for iron and platinum, which have almost identical thermal conductivities (see Table 5.1), should not differ appreciably from the data for sodium chloride. Instead, they show marked, but contrasting, differences. The two straight lines used to represent the data have the following equations:

$$F/D^2G = 9.85 \times 10^{-3}[1+0.27(\lambda/D)] \quad \text{dyn/cm}-°\text{K}, \quad D < 0.15 \quad \mu\text{m}$$
$$F/D^2G = 25.8 \times 10^{-3}(\lambda/D), \quad D > 0.15 \quad \mu\text{m}. \tag{5.5}$$

The straight line corresponding to the first equation was actually based on points out to $(2\cdot\lambda/D) = 4$, the maximum value for which data were available.

The terminal velocities of particles in a temperature gradient, according to the above equations, are given by

$$U/G = 5.6 \times 10^{-4}K_s\cdot(D+0.0176) \quad \text{cm}^2/°\text{K-sec}, \quad D < 0.15 \quad \mu\text{m}$$
$$U/G = 9.53 \times 10^{-5}\cdot K_s, \quad D > 0.15 \quad \mu\text{m}. \tag{5.6}$$

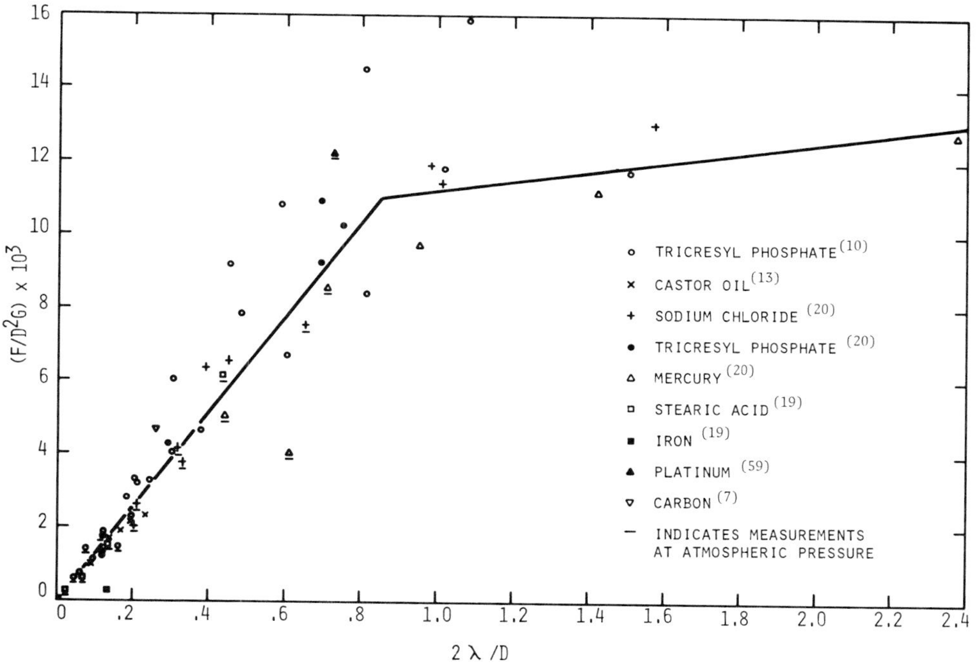

FIG. 5.4. Thermal forces per unit temperature gradient as a function of Knudsen number $(2\lambda/D)$.

TABLE 5.1

THERMAL CONDUCTIVITIES OF SEVERAL MATERIALS

Material	H_P (cal/cm–sec–°K)	Relative to air[a]	Ref.
Stearic acid	3.0×10^{-4}	5.3	[19]
Sodium chloride	1.55×10^{-2}	272	[19]
Iron	0.16	2,810	[18]
Mercury	0.02	351	[18]
Platinum	0.167	2,930	[18]
Silver	1.01	17,540	[18]
Asbestos	1.9×10^{-4}	3.3	[18]
Carbon	0.01	175	[18]
Granite	0.005	87.5	[18]
Glass	0.002	35	[18]
Magnesium oxide	$1.6\text{–}4.5 \times 10^{-4}$	2.8–7.9	[18]
Quartz	0.016, 0.030	281–526	[18]
Silica fused	0.0024	42	[18]
Tricresyl phosphate	4.8×10^{-4}	8.4	[20]
Glycerol	6.37×10^{-4}	11.2	[20]
Castor oil	4.32×10^{-4}	7.6	[13]
Paraffin oil	2.97×10^{-4}	5.2	[13]
Clay	0.0017	30	[18]
Air	5.7×10^{-5}		[17]

[a] $H_A = 4.67 \times 10^{-6} [T^{3/2}/(T+125)]$[18]. Ratios are for $T = 300$°K.

K_s is the particle's slip factor and D is its diameter in micrometers. For D greater than about 1 μm, the velocity per unit temperature gradient, normalized to 300°K, is very nearly constant and equal to 9.5×10^{-5} cm^2/°K-sec. The curve of U/G as a function of diameter and Waldman's theoretical curve have been included in Fig. 5.3. With the exception of Waldman's curve, the velocities were calculated by equating the thermal force to the Stokes–Cunningham resistance force acting on the particle. Some doubt has been expressed concerning the validity of this procedure [13]; however, Postma [57] has shown that the doubt is unfounded, at least for particle velocities described by the second equation of (5.6).

The empirical equations predict an almost constant velocity for $D > 1$ μm, in contrast to the theoretical predictions of a rapidly decreasing velocity, as particle size increases, for materials having large values of H. Few data are available on the force acting on large particles, but the deposition patterns observed in thermal precipitators do not suggest a rapidly changing thermophoretic velocity in the size range 1–10 μm. At larger sizes, however, the prediction of a constant velocity becomes increasingly inaccurate.

5-1.2 Thermal Precipitators

Aitken's investigations [4] included construction of a "thermic filter," identical in principle to modern thermal precipitators. Using his condensation nuclei counter, he was able to demonstrate that the thermic filter removed all particles from air passing through it. Lomax and Whytlaw-Gray [21] adopted the principle for the design of an instrument with which to collect smoke particles. Green and Watson [21] improved their design, developing the instrument shown schematically in Fig. 5.5. The success of this instrument, which is still widely used, has stimulated development of a number of other thermal precipitators, usually designed to overcome one or another of the shortcomings of the original. For a number of years, the Green and Watson precipitator has been the reference sampler for measuring dust concentrations in British mines and its sampling characteristics have been studied thoroughly. Since its use in mines was primarily to estimate the number of respirable particles per unit volume of air, the studies emphasized the size range ~0.5–10 μm. Many of the problems encountered with it, however, are common to other precipitators, particularly those having wires or metal ribbons as the hot element.

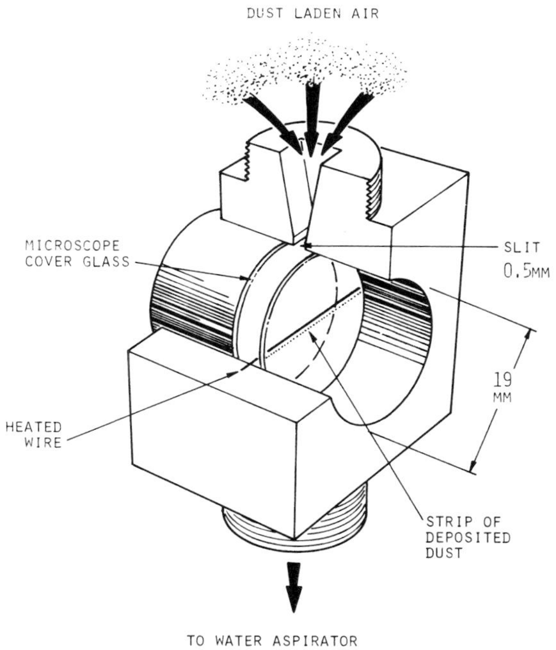

Fig. 5.5. Diagram of the standard thermal precipitator [58].

In the mines, the precipitator is operated with the inlet up so that particles cannot settle on the collecting surfaces. If the inlet were of constant cross-sectional area A, the number of particles available for deposition per unit time would be

$$C \cdot A \cdot (V_a + U_G), \tag{5.7}$$

where C is the concentration of particles in the air, V_a is the average velocity of air entering the duct, and U_G is the terminal settling velocity of the particle. The tapered inlet actually used has a volume of about 0.65 cm^3. The average transit time through it is about 6 sec, during which many particles settle on the inlet walls. However, Walton [22] has shown that the additional particles falling into the widened inlet exactly match the number lost to the walls and Eq. (5.7) is valid for the Green and Watson precipitator when A refers to the region between the collecting plates. Walton assumed velocity profiles characteristic of nonviscous flow, which is probably reasonable for such short converging ducts.

The settling velocity U_G that makes an excess number of particles available for deposition shortens the transit time through the deposition region, tending to reduce the efficiency of collection. The two effects combine to give an apparent collection efficiency that exceeds the real thermal deposition efficiency by a factor of 2 at $D = 30\ \mu m$ [23]. Above about 2.5 μm the real efficiency begins to decrease from 100%. Efficiency is often lowered by leakage, an undesirable effect that is especially difficult to avoid in precipitators that have moving deposition surfaces.

The empirical curves of Fig. 5.3 indicate that the precipitator deposit should show some segregation of particles on the basis of size, the smaller particles being deposited first and those in the range 0.5–10 μm showing roughly similar deposition patterns. This is borne out by the experimental deposition curves in Fig. 5.6, based on the histograms prepared by Ashford [24], and by the general observation that the small particles appear at the leading edge of the deposit [52]. Despite the overall similarity of the deposition patterns, it is not surprising that quite different size distributions can be obtained in different microscope fields along the deposit [25]. The studies of Dawes *et al.* [25] brought out the nonuniform pattern of deposition in the precipitator with respect to both number of particles per unit area and size distributions, and demonstrated that displacement or bowing of the wire, too slight to be readily detected, could cause marked differences in the deposition patterns on the two collecting surfaces (see Fig. 5.7).

The deposits in the thermal precipitator are consistently close to 0.175 cm in width, so that the volume of the region defined by the two deposits, and within which effective thermal forces occur, is 0.0084 cm^3, when allowance is made for the wire. At 6.75 cm^3/min, the average transit time of air passing through

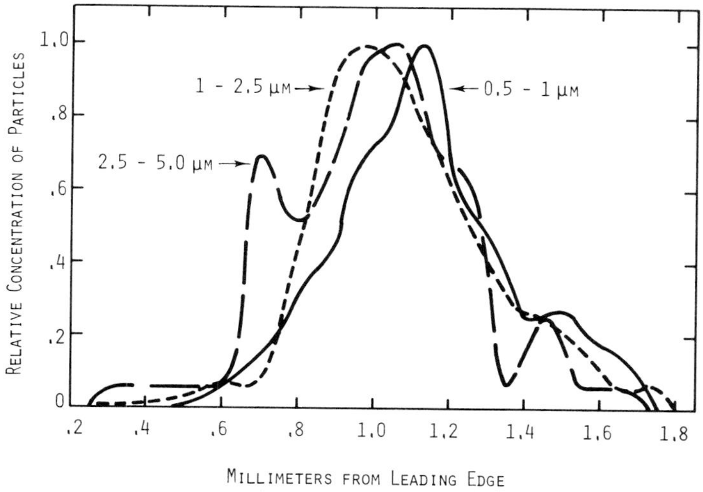

FIG. 5.6. Deposition pattern in thermal precipitator sample.

this volume is 0.075 sec. Since the upper size limit at which complete thermal deposition occurs at this flow rate is about 2.5 μm, the maximum deposition time of particles having no significant sedimentation velocity is

$$t_D = 0.075 = W_d/(9.5 \times 10^{-5} K_s G) = W_d W_e/(9.5 \times 10^{-5} K_s \Delta T), \quad (5.8)$$

where W_d is the maximum distance normal to the collecting surface that a

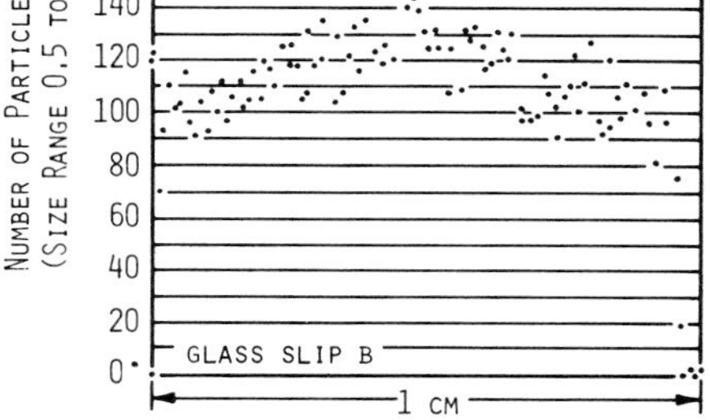

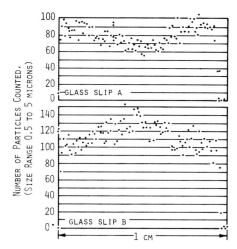

FIG. 5.7. Effect of bowed wire on sample deposits in standard thermal precipitator [25]. Each point is count of one traverse of 25 μm width.

particle must move owing to the thermal gradient, and W_e is the distance defined by the effective temperature gradient, $\Delta T/W_e$. Setting $W_d = \alpha_d W$ and $W_e = \alpha_e W$, where W is one-half the separation between the collecting plates, and taking $K_s = 1.07$, $\Delta T = 80°C$,

$$\alpha_d \cdot \alpha_e \cdot W^2 = 6.16 \times 10^{-4}$$

and

$$\alpha_d \cdot \alpha_e = 0.945.$$

Equation (5.8) and this relationship are used below to calculate the maximum time required to effect complete collection of particles that have sedimentation velocities much smaller than the velocity of air moving through the deposition region.

Pertinent data concerning some of the thermal precipitators that have been described in the literature are given in Table 5.2. Each of those included in the table has some design characteristic that distinguishes it from the others. Transit and deposition times are defined as given above, except that $W_d = W_e$ is the separation between the hot and cold surfaces which form the parallel plate precipitators. A comparison between the transit and deposition times suggests that the Lauterbach [26] and Walkenhorst [27] precipitators could be operated with greater confidence at lower flow rates and raises some question concerning the efficiency of the moving tape precipitator when operated under the specified conditions.

The precipitators having oscillating or rotating collecting surfaces were designed to reduce the effects of overlap and eliminate the problem of size segregation. Oscillating mechanisms are somewhat more elaborate than are rotating mechanisms. The Lauterbach precipitator increases the total deposition area by a factor of about 8 and reduces the maximum density of the deposit by about 12 times. In the precipitator described by Cember *et al.* [28], the rotating collecting surface is opposite a hot wire that is half embedded in the body of the precipitator. Since deposition always occurs beneath the wire, the fraction of incoming particles that is deposited in an element of length dR at a distance R from the midpoint of the wire is

$$f(R) = dR/R_0,$$

where R_0 is one-half the effective length of the wire. The rotation of the collector causes this fraction to be spread over an area of $2\pi R \cdot dR$, so that the relative density of particles at a distance R from the center of rotation is

$$\rho(R) = C(R)/N_T = (2\pi R R_0)^{-1} \qquad \text{cm}^{-2}.$$

$C(R)$ is the number of particles per unit area at R and N_T is the total number collected. The finite width of the deposit will cause some deviation from this

TABLE 5.2

CHARACTERISTICS OF VARIOUS THERMAL PRECIPITATORS

Ref.	Configuration	Surface separation (cm)	Temperature: Difference (°C)	Temperature: Effective gradient (°C/cm)	Flow rate (cm^3/min)	Flow rate (cm/sec)	Deposit: Length (cm)	Deposit: Width (cm)	Transit time (sec)	Deposition time (sec)	Remarks (FM)
[21]	0.0254 cm wire between glass cover slips	0.051[a]	80	3150	6.5	2.1–4.2	0.175	1.0	0.078	0.075	(0.32)
[26]	0.0318 cm wire	0.102[a]	230	4530	15	1.5–2.2	0.175	1.6	0.11	0.11	Collection surfaces oscillate (0.45)
[27]	0.005 × 0.15 cm ribbon between two glass cover slips	0.026[a]	56	5600	10	6.5–7.7	0.15	1.0	0.018	0.018	(0.56)
[28]	0.0254 cm wire half embedded opposite cover slip	0.0254	—	—	10	6.5–13	—	—	—	—	Rotating collecting surface
[29]	Truncated cone in 2.54 cm diameter cylinder, 8.25 cm long	0.143– 0.016	130 150	900–8100 1040–9350	255 1600	4.1–42 25.8–264	~8.25 ~8.25	— —	1.12 0.184	0.10 0.086	— —
[30]	Parallel circular plates, air enters through 0.9 cm diameter hole in center of hotplate of 7.62 cm diameter.	0.038	100	2640	300	5.6–46	~3.8 available	—	0.34	0.15	(0.25)

TABLE 5.2 (Continued)

Ref.	Configuration	Surface separation (cm)	Temperature: Difference (°C)	Temperature: Effective gradient (°C/cm)	Flow rate (cm^3/min)	Flow rate (cm/sec)	Deposit: Length (cm)	Deposit: Width (cm)	Transit time (sec)	Deposition time (sec)	Remarks (FM)
[31]	Parallel plate, 1 cm diameter hot plate with 0.16 cm inlet hole	0.038	125	3290	12		~0.5 cm available	—	0.15	0.12	Cold plate rotates with periphery under hot plate
[32]	Parallel plate, hot plate 7.62 cm diameter with ~1 cm hole. Air enters around periphery.	0.0508	48	956	1000	14–114	~3.8 available	—	0.136	0.51	Moving tape serves as cold plate

[a] Separation between collecting surfaces.

relationship, but near the periphery the deviation should be small. This method of collection should have some real advantages for studying overlap.

The parallel plate precipitators having stationary deposition surfaces show a segregation of particles according to size as does the Green and Watson precipitator [33]. Neither the moving tape collector nor the collector rotating off-axis can provide a deposition that is uniform with respect to size over the whole area of collection; however, adequate size distributions should be obtained along the center line of the deposit. They have the advantage of collecting samples large enough for accurate chemical or gravimetric analysis.

5-2 Electrostatic Precipitation

Electrostatic precipitators that deposit a sample over a large area have been found unsuitable as a means of collecting samples for particle size determination. With the standard concentric cylinder arrangement described in Chapter 4, it was deemed necessary to use a total of 12 electron microscope grids along the length of the deposit in order to get an adequate sample for sizing [2]. Fraser [34] used four grids in a precipitator designed especially for microscopic examination of the sample and determination of particle size distributions [35]. Each of the four grids, when examined by electron microscopy, provided a different size distribution and all of them differed from the distribution obtained from a membrane filter sample of the same aerosol. When the grids were examined in the optical microscope, however, the median diameters of the particles on the various grids were in good agreement, leading Fraser to conclude that the poor agreement observed when electron microscopy was used resulted from rapid deposition of submicron particles at the leading edge of the deposit.

The effect that Fraser observed is due to the charging characteristics of small particles (see Section 4-2.3). It cannot be eliminated by adopting different electrode configurations; however, the problem of microscopic examination can be greatly reduced by collecting the entire sample on a single electron microscope grid. To do this, it is necessary to restrict the charging and precipitation to a very small volume while maintaining a large ion density and a strong precipitating field. Instruments in which a unipolar ion field is produced by a point-to-plane corona discharge or by a radioactive isotope have proved successful for this purpose.

5-2.1 Point-to-Plane Precipitators

Two electrode configurations for this type of precipitator are shown in Fig. 5.8. The one in Fig. 5.8(A) was used by Baum [1], who was apparently the first to adapt the point-to-plane corona to this purpose. In this model, the air entered

(A)

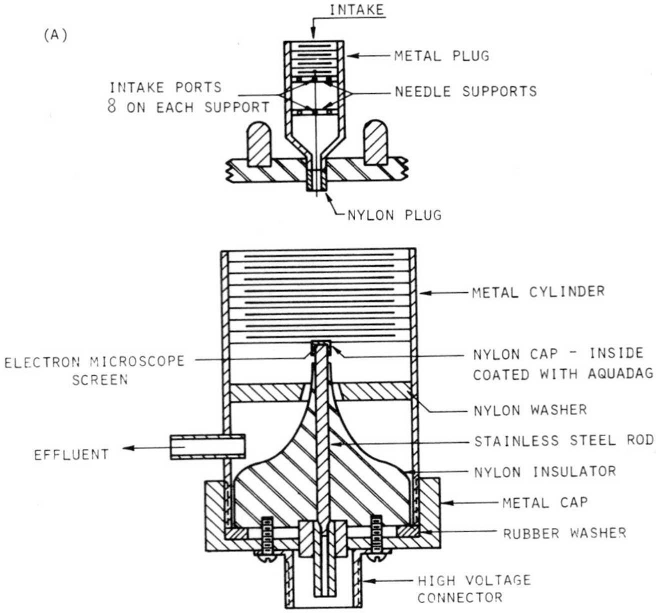
INTAKE
METAL PLUG
INTAKE PORTS
8 ON EACH SUPPORT
NEEDLE SUPPORTS
NYLON PLUG
METAL CYLINDER
ELECTRON MICROSCOPE
SCREEN
NYLON CAP - INSIDE
COATED WITH AQUADAG
NYLON WASHER
EFFLUENT
STAINLESS STEEL ROD
NYLON INSULATOR
METAL CAP
RUBBER WASHER
HIGH VOLTAGE
CONNECTOR

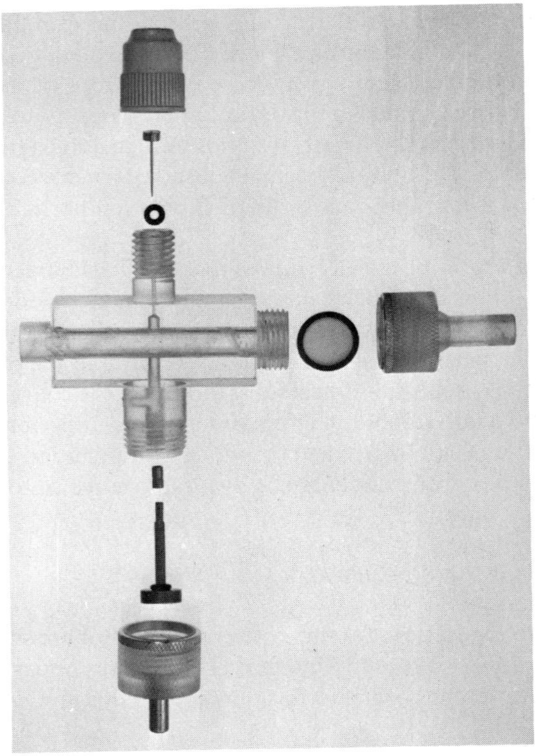

around the high voltage electrode in an annulus having an outside diameter equal to the diameter of the collecting electrode. At a flow rate of 2.8 liters/min, the collection efficiency was 37% at 4 kV on the point and 25 μA, and 41% at 5 kV and 100 μA. A negative corona was used.

Several precipitators similar in design to that shown in Fig. 5.8(B) have been described [36–38], probably the most widely used being those developed at Harvard [36] and at Rochester [38]. The Harvard instrument can operate with an ac high voltage but dc voltages, usually negative, are more common. Operating voltages range from about 4 to 15 kV and the average linear flow rates range from about 1.5 to 70 cm/sec. In the prototype of the present Rochester instrument [40], all of the sampled air passed over the collecting electrode and approximately 65% of the available aerosol mass was collected on the electron microscope grid. In subsequent models, and in all the other point-to-plane precipitators, only a fraction (sometimes very small) of the incoming air passes between the electrodes. Characteristics of these precipitators, and others that are described below, are given in Table 5.3.

Particles acquire a charge and undergo precipitation in point-to-plane precipitators in the same manner as in the concentric cylinder precipitators, but it is more difficult to predict the electric fields and ion densities in the former. The voltage–current characteristics of the two types of precipitators are similar in shape, and the differences between the negative and positive coronas are similar also, as the curves of Fig. 5.9 show. The positive corona usually begins at a voltage higher by about 30 to 40% than that required to start a negative corona [41]. The electric field between the point and the plane, before there is any appreciable flow of current, gives rise to electric lines of force such as those shown in Fig. 5.10. The potential decreases with distance from the point in the manner shown in the same figure. When current flows, the space charge within the gap raises the potential at all points between the electrodes.

It is apparent from Fig. 5.10 that the region between the electrodes within which particles will be subjected to charging and precipitation on the grid is greater in area than a triangle having the sample area as its base and the distance from grid to point as its altitude, but is less than a rectangle having the same dimensions as its sides. Therefore, half the sum of these two areas has been taken as the region within which particles will be subject to deposition in the sample area. The ratio of this area to the area of the tube normal to the direction of air flow has been taken as the geometric efficiency and has been used with Eq. (5.1) to calculate the tabulated figures of merit.

FIG. 5.8. Point-to-plane precipitators. (A) Baum [1] precipitator with annular flow of air past point; (B) exploded view of U.R. [38] precipitator with inlet directed left, point assembly above, and electron microscope grid holder below. Courtesy of American Industrial Hygiene Association.

TABLE 5.3

CHARACTERISTICS OF SEVERAL ELECTROSTATIC PRECIPITATORS

Type	High voltage (kV)	Current (μA)	Electrode separation (cm)	Volumetric flow rate	Re	Average linear velocity (cm/sec)	Effective area of deposit (cm^2)	Geometric efficiency	Figure of merit (cm/sec)	Transit time to sample zone (sec)	Ref.
Point-to-plane annular flow	4.5 neg.	25–100	0.3	2.8 liters/min	1240	590.0	0.031	—	5–600	—	[1]
Point-to-plane normal flow	5 neg.	~5	~0.6	70 cm^3/min	24	11.5	0.044	0.57	15	0.3	[40]
Point-to-plane normal flow	10–15 a.c.	—	3.8	5.0 liters/min	185	7.3	0.044	0.058	110	0.87	[36]
Point-to-plane normal flow	7 neg	3.5	1.0	70 cm^3/min	10	1.6	0.044	0.25	7	2.5	[38]
H^3 ionization normal flow	2 neg.	0.0022	0.16	7 cm^3/min	6	5.8	0.026	1.0	4.5	0.43	[43]
Pulsed precipitator	2.3 neg. to charge; 4.2 neg. to precipitate	2	0.77	2.6 liters/min	368 98	55.3 (charge); 10.9 (precipitate)	102	—	0.17[a]	1.8	[44]
Point-to-plane normal flow	7 a.c.	—	0.75	0.38–5.3 liters/min	43–590	5–70	0.044	0.047	6.7–94.5	—	[39]

[a] Based on volume sampled onto uniform deposit area.

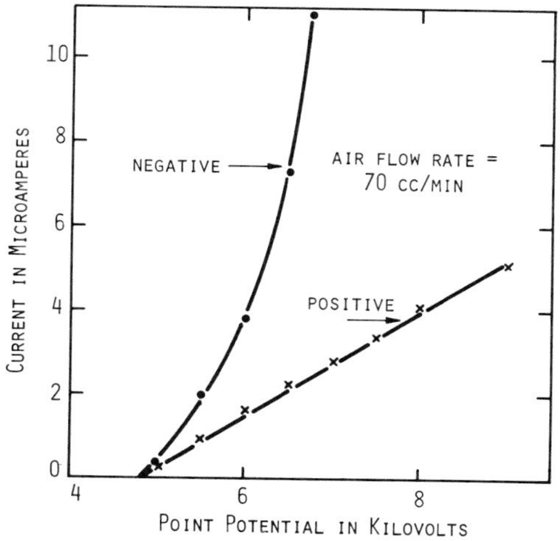

FIG. 5.9. Current-voltage characteristics for U.R. point-to-plane precipitator.

Because all of these instruments were designed to be used to obtain particle size distributions, their gross collection efficiencies have not been studied to any great extent. McGreevy [42] studied the efficiency of the Harvard precipitator for particles of 0.028 to 0.19 μm diameter, using a condensation nuclei counter to determine the fraction of incoming particles that was deposited in the precipitator. He found a considerable variation in efficiency with particle size.

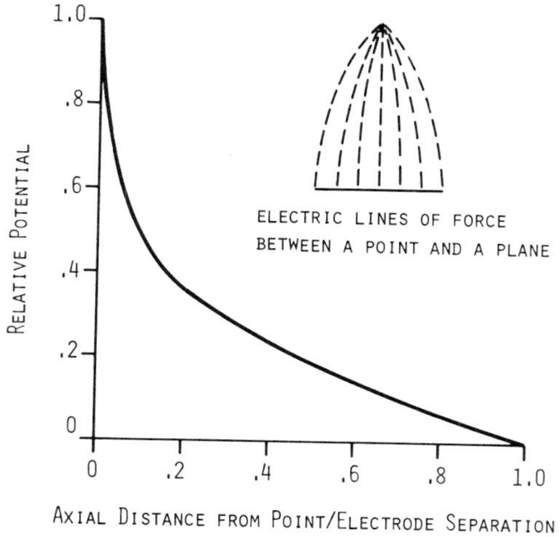

FIG. 5.10. Potential drop between electrodes of point-to-plane precipitator.

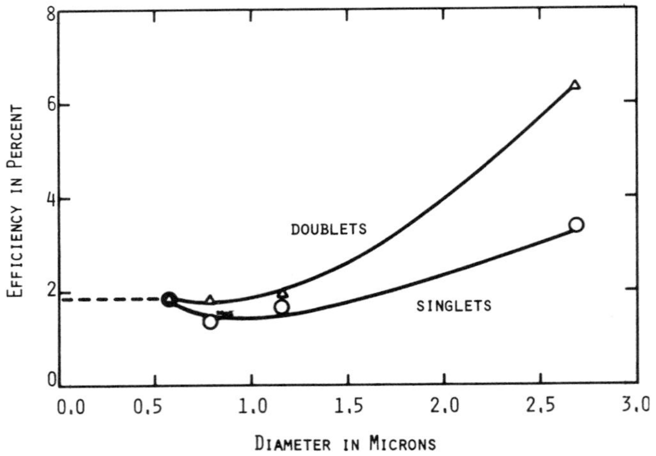

FIG. 5.11. Efficiency as a function of particle size for point-to-plane precipitator [39].

Since only a small fraction of the incoming aerosol passes between the electrodes of this instrument, which is normally operated until deposited aerosol is visible on the glass walls near the electrodes [36], McGreevy's observations are largely irrelevant to its use in size distribution studies. A more useful study of efficiency was made by Reist [39], using a precipitator of similar design and operating characteristics, but having somewhat smaller dimensions. He compared the number of particles actually deposited in the sampling area of the collecting grid with the total number entering the precipitator. The particles were produced by nebulization of suspensions of polystyrene latexes. Collection efficiencies for singlets and doublets were determined separately. For linear air velocities between 5 and 70 cm/sec, Reist found efficiencies somewhat less than those expected on the basis of geometry (see Fig. 5.11). He found an increased efficiency at larger sizes but felt there was no effect due to flow rate.

The Lovelace ESP (Fig. 5.12) [43] was designed to provide a deposit adequate for particle size analysis from a sample of a very small volume of air highly contaminated with radioactivity. The point-to-plane precipitators require higher voltages to produce the corona than are needed in the precipitation process itself, and the corona is not always stable, leading, at times, to an "arcing over" effect that destroys the collecting surface of the electron microscope grid. To avoid these disadvantages, a tritium source was used to provide the necessary ionization. The relatively long half-life of tritium (about 12 years) prevents any rapid decay of the strength of ionization and the short range ($\sim$0.3 cm) of its beta particle in air makes it possible to confine the region of bipolar ionization to a very small volume. This

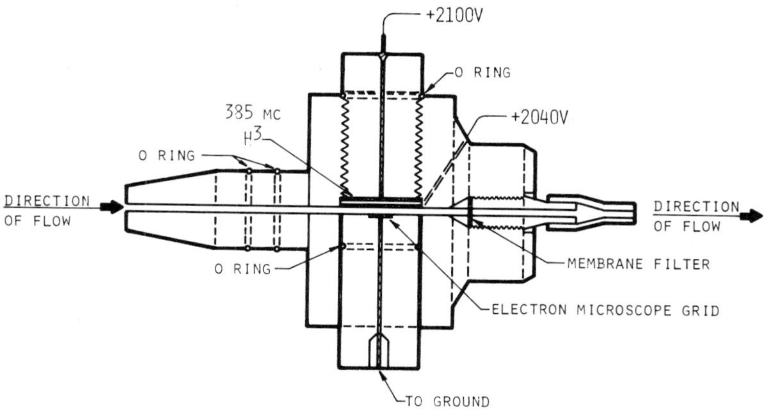

FIG. 5.12. Diagram of the Lovelace electrostatic precipitator [43].

volume is contained between the tritium source, which is deposited on a stainless steel foil, and a thin brass disk of the same size that separates the ionization space from the air flow channel. When the tritium source is held at 2100 V and the separator at 2040 V, there is a flow of ions through a 1/16 in. square hole in the center of the separator to the electron microscope grid serving as the ground plate. Under normal operating conditions, the ion current is 0.086 $\mu A/cm^2$. If it is assumed that the field strength between the separator disk and the grid is given by V_s/h, where V_s is the voltage on the separator disk and h is its height above the grid, then the ion density can be calculated as

$$N = 6.3 \times 10^{18} i \cdot h/\mu \cdot V_s = 2.6 \times 10^7 \quad \text{cm}^{-3}$$

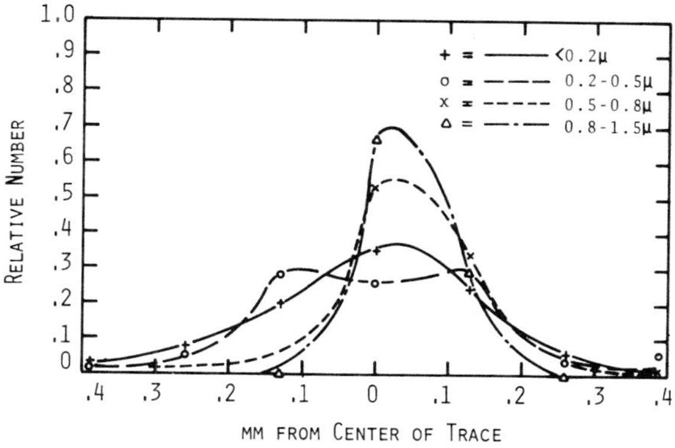

FIG. 5.13. Deposition pattern of particles collected in the Lovelace precipitator.

for $h = 0.16$ cm, $\mu = 1.6$ cm^2/V-sec for positive ions, and current (i) and voltage as given above. At 7 cm^3/min, the time required for the air stream to traverse the ion current region is $t = 0.035$ sec. The minimum value of Nt is then 9.1×10^5 sec/cm^3, which should lead to adequate charging of the particles. Despite the short distance within which precipitation occurs, there is a size-segregation effect that makes it necessary to examine electron micrographs of several positions along the deposit in order to get a representative sample (see Fig. 5.13). A comparison of this figure with that for the thermal precipitator indicates that the segregation effect in the two instruments is very similar. The effect probably occurs with the point-to-plane precipitators, also, unless turbulent conditions occur in the precipitating region, but it does not appear to have been studied.

The size segregation effect has been very nearly overcome by the use of a pulsed precipitating voltage [44]. A schematic diagram of an instrument that operates in this manner is shown in Fig. 5.14. The necessary corona is produced

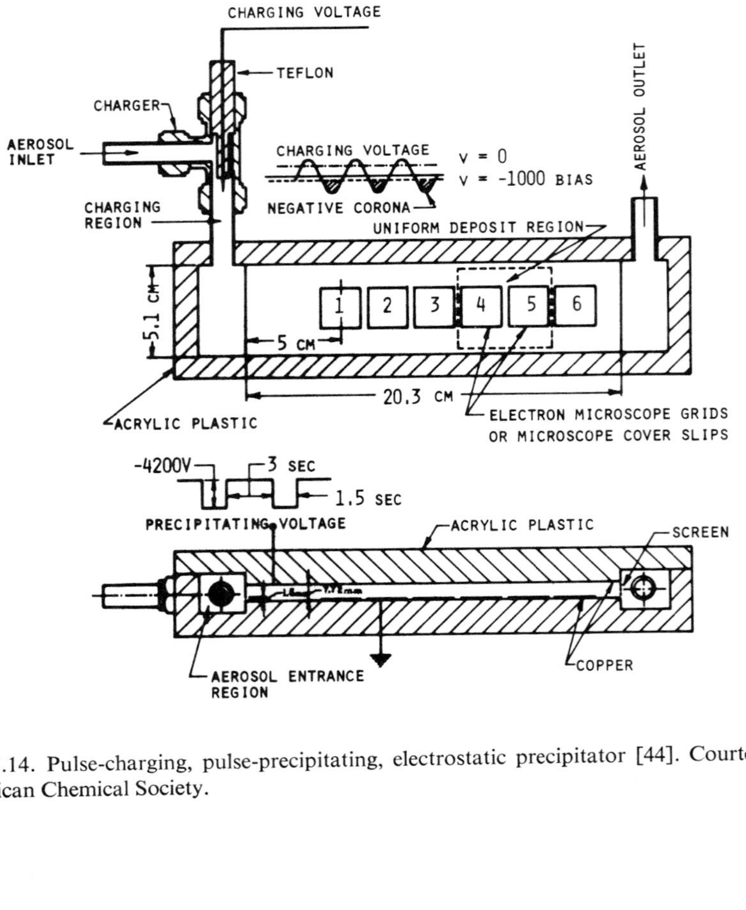

FIG. 5.14. Pulse-charging, pulse-precipitating, electrostatic precipitator [44]. Courtesy of American Chemical Society.

by impressing 1.4 kV ac voltage on a point having a negative 1 kV bias voltage on it. A 2 μA corona current is obtained, which provides a good charge on the particles. The alternating voltage tends to prevent deposition of particles in the charging region, and they pass into the precipitation region between two plane parallel electrodes. The precipitating voltage is on one-third of the time for periods of 1.5 sec. During the intervening 3-sec periods, the precipitating region is flushed with fresh aerosol. Since the air flow continues during the on-interval, there is a tendency for size segregation to occur, but in the areas marked in Fig. 5.14 the effect is minimal and the relative collection efficiency varies between 0.54 at 0.16 μm and 0.81 at 1.8 μm.

Several comparisons have been made between size distributions obtained from electrostatic precipitator samples and those obtained from thermal precipitator samples when the instruments were sampling the same aerosol. Agreement between the two methods has generally been good, but the comparisons have been limited to size distributions having count median diameters below 1 μm. Figure 5.15 presents the pertinent data from several comparisons

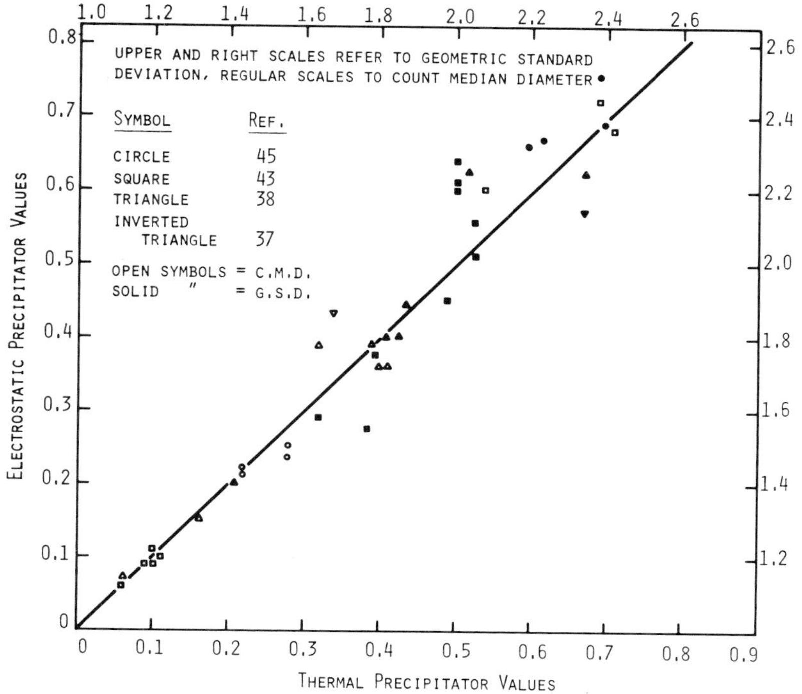

FIG. 5.15. Comparison of size distribution statistics according to electrostatic and thermal precipitators. (Line represents exact agreement.)

of this sort. The study by Ettinger and Posner [45] brought out the fact that variations of $\pm 10\%$ in count median diameters were overshadowed by variations of $\pm 15\%$ in geometric standard deviation, with the result that estimates of the mass median diameter differed by as much as 2.7 times for any one sampler and 6.7 times for all samplers.

5-2.2 Errors Associated with Particle Deposition

The process of particle deposition on a sampling surface may lead to errors both in size measurement and in the estimation of particle concentration. Fraser [34] observed that electron microscope grids having conductive films were required for proper results with electrostatic precipitators, an observation confirmed by others [36]. Otherwise, the particles were preferentially deposited on the grid wires where they could not contribute to the useful sample. A similar preferential deposition of particles on grid wires occurs in the thermal precipitator [46–48], but it cannot be offset using metal-coated or carbon films. Zebel [48] found that the grid openings receive, on the average, less than two-thirds of their proper share of particles. He did not mention any size discrimination, but the possibility of sample bias cannot be ignored. Cartwright [46] recommends that thermal precipitator samples be collected on formvar-coated cover slips from which the formvar film can be floated after the collection of a sample and transferred to an electron microscope grid. When the sample is one of soluble particles, the collecting slide can be shadowed and replicated.

Particles may be deposited on particles that were collected earlier, giving rise to spurious aggregates that cause concentrations to be underestimated and sizes to be overestimated. The problem of overlap was treated theoretically by Irwin *et al.* [49], who considered the particles to be circular laminae, all of the same diameter D. They found that if N laminae are deposited in an area A, and C particles are counted in the same area when overlapping laminae are treated as single units, then

$$N = (-A/\pi D^2) \cdot \ln[1 - C \cdot (\pi D^2/A)]. \tag{5.9}$$

Later, Armitage [50] showed that this equation is valid, also, for a distribution of sizes, if D^2 is replaced with

$$(v + 2\bar{D}^2)/2,$$

where v is the variance and $\bar{D}$ the average diameter of the distribution. For a lognormal distribution of median diameter D_g and geometric standard deviation $\sigma_g = e^{\sigma}$,

$$(v+2\bar{D}^2) = D_g^{\ 2}\cdot(e^{2\sigma^2}+e^{\sigma^2}) = D_g^{\ 2}\cdot e^{2\sigma^2}\cdot(1+e^{-\sigma^2}). \qquad (5.10)$$

The use of Eq. (5.9) with D^2 replaced by the *average squared diameter*, $\overline{D^2} = D_g^{\ 2}\cdot e^{2\sigma^2}$, is accurate only for small values of σ. The total particle area is 4–5% of the sampling area at an overlap error of 10%.

Roach [51] made an experimental study of overlap in thermal precipitator samples of coal dust, fitting his experimental data to the equation

$$C = N\cdot e^{-aN}, \qquad (5.11)$$

where C and N are, respectively, the number of units counted and the number of particles deposited in a 20 μm wide by ~1 mm long traverse across the deposit. For typical samples, he found $a = 0.0026$ when all sizes were included and 0.0048 when only sizes between 1 and 5 μm were considered. For comparison, Eq. (5.9) can be rearranged to give

$$C = (A/\pi D^2)\cdot(1-e^{-m}),$$

where $m = N\pi D^2/A$. If we set $2a = \pi D^2/A$, these two equations give results

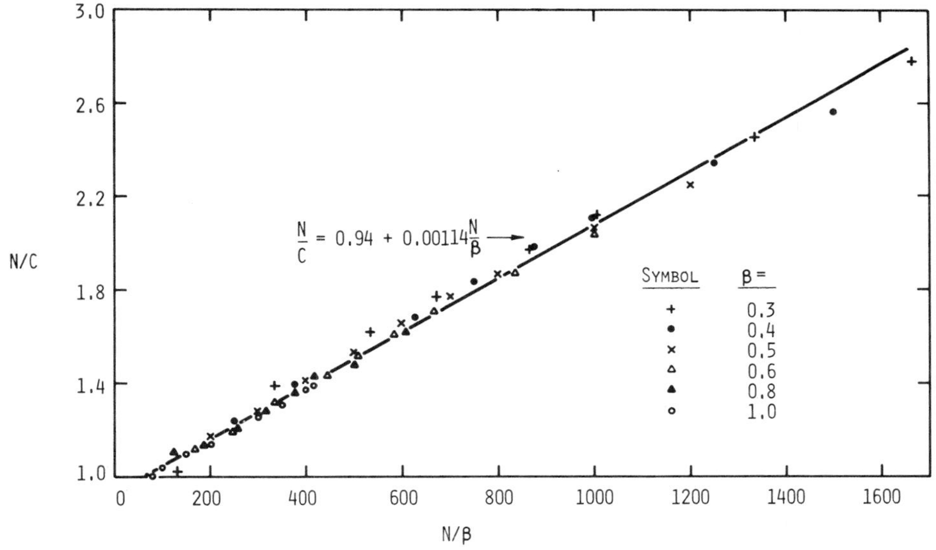

FIG. 5.16. Correlation of overlap data for various values of size parameter of a truncated exponential distribution. (1μm ⩽ D ⩽ 5 μm).

that differ by less than 4% for $m \leqslant 0.8$. Thus, the theoretical and experimental results agree in form. However, when a is calculated from the experimental values of D and A, the result is not in good agreement with the experimental value.

Ashford *et al.* [52] made an extensive investigation of overlap in samples of the type studied by Roach. They expressed size distributions in a truncated exponential form:

$$f(D) = \beta e^{\beta(1-D)}.$$

This is Eq. (3.45) applied to a population of particles from which all particles smaller than 1 μm have been eliminated. Normalization of the frequency distribution function introduced a constant term that includes e^{β}. They condensed their overlap data into a family of curves showing N as a function of C, with β as the parameter distinguishing the curves. The traverses across the precipitator deposit, to which C and N were referred, were 30 μm wide by 1 mm long. Their family of curves can be consolidated reasonably well if the correction factor, N/C, is plotted against N/β, as shown in Fig. 5.16. The equation for the straight line in that figure can be rearranged to give

$$N = 0.94C/(1-0.00114C/\beta). \tag{5.12}$$

Since all of the data plotted in Fig. 5.16 refer to a constant sampling area, the relationship should be valid for similar size distributions under other sampling conditions, if proper allowance is made for differences in sampling area. That this is, unfortunately, not true is apparent from Fig. 5.17 in which curves for Roach's empirical equation and Armitage's theoretical equations have been plotted for comparison with the straight line of Fig. 5.16. From Roach's data, $A = 2\times 10^4\ \mu\text{m}^2$ and $\beta \simeq 0.45$ [52], while $A = 3\times 10^4\ \mu\text{m}^2$ for Ashford's data. The value of $D^2 = (v+2\bar{D}^2)/2$ used with Armitage's equation was calculated for a truncated exponential distribution having an upper size limit of 5 μm. Apparently, factors other than those considered in the theoretical development must have a significant effect on overlap.

In discussing his results, Roach pointed out that the conditions assumed theoretically are essentially those that apply to sedimentation cells in which particles follow a straight vertical path to the collecting surface. In thermal precipitators, however, the particle approaches the collector along a slanting trajectory as shown in Fig. 5.18 and may be intercepted by another particle before it reaches the site at which it would have deposited if the second particle had not been there. When thermal precipitators are operated with their inlets up, larger particles have sedimentation velocities equal to or greater than their thermal velocities and their trajectories may be quite flat. The trajectories of particles having negligible sedimentation velocities may also deviate

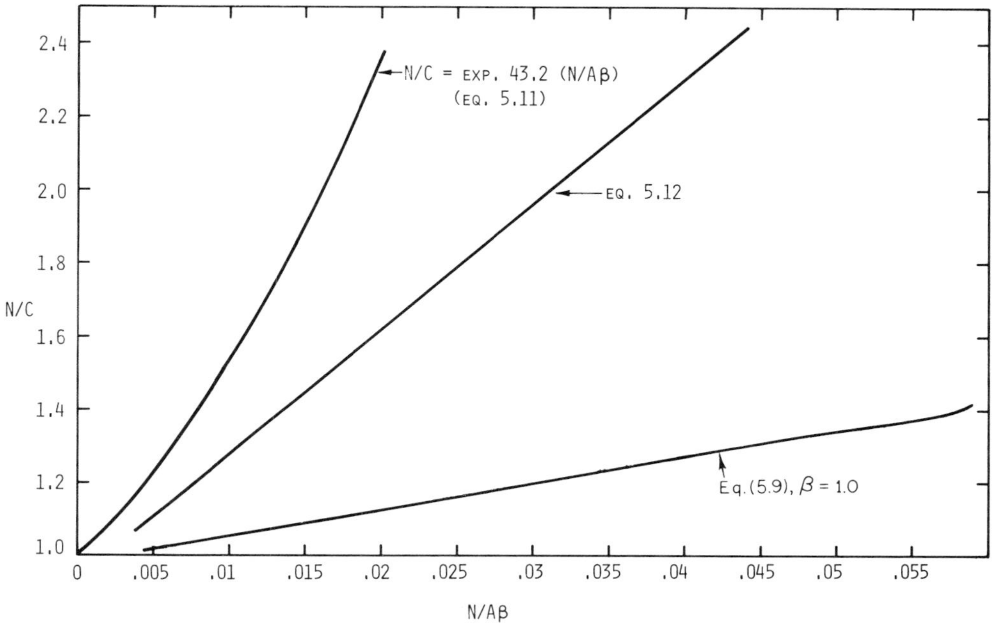

FIG. 5.17. Comparison of overlap predictions for truncated (1 μm $\leqslant D \leqslant$ 5 μm) exponential distributions.

significantly from the normal to the surface because of their flow velocity parallel to the surface.

In electrostatic precipitation, particles will approach the collecting surface along a trajectory more nearly normal to the surface, because of the large velocities acquired by particles in an electrostatic field. However, other effects become important in the problem of overlap. If the deposited particle

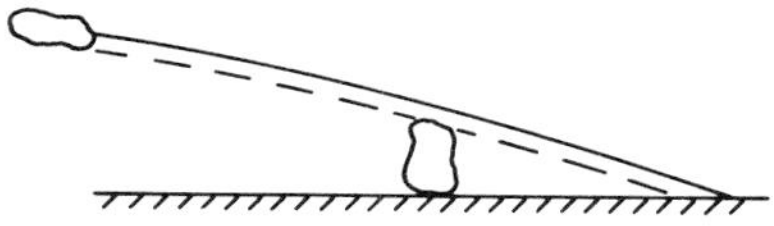

FIG. 5.18. Particle trajectories in thermal precipitators.

readily gives up its charge and assumes the potential of the collecting surface, the electric field will be distorted in its vicinity, tending to deposit incoming particles on it. On the other hand, if the deposited particle retains its charge, it will tend to repel incoming particles, giving rise to dense deposits with very little overlap, as shown in Fig. 5.19.

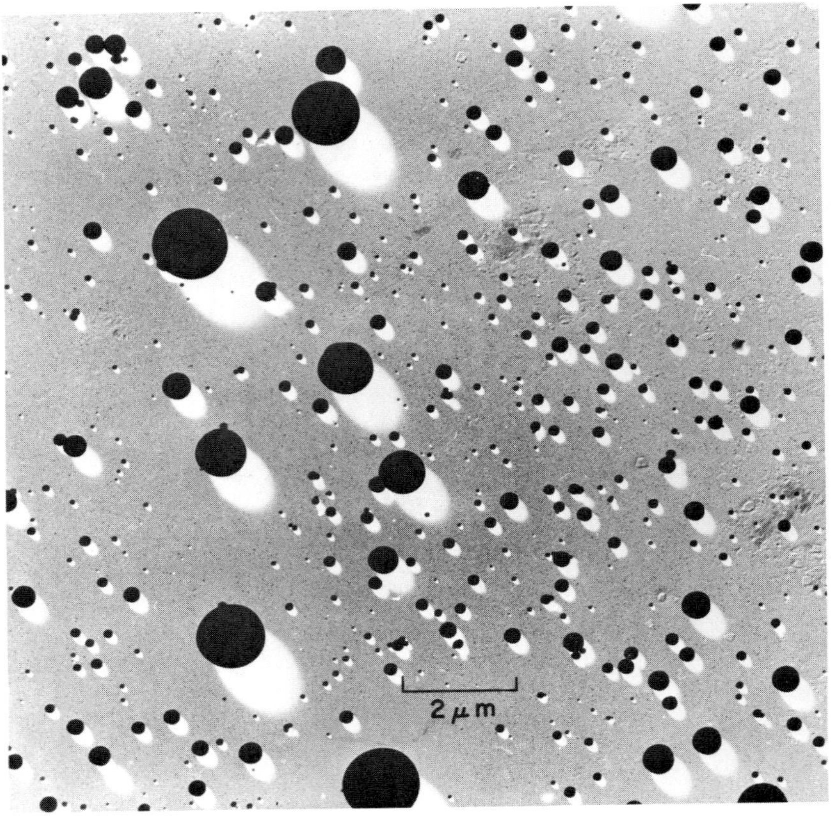

FIG. 5.19. Electrostatic precipitator sample of fused clay particles showing negligible overlap [61]. Courtesy of Pergamon Press, Ltd.

5-3 Membrane Filters

Membrane filter samples of particles having an appropriate index of refraction can be viewed directly in the optical microscope if the interstices of the filter are filled with a fluid having an index of refraction close to that of the filter material. Crossman [53] recommends a fluid having an index of refraction of 1.507, although somewhat different values may be applicable, depending on the filter manufacturer [54]. Hosey *et al.* [54] mixed light mineral oil and alpha chloronaphthalene in a ratio of 3:1, adding small amounts of one or the other until the desired index of refraction was obtained, and found that the sample showed very little evidence of filter structure. The technique cannot be used with particles having an index of refraction near 1.5. Small particles having a refractive index as large as 1.54 or as small as 1.43 are likely to be overlooked when counting or sizing [55] unless phase contrast optics are used.

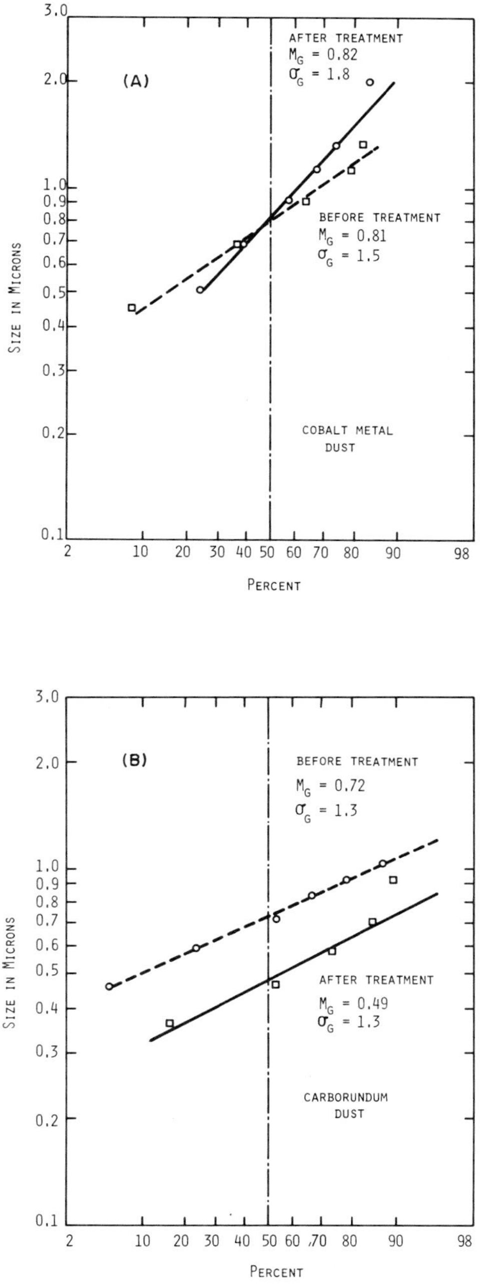

FIG. 5.20. Size distributions of particles collected on membrane filters and measured before and after dissolution of filter with ethyl acetate [60]. Courtesy of American Medical Association.

Samples collected on membrane filters can also be processed for examination by electron microscopy. The more widely used techniques are those in which a small square of the filter sample is placed on a grid covered with a formvar film and the filter material slowly dissolved away by a solvent such as acetone [56] or ethyl acetate [55, 60]. The validity of the method for determining particle size distributions is open to question. Fraser [60] made optical microscope measurements on samples from the same filter before and after dissolution of the filter in ethyl acetate. His results for samples of cobalt metal and carborundum dust are shown in Fig. 5.20A, B. He concluded that changes in the size distribution were related to the composition of the aerosol and should be determined for each substance investigated. Moreover, Ettinger and Posner [45] found that membrane filter samples treated for electron microscopy gave count median diameters 30–75% larger than those samples from the same aerosol collected with an electrostatic precipitator. Poor agreement between membrane filter samples was also observed.

References

1. J. W. Baum, *Electrical Precipitator for Aerosol Collection on Electron Microscope Grids*, USAEC Rep. HW–39129, 1955.
2. K. E. Lauterbach *et al.*, Efficiency studies of the electrostatic precipitator, *AMA Arch. Ind. Hyg. Occup. Med.*, *9:* 69–75 (1954).
3. J. Tyndall, On Dust and Disease, *Proc. Roy. Inst.*, *6:* 1–14 (1870).
4. J. Aitken, On the Formation of Small Clear Spaces in Dusty Air, *Trans. Roy. Soc.* (*Edinburgh*) *32:* 239–272 (1894).
5. O. J. Lodge and J. W. Clark, On the Phenomena Exhibited by Dusty Air in the Neighbourhood of Strongly Illuminated Bodies, *Phil. Mag. Ser. 5*, *17:* 214–239 (1884).
6. H. H. Watson, The Dust-Free Space Surrounding Hot Bodies, *Trans. Faraday Soc.* *32:* 1073–1081 (1936).
7. M. K. Paranjpe, The Convection and Variation of Temperature Near a Hot Surface, *Proc. Ind. Acad. Sci.*, *4:* 423–435 and 639–651 (1936).
8. L. Waldman, Über die Kraft eines inhomogenon Gases auf kleine suspendierte Kugeln, *Z. Naturforsch.*, *14a:* 589–599 (1959).
9. W. Cawood, The Movement of Dust or Smoke Particles in a Temperature Gradient, *Trans. Faraday Soc.*, *32:* 1068–1073 (1936).
10. P. Rosenblatt and V. K. LaMer, Motion of a Particle in a Temperature Gradient; Thermal Repulsion as a Radiometer Phenomenon, *Phys. Rev.*, *70:* 385–395 (1946).
11. J. C. Maxwell, On Stresses in Rarified Gases Arising from Inequalities of Temperature, *Trans. Roy. Soc.* (*London*), *170:* 231–245 (1879).
12. P. Epstein, Zur Theorie des Radiometers, *Z. Phys.*, *54:* 537–563 (1929).
13. R. L. Saxton and W. E. Ranz, Thermal Force on an Aerosol Particle in a Temperature Gradient, *J. Appl. Phys.*, *23:* 917–923 (1952).
14. C. F. Schadt and R. D. Cadle, Critical Comparison of Collection Efficiencies of Commonly Used Aerosol Sampling Devices, *Anal. Chem.*, *29:* 864–868 (1957).
15. J. R. Brock, On the Theory of Thermal Forces Acting on Aerosol Particles, *J. Colloid Sci.*, *17:* 768–780 (1962).

16. J. R. Brock, Forces on Aerosols in Gas Mixtures, *J. Colloid Sci., 18:* 489–501 (1963).
17. *Handbook of Chemistry and Physics*, 37th ed., Chem. Rubber Publ. Cleveland, Ohio, 1955.
18. *Mechanical Engineers Handbook*, 4th ed., McGraw-Hill, New York, 1941.
19. C. F. Schadt and R. D. Cadle, Thermal Forces on Aerosol Particles in a Thermal Precipitator, *J. Colloid Sci., 12:* 356–362 (1957).
20. C. F. Schadt and R. D. Cadle, Thermal Forces on Aerosol Particles, *J. Phys. Chem., 65:* 1689–1694 (1961).
21. H. L. Green and H. H. Watson, *Physical Methods for the Estimation of the Dust Hazard in Industry*, Med. Res. Council Spec. Rep. Ser. No. 199, 1935.
22. H. Walton, *Symposium on Particle Size Analysis*, p. 140, Inst. Chem. Eng. and Soc. Chem. Ind., 1947.
23. H. H. Watson, The Sampling Efficiency of the Thermal Precipitator, *Brit. J. Appl. Phys., 9:* 78–79 (1958).
24. J. R. Ashford, Some Statistical Aspects of Dust Counting, *Brit. J. Appl. Phys., 11:* 13–21 (1960).
25. J. G. Dawes *et al.*, *The Thermal Precipitator and the PRU Handpump: A Critical Study*, S.M.R.E. Res. Rep. No. 187, 1960.
26. K. E. Lauterbach *et al.*, *Design of an Oscillating Thermal Precipitator*, USAEC Rep. UR–199, 1952.
27. W. Walkenhorst, Ein neuer Thermalpräzipitator mit Heizband und seine Leistung, *Staub, 22:* 103–105 (1962).
28. H. Cember, T. Hatch and J. A. Watson, Dust Sampling with a Rotating Thermal Precipitator, *Amer. Hyg. Ass. Quart., 14:* 191–194 (1953).
29. J. Bredl and T. W. Grieve, A Thermal Precipitator for the Gravimetric Estimation of Solid Particles in Flue Gases, *J. Sci. Instrum., 28:* 21–23 (1951).
30. C. Orr, Jr., M. T. Gordon, and M. C. Kordecki, Thermal Precipitation for Sampling Airborne Microorganisms, *J. Appl. Microbiol., 4:* 116–118 (1950).
31. W. P. Hendrix and C. Orr, Jr., Thermal Precipitation, *Rev. Sci. Instrum., 35:* 1373–1374 (1964).
32. C. Orr, Jr. and R. A. Martin, Thermal Precipitator for Continuous Aerosol Sampling, *Rev. Sci. Instrum., 29:* 129–130 (1958).
33. O. R. Moss, *Shape Factors for Airborne Particles*, Master of Science Thesis, Univ. of Rochester, 1969.
34. D. A. Fraser, The Collection of Submicron Particles by Electrostatic Precipitation, *Amer. Ind. Hyg. Ass. Quart., 17:* 75–79 (1956).
35. A. D. Hosey and H. H. Jones, Portable Electrostatic Precipitator Operating from 110 Volts A.C. or 6 Volts D.C., *AMA Arch. Ind. Hyg. Occup. Med., 7:* 49–57 (1953).
36. C. E. Billings and L. Silverman, Aerosol Sampling for Electron Microscopy, *J.A.P.C.A., 12:* 586–590 (1962).
37. M. Arnold, P. E. Morrow and W. Stöber, Vergleichende Untersuchung über die Bestimmung der Korngrössenverteilung fester Stäube mit Hilfe eines Hochspannungsabscheiders und des Elektronenmikroscops, *Kolloid-Z., 181:* 59–65 (1962).
38. P. E. Morrow and T. T. Mercer, A Point-to-Plane Electrostatic Precipitator for Particle Size Sampling, *Amer. Ind. Hyg. Ass. J., 25:* 8–14 (1964).
39. P. C. Reist, Size Distribution Sampling Errors Introduced by the Point-Plane Electrostatic Precipitator Sampling Device, in *Proc. A.E.C. Air Cleaning Conf. 9th*, CONF–660904, p. 613, 1966.
40. T. T. Mercer, *A Study of Some Physical Properties of an Aerosol in Relation to Airborne Decay Products of Radon*, USAEC Rep. UR–474, 1957.

41. E. Warburg, Über die Stille Entladung in Gasen, *Handbuch der Physik*, Vol. 14, Chapter 4, Springer–Verlag, Berlin, 1927.
42. G. McGreevy, The Evaluation of the Performance of an Electrostatic Precipitator Using a Pollak-Nolan Nucleus Counter, *Atmos. Environ.*, *1:* 87–95 (1967).
43. T. T. Mercer, M. I. Tillery, and M. A. Flores, *An Electrostatic Precipitator for the Collection of Aerosol Samples for Particle Size Analyses*, USAEC Rep. LF–7, 1963.
44. B.Y.H. Liu and A. C. Verma, A Pulse-Charging, Pulse-Precipitating Electrostatic Aerosol Sampler, *Anal. Chem.*, *40:* 843–847 (1968).
45. H. Ettinger and S. Posner, Evaluation of Particle Sizing and Aerosol Sampling Techniques, *Amer. Ind. Hyg. Ass. J.*, *26:* 17–25 (1965).
46. J. Cartwright, The Electron Microscopy of Airborne Dust, *Brit. J. Appl. Phys., Suppl.*, *3:* S109–S120 (1954).
47. C. E. Billings, W. J. Megaw, and R. D. Wiffen, Sampling of Sub-micron Particles for Electron Microscopy, *Nature* (*London*), *189:* 336 (1961).
48. G. Zebel, Vergleichende Messungen an feinteiligen Aerosolen zwischen Thermal-präzipitator und Spaltultramikroscop, *Staub*, *39:* 21–29 (1955).
49. J. O. Irwin, P. Armitage, and C. N. Davies, The Overlapping of Dust Particles on a Sampling Plate, *Nature* (*London*), *163:* 809 (1949).
50. P. Armitage, An Overlap Problem Arising in Particle Counting, *Biometrika*, *36:* 257–266 (1949).
51. S. A. Roach, Counting Errors Due to Overlapping Particles in Thermal Precipitator Samples, *Brit. J. Ind. Med.*, *15:* 250–257 (1958).
52. J. R. Ashford *et al.*, The Effect of Particle Size on the Errors Associated with the Overlapping of Particles on Thermal Precipitator Samples, *Ann. Occup. Hyg.*, *6:* 201–222 (1963).
53. G. C. Crossman, The Value of Phase Microscopy for the Examination of Particulate Material Collected on Membrane Filters, *Amer. Ind. Hyg. Ass. J.*, *20:* 190–193 (1959).
54. A. D. Hosey, H. H. Jones, and H. E. Ayer, Evaluation of an Aerosol Photometer for Dust Counting and Sizing, *Amer. Ind. Hyg. Ass. J.*, *21:* 491–501 (1960).
55. H. J. Paulus *et al.*, Use of Membrane Filters in Air Sampling, *Amer. Ind. Hyg. Ass. J.*, *18:* 267–273 (1957).
56. E. H. Kalmus, Preparation of Aerosols for Electron Microscopy, *J. Appl. Phys.*, *25:* 87 (1954).
57. A. K. Postma, *Thermophoretic Velocity of Large Aerosol Particles*, USAEC Rep. BNWL–163, 1965.
58. J. R. Hodkinson, Thermal Precipitators, in *Air Sampling Instruments*, Amer. Conf. Gov. Ind. Hygsts, Cincinnati, 1962.
59. P. Goldsmith and P. G. May, Diffusiophoresis and Thermophoresis in Water Vapour Systems, in C. N. Davies (Ed.), *Aerosol Science*, Academic Press, New York and London, 1966.
60. D. A. Fraser, Absolute Method of Sampling and Measurement of Solid Air-Borne Particulates, *A.M.A. Arch. Ind. Hyg. Occup. Med.*, *8:* 412–419 (1953).
61. G. M. Kanapilly, O. Raabe, and G. J. Newton, A New Method for the Generation of Aerosols of Insoluble Particles, *Aerosol Sci. 1:* 313–324 (1970).

6

Measurement of Aerodynamic Diameter

If the air resistance acting on a particle is proportional to the relative velocity between it and air, then both the terminal settling velocity U_G and the Stokes number of Stk of the particle are proportional to the product mZ of the particle's mass and mobility. Measurement of U_G or Stk makes it possible to express the size of a particle of irregular shape as an aerodynamic diameter as tabulated below. D_A is, by definition, the aerodynamic diameter of the

Irregular particle		Measured quantity		Unit density sphere
mZg	=	U_G	=	$\rho_0 D_A^2 K_{sA} g/18\eta$
$2mZU_0/W$	=	Stk	=	$\rho_0 D_A^2 K_{sA} U_0/9\eta W$

particle, K_{sA} is the slip factor for D_A, η is the viscosity of air, ρ_0 is the density (unity in cgs system) of the sphere of diameter D_A, g is the acceleration due to gravity, and U_0 is an initial velocity of the particle relative to some fixed coordinate system having a characteristic length W.

Although instruments such as the slit ultramicroscope can be used to determine the aerodynamic diameters of individual particles by direct measurement of their settling velocities, it is more useful and convenient to employ instruments that distribute particles in a continuous size spectrum or sort them into discrete size intervals on the basis of their aerodynamic diameters. The collected particles can be analyzed thereafter to determine the distribution of activity as a function of aerodynamic diameter. Instruments that are most suited for this application operate on the principles of air elutriation, centrifugation, or impaction, as described below.

6-1 Air Elutriation

Air elutriation is a process in which particles are separated on the basis of size by pitting their settling velocity against the velocity of a current of air with which they move. For vertical elutriation, particles are carried upward in a diverging air stream until they reach a level at which their settling velocity equals the vertical component of the diminishing air velocity. For horizontal elutriation, the particle's settling velocity is normal to the velocity of the air stream, which usually is confined within a duct of rectangular cross section. Vertical elutriators have been used successfully for the size-fractionation of powders [1], but are not often used for air sampling.* Horizontal elutriators, on the other hand, have proved valuable for respirable activity sampling and in the study of shape factors. Moreover, any duct that is not perfectly vertical will have some characteristics of a horizontal elutriator, so it is useful to consider their operation in some detail.

6-1.1 Theoretical Considerations

Walton [2] has advanced the following two theorems, valid for laminar flow, which are of special interest in elutriation:

(a) A change in air movement (due, for example, to the introduction of a sampling flow) within a cloud of sedimenting inertia-less particles whose concentration is initially uniform will not give rise to a change in particle concentration at any point within the cloud; and

(b) The number of particles passing per unit time through any given area within a cloud of sedimenting inertia-less particles is equal to the number that would fall through if the air were stationary plus the product of the particle concentration and the flux of air through the area.

* See Chapter 8 for a recent exception to this statement.

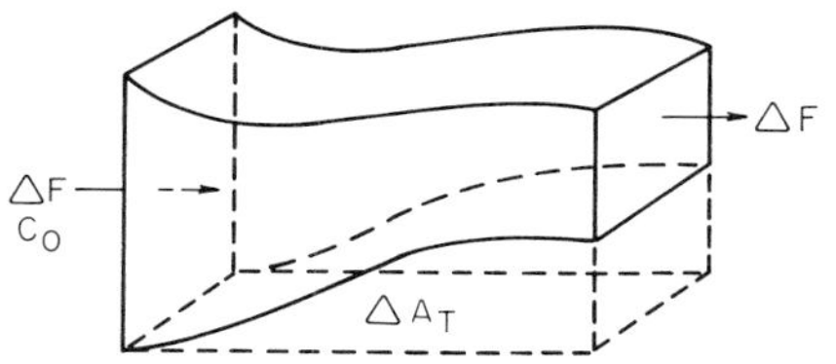

FIG. 6.1. Elementary tube of airflow.

The second of these was used by Walton to establish the validity of Eq. (5.7) when applied to the thermal precipitator. By introducing the concept of particle "tubes of flow," which are elemental volumes bounded by the trajectories of particles of the same settling velocity, Walton was able to derive an important corollary to Theorem (a): If air entering a duct has a uniform concentration C_0 of particles of a given settling velocity, then the concentration of such particles in any volume element within the duct is either C_0 or zero.

Consider an elementary tube of air flow having vertical sides and bounded by upper and lower walls (see Fig. 6.1). The tube width is small but its height may have any value and either may vary along the length. The air velocity at any point is a function of both longitudinal and horizontal position. Air enters at a constant volumetric flow rate ΔF, carrying C_0 particles per unit volume that have terminal settling velocity U_G. Particles begin depositing on the floor at its leading edge at a rate equal to $C_0 \cdot U_G$ per unit of horizontal projected area per unit time and continue depositing at that rate until either the concentration at floor level abruptly changes to zero, or the end of the tube is reached. The first condition occurs when the limiting trajectory of the particles that enter at the upper boundary of the inlet intersects the tube floor (see Fig. 6.2). In this case, all of the incoming particles will be deposited on an area of the floor equivalent to a projected area ΔA_{m}. Therefore,

$$C_0 \cdot \Delta F = C_0 \cdot U_G \cdot \Delta A_{\mathrm{m}} \quad \text{and} \quad \Delta A_{\mathrm{m}} = \Delta F / U_G. \tag{6.1}$$

After a sampling time t the concentration of particles per unit projected area is

$$C_A = C_0 \cdot \Delta F \cdot t / \Delta A_{\mathrm{m}} = C_0 \cdot U_G \cdot t. \tag{6.2}$$

If $\Delta F / U_G \geqslant \Delta A_T$, then the entire floor is covered with particles at a constant

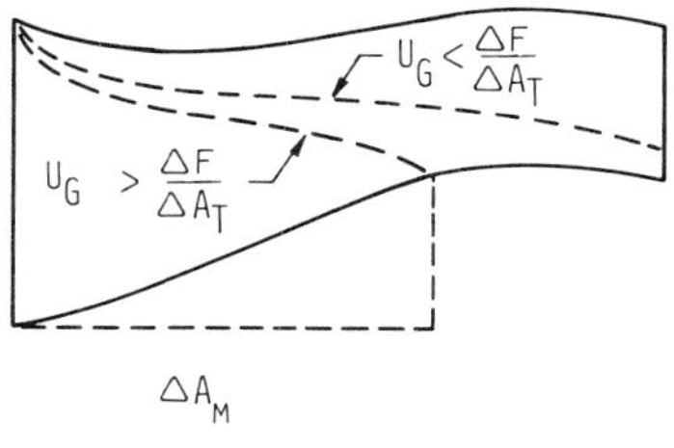

FIG. 6.2. Limiting particle trajectory intersecting tube floor.

number per unit projected area:

$$C_A = C_0 U_G \cdot t \tag{6.2a}$$

The fraction penetrating the tube is

$$P = [C_0 \cdot \Delta F - C_0 U_G \cdot \Delta A_T]/C_0 \Delta F = 1 - U_G \cdot (\Delta A_T/\Delta F). \tag{6.3}$$

If the flow pattern within a duct is such that the duct can be divided into similar flow tubes satisfying the conditions given above, then the penetration of particles through the duct will be

$$P = 1 - U_G \cdot (\textstyle\sum \Delta A / \sum \Delta F) = 1 - U_G \cdot A/F, \tag{6.4}$$

where A is the horizontal projection of the total floor area and F is the total flow rate.

If the duct has a constant height H_0, a length L, and a transverse cross section that is rectangular, but not necessarily of constant width, and if its sides are straight, then

$$P = 1 - [U_G \cdot L(W+W_0)/2F], \tag{6.5}$$

where W_0 and W are the entrance and exit widths, respectively, and $U_G < 2F/L(W+W_0)$. For larger values of U_G, $P = 0$, and the particles are deposited over a length, l_D, in an area

$$A = [(W-W_0)/2L]\, l_D^2 + l_D W_0 = F/U_G,$$

from which

$$l_D = \left\{ \frac{2FL}{U_G(W-W_0)} \cdot \left[1 + \frac{LW_0^2 U_G}{2F(W-W_0)} \right] \right\}^{1/2} - \frac{LW_0}{W-W_0}. \tag{6.6}$$

For a duct of constant cross section, $W = W_0$, and Eq. (6.6) reduces in the limit to

$$l_D = \frac{F}{U_G W_0} = \frac{\bar{V} H_0}{U_G}, \tag{6.6a}$$

where $\bar{V}$, is the average linear air speed in the duct. The deposition length is inversely proportional to the terminal settling velocity, U_G. If $W \gg W_0$, then except at large values of LU_G/F,

$$l_D \simeq \left[\frac{2FL}{U_G(W-W_0)} \right]^{1/2} - \frac{LW_0}{W-W_0} \simeq \left(\frac{2FL}{U_G W} \right)^{1/2} - \frac{LW_0}{W}, \tag{6.6b}$$

and the deposition length is almost inversely proportional to $U_G^{1/2}$.

For most particle sizes for which horizontal elutriation is useful, U_G can be assumed to be proportional to D_A^2 and $P = 1 - k^2 D_A^2$, where

$$k^2 = \frac{L(W+W_0)}{2F} \cdot \frac{\rho_0 g}{18\eta}. \tag{6.7}$$

Penetration varies with aerodynamic diameter as shown in Fig. 6.3 and equals

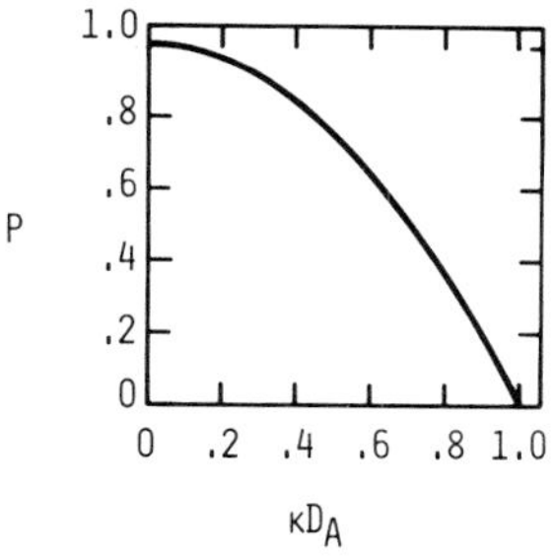

FIG. 6.3. Variation of penetration with aerodynamic diameter.

50% when $kD_A = 0.707$. By choosing k properly, the aerodynamic diameter at 50% penetration can be fixed at any desired value within the limits of validity of the assumption $U_G \propto D_A{}^2$. For respirable activity samplers (see Chapter 8), $k^2 = 0.02\ \mu\text{m}^{-2}$ when D_A is in micrometers.

While laminar flow is necessary for these results to be valid, no specific velocity profile is required, and the shapes of the duct walls are limited only by the requirement that $\Delta F/\Delta A$ be the same for all longitudinal elements of the duct. Walton extended his calculations to the situation in which the latter requirement is not satisfied. If the deposition velocity, $\Delta F/\Delta A$, is represented by U', and if the horizontal projection, dA, of the total floor area of all tubes having deposition velocities between U' and $U'+dU'$ is represented by

$$dA = R(U')\,dU',$$

then the number of particles deposited in the elutriator per unit time is

$$N = C_0 \int_0^{U_G} U'R(U')\,dU' + C_0\,U_G \int_{U_G}^{U_m'} R(U')\,dU', \tag{6.8}$$

where U_m' is the maximum value of $\Delta F/\Delta A$ that can occur in the system under consideration and C_0 is the input concentration of particles having a common settling velocity U_G. The first term on the right gives the total number of particles entering the elutriator per unit time in flow tubes for which $\Delta F/\Delta A < U_G$; the integral itself is the total air flow rate through those tubes. The second term on the right is the total number of particles deposited in the elutriator from flow tubes for which $\Delta F/\Delta A > U_G$. The fraction penetrating the elutriator is

$$P = \frac{\int_{U_G}^{U_m'} (U' - U_G) \cdot R(U') \cdot dU'}{\int_0^{U_m'} U' \cdot R(U') \cdot dU'}. \tag{6.8a}$$

Walton used this result to determine the penetration of particles through a horizontal duct having a circular cross section of radius R_0 (see Fig. 6.4) and a length L. For laminar, viscous flow in the duct, the air velocity at any distance r from the axis is

$$V(r) = 2\overline{V}\,[1-(r^2/R_0{}^2)],$$

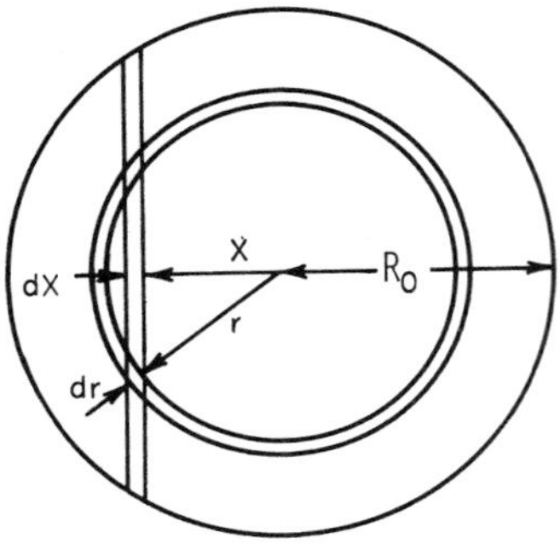

FIG. 6.4. Cross section of horizontal circular duct.

where $\overline{V} = F/\pi R_0^2$ is the average air speed in the elutriator. For a flow tube of length L and width dx at a distance x from the axis, $\Delta A = L\,dx$, and

$$U' = \frac{\Delta F}{\Delta A} = \frac{2}{L} \cdot \int \frac{V(r) \cdot dr}{[1-(x^2/r^2)]^{1/2}} = \frac{8\overline{V}R_0}{3L}\left[1 - \frac{x^2}{R_0^2}\right]^{3/2}, \tag{6.9}$$

where the integral is evaluated between x and R_0. Introducing this into Eq. (6.8a), and setting $R(U')\,dU' = dA = 2L\,dx$, gives

$$P = \frac{\int_{x(U_G)}^{0} \{[1-(x^2/R_0^2)]^{3/2} - (3U_G L/8\overline{V}R_0)\}\,dx}{\int_{R_0}^{0} [1-(x^2/R_0^2)]^{3/2}\,dx}$$

$$= \frac{2}{\pi}\{Y\cdot(1-Y^2)^{1/2} - 2Y\cdot(1-Y^2)^{3/2} + \sin^{-1}Y\}, \tag{6.10}$$

where

$$Y^2 = 1 - [3U_G L/8\overline{V}R_0]^{2/3}.$$

$P = 0$, when $Y^2 \leqslant 0$. If the duct makes an angle θ with the horizontal, Eq. (6.10) must be corrected by multiplying U_G by $\cos\theta$. The equation is used to estimate sedimentation losses in the lung. Other forms of this equation have been derived by Natanson and by Thomas [3].

The horizontal elutriator can be used to determine the distribution of activity of an aerosol as a function of aerodynamic diameter [4]. Although the derivations given above were related to the number of particles per unit volume, they are valid for the general description of C as activity per unit volume of air. To generalize to particles of more than one settling velocity, let dC, the amount of activity per unit volume associated with particles having settling velocities in the increment U to $U+dU$, be given by

$$dC/dU = C_0\,f(U),$$

where

$$\int_{U(\min)}^{U(\max)} f(U)\,dU = 1.$$

At a distance from the elutriator inlet equivalent to a floor area A, the change in activity per unit area due to the cutoff of activity associated with particles having settling velocities, $U = F/A$, is given by Eq. (6.2) as

$$dC_A = C_0 \cdot f(U)\, dU \cdot U \cdot t, \tag{6.11}$$

from which

$$f(U) = \frac{1}{C_0 tU} \cdot \frac{dC_A}{dU} = \frac{-A^3}{F^2 C_0 t} \cdot \frac{dC_A}{dA}. \tag{6.12}$$

Assuming $U = \alpha^2 D_A{}^2$, $2\alpha^2 D_A f(U) = f(D_A)$ and

$$f(D_A) = -2\left(\frac{A}{F}\right)^{3/2} \cdot \frac{\alpha A}{C_0 t} \cdot \frac{dC_A}{dA}, \tag{6.13}$$

where $\alpha^2 = \rho_0 g/18\eta$. If a curve is plotted showing the measured activity per unit area as a function of the area from the elutriator inlet, then for any value of A, dC_A/dA is the tangent to that curve at A, and D_A is the aerodynamic diameter of particles having a terminal setting velocity $U = F/A$. From Eq. (6.11) the total activity per unit volume of air is

$$C_0 \int_{U(\min)}^{U(\max)} f(U)\, dU = C_0 = -\int_{C_A(\max)}^{0} dC_A/Ut \tag{6.13a}$$

$$= (-1/Ft) \cdot \int_{C_A(\max)}^{0} A \cdot dC_A = M/Ft,$$

where M is the total activity deposited in the elutriator. Practical considerations may limit the elutriator length so that some activity escapes the elutriator to be collected, presumably, on a back-up filter. M is then the sum of the activities collected in the elutriator and on the back-up filter.

6-1.2 Elutriation Spectrometers

In a horizontal elutriator having a rectangular transverse cross section of constant area, particles of terminal settling velocity U will deposit in uniform concentration out to a cut-off distance l_D from the inlet. The fraction of these particles deposited out to a distance $l < l_D$ is

$$\frac{l}{l_D} = \frac{lU}{\bar{V}H_0}.$$

All of these particles entered the elutriator at heights equal to or less than H, which is defined by

$$lU/\bar{V}H_0 = (1/F) \cdot \int_0^{F(H)} dF. \tag{6.14}$$

For laminar viscous flow in a duct having a width W very much greater than its depth H_0,

$$dF = (\bar{V}W/H_0) \cdot 6[H - (H^2/H_0)]\, dH,$$

and integration of Eq. (6.14) yields

$$l = \bar{V}H^2[3 - (2H/H_0)]/UH_0. \tag{6.15}$$

Particles entering in a layer of thickness δH centered at H, are deposited over a length δl, centered at l, given by

$$\delta l = 6\bar{V}H/UH_0{}^2 \cdot (H_0 - H) \cdot \delta H. \tag{6.16}$$

For aerosol spectrometers operating by horizontal elutriation, the particles are confined in this way to a thin layer surrounded by filtered air. Particles of a given size are deposited in a narrow band across the elutriator floor. The finite length, δl of the deposit limits the accuracy with which the diameter can be measured (see Fig. 6.5). The resolution of a spectrometer is usually defined as $\delta D/D$, where δD is the error in D due to the length of the deposit. In the following discussion, however, the ratio, $\delta l/l$, will be treated as a measure of resolution. Later, it will be converted to $\delta D/D$ for some specific instruments. In the present case, the resolution of the instrument is related to

$$\frac{\delta l}{l} = \frac{6(H_0 - H)}{3H_0 - 2H} \cdot \frac{\delta H}{H}. \tag{6.17}$$

If the layer of particles enters at the duct midline, then $H = H_0/2$, and

$$\frac{\delta l}{l} = 3\frac{\delta H}{H_0}. \tag{6.17a}$$

If the layer enters at the upper boundary, then $H = H_0 - \delta H/2$, δH is small compared to H_0, and

$$\frac{\delta l}{l} = 3\left(\frac{\delta H}{H_0}\right)^2. \tag{6.17b}$$

For a given value of δH, introducing the aerosol at the upper boundary provides much better resolution; however, the aerosol flow rate is much smaller.

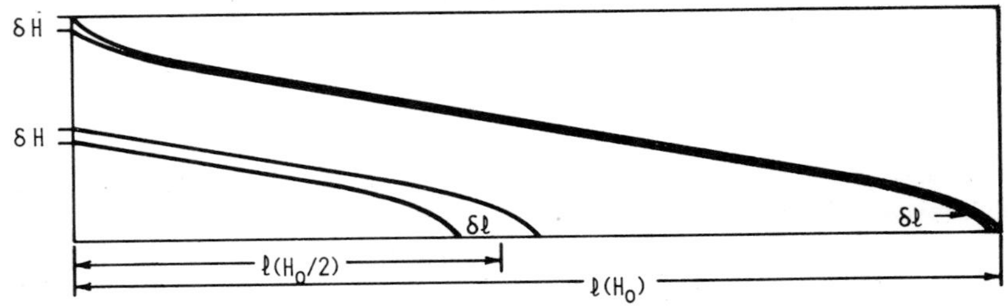

FIG. 6.5. Trajectories of particles entering horizontal elutriator at different heights.

Moreover, to insure the collection of particles of a given minimum diameter, a spectrometer having an upper boundary input must have twice the length, or operate at half the flow rate, required for a spectrometer having midline input. To provide the same aerosol sampling rates for spectrometers of the same dimensions, the layer thicknesses must be related by

$$\delta H_{\mathrm{m}} \simeq (2\bar{V}_{\mathrm{u}}/\bar{V}_{\mathrm{m}}) \cdot (\delta H_{\mathrm{u}}{}^2/H_0),$$

where m and u identify, respectively, the midline input and the upper boundary input. To collect all particles in a size range having the same minimum value for both types of spectrometers, $\bar{V}_{\mathrm{m}} = 2\bar{V}_{\mathrm{u}}$, and

$$\frac{\delta H_{\mathrm{m}}}{H_0} \simeq \frac{\delta H_{\mathrm{u}}{}^2}{H_0{}^2}.$$

Except for a negligible term of higher order, the two types of instruments have the same resolution for the same aerosol sampling conditions.

The resolution of the instrument can be altered by changing the ratio $\delta H/H_0$. Although the dimensions of the aerosol layer are initially defined by the fixed walls of an inlet duct, the thickness of the layer must be compatible with the conditions necessary to maintain the proper velocity profile for laminar flow. The resolution can be adjusted, therefore, by altering the ratio of clean air flow to aerosol flow; larger ratios providing better resolution, within limits. In general, $\delta l/l = \delta F/F_{\mathrm{H}}$, where δF is the aerosol flow rate and F_{H} is the total flow rate between the midline of the aerosol inlet and the floor of the channel.

Resolution is also affected to some extent by a particle's diffusivity Δ, which causes it to deviate from its theoretical trajectory by an amount related to its root-mean-square displacement in the time, $t = H/U$, necessary for a particle of mass m to settle out after entering at a height H:

$$\begin{aligned}(\overline{X^2})^{1/2} &= (2\Delta H/U)^{1/2} = (2HkT/mg)^{1/2} \\ &= 9.1 \times 10^{-9}(H/m)^{1/2} \quad \text{cm} \qquad \text{at} \quad T = 294°\text{K}.\end{aligned}$$

H is in centimeters and m in grams. The significance of this displacement depends on one's point of view. Zebel [5] interprets it as the amount by which the deposition length l_{D} is increased, ignoring the additional effect as the particle is displaced to another trajectory. Stöber [4], on the other hand, felt that the increase in deposition length was $l_{\mathrm{D}}[(\overline{X^2})^{1/2}/H]$, as it would be if all the displacement occurred at the entrance. In this case, the resolution is increased by the factor

$$1 + [2(\overline{X^2})^{1/2}/\delta H] \tag{6.18}$$

for a midline input. A middle-of-the-road interpretation is that the displacement averaged over the sedimentation time provides an effective velocity,

$$U' = U \pm (2U\Delta/H)^{1/2},$$

which can be used with Eq. (6.15) to calculate the deposition length, allowing for the effect of diffusion. The resolution is increased by the factor

$$1 + (2H_0/3\delta H) \cdot \{[(H/2) \cdot (mg/kT)]^{1/2} - 1\}^{-1}. \tag{6.19}$$

For $H_0 = 2H = 1$ cm, $\delta H = 0.05$ cm, and $m = 10^{-14}$ gm, Eqs. (6.18) and (6.19) give factors of 3.56 and 2.97, respectively; for $m = 10^{-12}$ gm, the factors are 1.25 and 1.17. Factors based on Zebel's interpretation do not differ significantly from unity.

6-1.3 Applications of Horizontal Elutriators

Aside from their use as respirable mass samplers, horizontal elutriators have been used effectively in the determination of shape factors relating aerodynamic diameters (D_A) to projected area diameters (D_P) for particles of irregular shape. Stein *et al.* [6] used nonspectrometric elutriation to measure D_A/D_P for particulate air pollutants. Their calibration of the elutriator, shown in Fig. 6.6, brings out the sharpness of cutoff that is obtained with elutriators. They observed ratios of D_P/D_A ranging from 0.5 to 6.5 with most values less than 2.8. Their work emphasizes the difficulty in predicting the aerodynamic behavior of particles of unknown source and composition from arbitrary measurements of diameter. Thomas and Knuth [7] used a horizontal elutriator of circular cross section to measure the settling velocities of monodisperse particles.

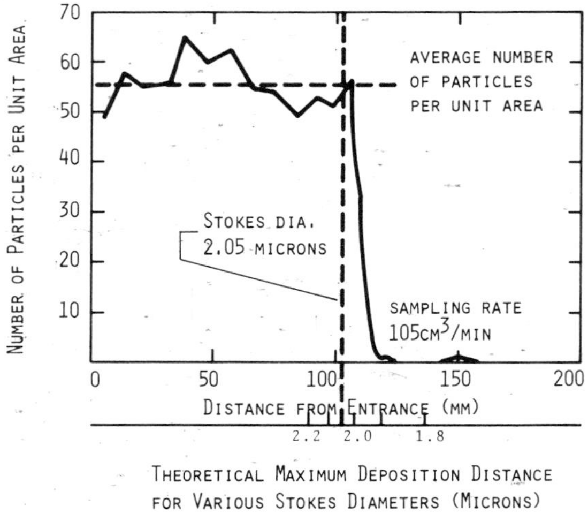

Fig. 6.6. Calibration of a horizontal elutriator for 2.05 μm polystyrene particles [6]. Courtesy of American Industrial Hygiene Association.

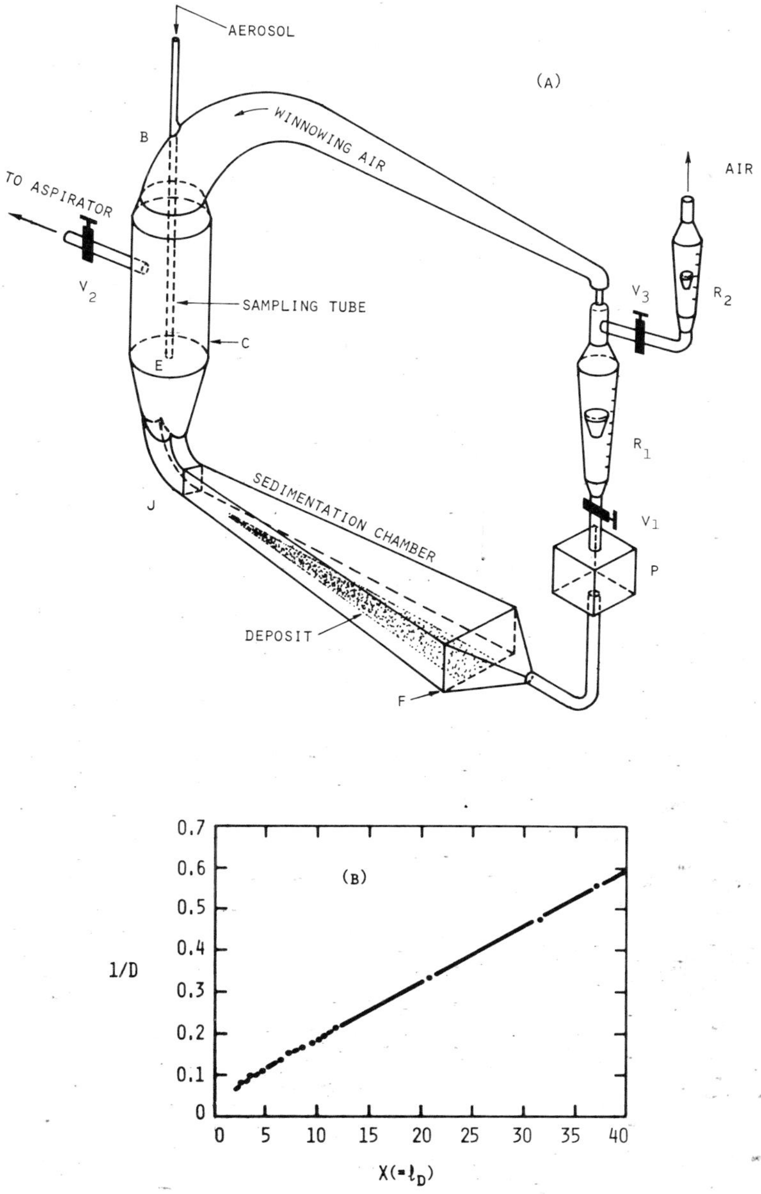

FIG. 6.7. (A) The Timbrell aerosol spectrometer; (B) calibration of Timbrell spectrometer in terms of reciprocal diameter as a function of distance along elutriator floor [8]. Courtesy of Charles C. Thomas, Publisher.

Timbrell [8] developed the spectrometer shown in Fig. 6.7(A). Since $W \gg W_0$, the deposition length [Eq. (6.6b)] is inversely proportional to D_A:

$$l_D = \frac{1}{D_A}\left[\frac{FL}{(W-W_0)\,\alpha^2}\right]^{1/2} - \frac{LW_0}{W-W_0}.$$

The factor 2 has been omitted from the radical because the aerosol layer enters at the midline. For $F = 1.67\ \text{cm}^3/\text{sec}$, $L = 40$ cm, $W = 7.62$ cm, $W_0 = 0.635$ cm, $\alpha^2 = 2.99 \times 10^{-3}\ \text{cm}/\mu\text{m}^2\text{-sec}$, and D_A in micrometers,

$$l_D = (56.7/D_A) - 3.64.$$

α^2 is defined with Eq. (6.13). Timbrell's calibration curve (Fig. 6.7B) yields a slope of 74.8 and an intercept of -3.87. The experimental slope is larger because the instrument's dimensions do not satisfy the assumption $W \gg H_0$. For this instrument, $H_0 = 1.27$ cm, δH has an effective value of 0.09, and the resolution is

$$\frac{\delta D}{D} = \frac{3}{2}\frac{\delta H}{H_0} = 0.11.$$

Timbrell [8] has studied a variety of aerosols of a hazardous nature. He has made an extensive study of the aerodynamic properties of fibers, particularly those of asbestos, showing that a fiber's aerodynamic diameter depends mostly on its real diameter, with relatively little effect due to fiber length (Fig. 6.8). In addition, he has used the instrument to study aerodynamic size distributions and the random clumping of particles during deposition. Much of this work is summarized in the reference cited.

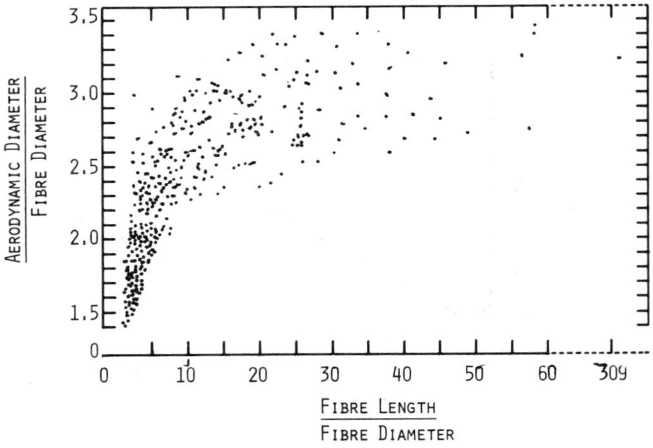

FIG. 6.8. Relationship between the aerodynamic diameter of a fiber and its length [8]. Courtesy of Charles C. Thomas, Publisher.

6-2 Aerosol Centrifuges

The operation of an aerosol centrifuge is similar in principle to that of a horizontal elutriator, with the force of gravity being replaced by a centrifugal force. Consider a duct of constant rectangular cross section bent into an annulus of outer radius R_0, width H_0, and depth W, as shown in Fig. 6.9, and rotated at an angular velocity, ω, about an axis at 0. Both R_0 and W are very much larger than H_0. Assume air enters uniformly over the duct cross section at I, flowing through the duct at a volumetric flow rate F to disappear uniformly over the duct cross section at E. Particles of mass m and mobility Z at a distance R from the axis of rotation are subjected to a centrifugal force $m\omega^2 R$ directed outward from the axis. For the particle size range of interest here, their radial velocity at R will be

$$U(R) = mZ\omega^2 R, \tag{6.20}$$

when $\omega R \gg \overline{V}$, the average air velocity in the duct. Assuming the velocity profile in the annulus is the same as that in a horizontal elutriator of similar cross section, particles distributed uniformly throughout the air at the entrance will deposit over a length of the outer wall given by

$$l_D = [\overline{V}H_0/U(R_i)] \cdot [1-(H_0/2R_i)], \tag{6.21}$$

where $R_i = R_0 - H_0$ and terms in H_0/R_i higher than the third power have been neglected.

If the particles are confined to a thin layer δR at R_i, then l_D is the position of the leading edge of a deposit of length

$$\delta l \simeq 3(\delta R^2/H_0{}^2)[1-(2\delta R/3R_i)]l_D. \tag{6.22}$$

The resolution of the system is

$$\frac{\delta D}{D} \simeq \frac{\delta l}{2l} \simeq \frac{3}{2}\frac{\delta R^2}{H_0{}^2}\left(1 - \frac{2\delta R}{3R_i}\right). \tag{6.22a}$$

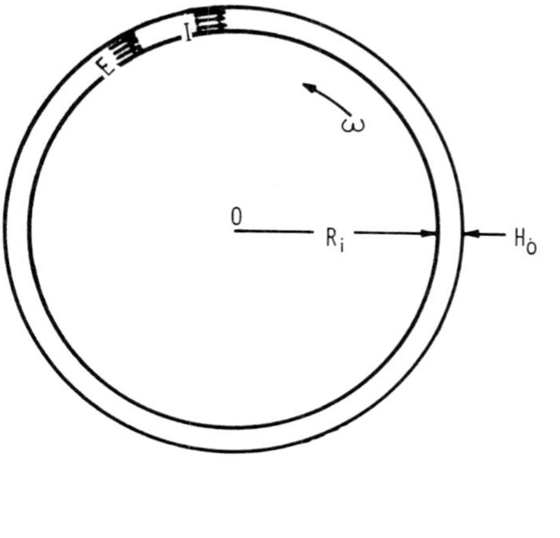

FIG. 6.9. Annular centrifuge duct.

This simple extension of the principles of horizontal elutriation to the centrifugation of aerosols has not been achieved in practice because no completely satisfactory method of introducing air and aerosol into the duct has been devised. In the first successful centrifugal aerosol spectrometer [9], the aerosol, surrounded with a sheath of clean air, entered near the apex of a conical annulus formed by two cones rotating about a common axis. In other instruments, rotating spiral ducts, either winding around a conical surface [10] or coiled in a plane [11] have been used. Some of these devices are described below.

6-2.1 Nonspectrometric Aerosol Centrifuges

(a) *The Goetz Aerosol "Spectrometer."* A cutaway view of this sampler is shown in Fig. 6.10. An aluminum cone, grooved with two independent

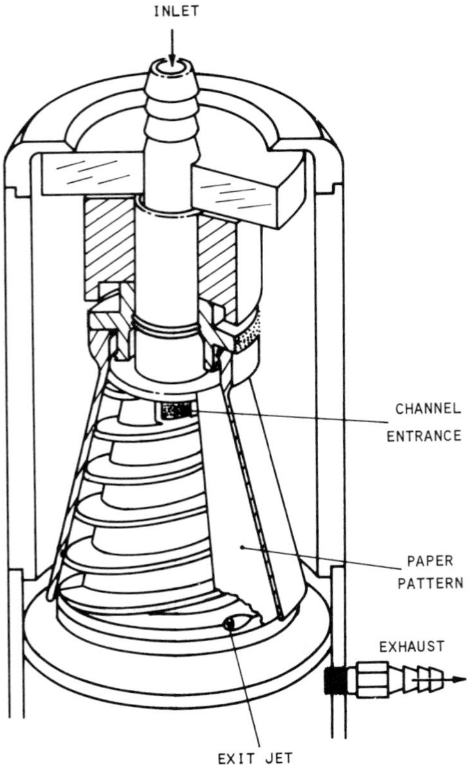

Fig. 6.10. Cutaway view of Goetz spectrometer [12]. Courtesy of International Atomic Energy Agency.

helical channels and covered with a close-fitting conical shell, is rotated at speeds up to 24,000 rpm. The direction of rotation is counterclockwise, as seen from above, and the channel spirals form right-hand screws, causing an impeller action that pumps air downward along the channels. Air is drawn through the stationary inlet tube and into the channels through ports at the top. It traverses the length of the channel and is discharged through a jet orifice, the diameter of which controls the rate of air flow through the channel. Airborne particles moving through the channels are subjected to a constantly increasing centrifugal acceleration that causes them to deposit on the channel floor. The floor is formed by a thin, removable foil covering the inner surface of the outer cone (see Fig. 6.10). Since incoming particles are distributed over the channel cross section at the entrance, particles of all sizes begin depositing at the inlet and continue to do so over a channel length that is a function of the operating characteristics of the instrument and the aerodynamic diameter of the particle.

When the collecting foil is laid out flat, the channel floors describe Archimedean spirals starting at a common origin but having their axes rotated to maintain a constant width between the boundaries of the channel floor. A similar spiral, always at a constant distance from a channel boundary, can be drawn through any point on the channel floor. The length of the spiral segment intercepted by arcs of radius R and R_0 (see Fig. 6.11) is given very nearly by [13]

$$l = \left(\frac{\pi \sin \gamma}{a}\right)(R^2 - R_0{}^2). \tag{6.23}$$

Since all spirals enter a channel at R_0, they all intercept the arc of radius R at the same distance from the inlet. With this relationship, the length of channel traversed by a particle before deposition can be calculated after measuring the radial distance from the origin to the point of deposition. Accessory equipment is available which facilitates the measurement of R [14]. The measurement can also be simplified by the use of gridded collecting foils [12].

Stöber and Zessack [13] made a theoretical study of the deposition of particles in the Goetz centrifuge. They assumed that the zone of turbulence, caused by the acceleration of the air as it moved from the stationary inlet tube into the rotating channels, extended down the channel for a distance l_a; that no particles were lost in the turbulent zone; and that particles were uniformly distributed over the channel cross section at l_a, where laminar flow began. With the additional assumption that there was no velocity gradient between channel floor and ceiling, they were able to calculate the trajectory of a particle of a given aerodynamic diameter for a given set of operating conditions. They found that deposition, at the midline of the channel floor, of spherical particles

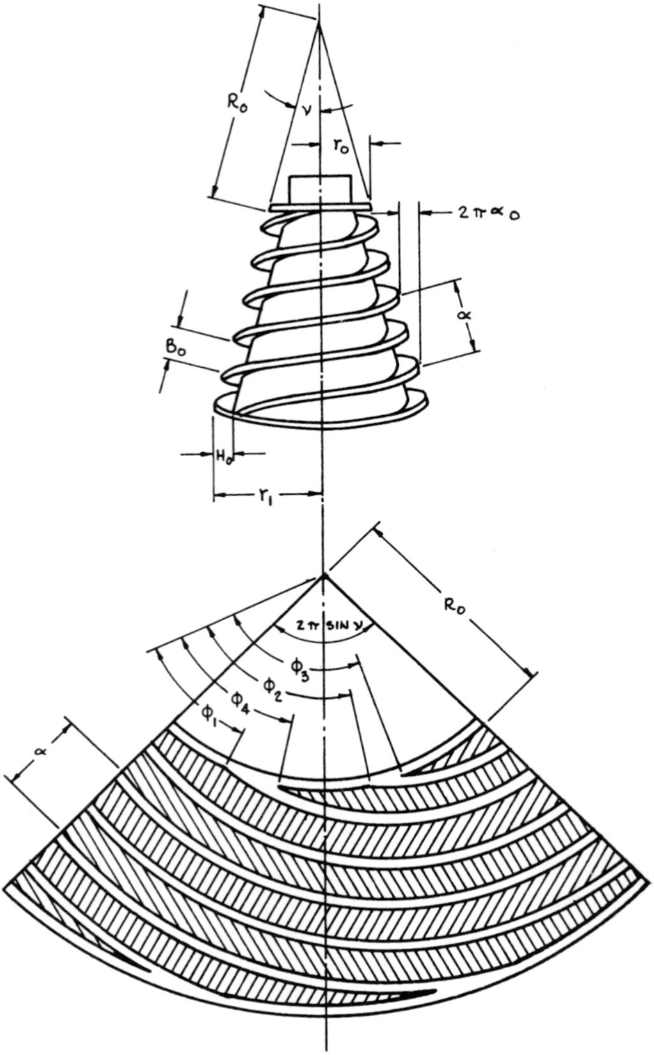

FIG. 6.11. Rotor and foil geometry in Gcetz spectrometer [13]. Couresty of Staub-Reinhaltung der Luft.

of diameter D would occur over a length l_D from the inlet given by

$$l_D = [(2.29\eta F/\rho K_s D^2 N^2) + (l_a + \lambda)^{3/2}]^{2/3} - \lambda, \tag{6.24}$$

where ρ and K_s are the particle's density and slip factor, respectively, η is the

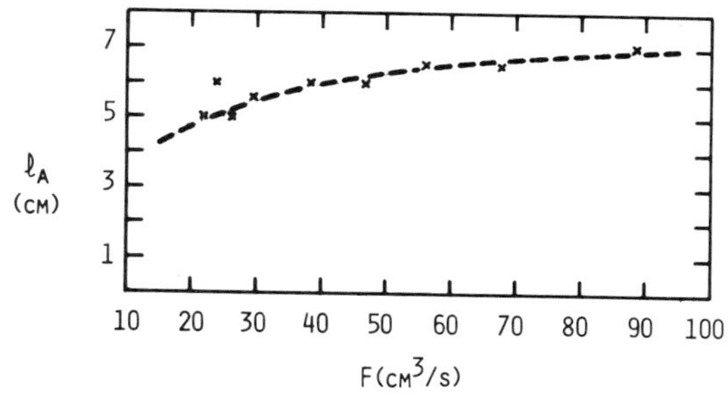

FIG. 6.12. Length of inlet turbulent zone as a function of flow rate [13]. Courtesy of Staub Reinhaltung der Luft.

viscosity of air, F is the volumetric flow rate, N is the velocity of rotation in revolutions per second, and $\lambda = (\pi R_0^2 \sin\gamma)/a = 16.5$ cm (see Fig. 6.11). All dimensions must be in cgs units, because the numerical factor includes values of fixed channel dimensions. They made experimental measurements of l_D under a variety of conditions and by matching their data against Eq. (6.24) were able to establish the relationship between flow rate and the length l_a of the turbulent zone, as shown in Fig. 6.12.

Baust [15], operating the centrifuge at 22,000 rpm, measured the surface concentration of monodisperse particles at the channel center as a function of deposition length. His results, with the corresponding theoretical curves according to Stöber and Zessack, are shown in Fig. 6.13(A),(B). The sharp cut-off in deposition, which is predicted theoretically, does not occur experimentally, a fact that was also observed by Stöber and Zessack and others [16, 17]. In addition, Baust verified (Fig. 6.13C) the skewed concentration profile encountered by Stöber and Zessack.

The relationship between aerodynamic diameter and deposition length depends on the rotational speed N and the air flow rate F, which is itself a function of N and of the diameter of the jet orifices. Some experimental data for two jet diameters at different values of N are shown in Fig. 6.14. For a given jet, an increase in N does not greatly alter the range of diameters that can be collected, the larger deposition velocities being partially offset by increased flow rate. A much greater effect is observed when N is fixed and the jet orifice diameter is changed. The double channel arrangement permits simultaneous operation at two orifice settings. The significance of flow rate makes it essential that it be reproducible for a given set of operating conditions. Ludwig and Robinson [17] and Raabe [12] have pointed out, however, that flow rate is quite sensitive to conditions external to the instrument. This view is supported by reported measurements of flow rate as a function of operating conditions. Goetz *et al.* [18] expected a linear relationship between F and the product,

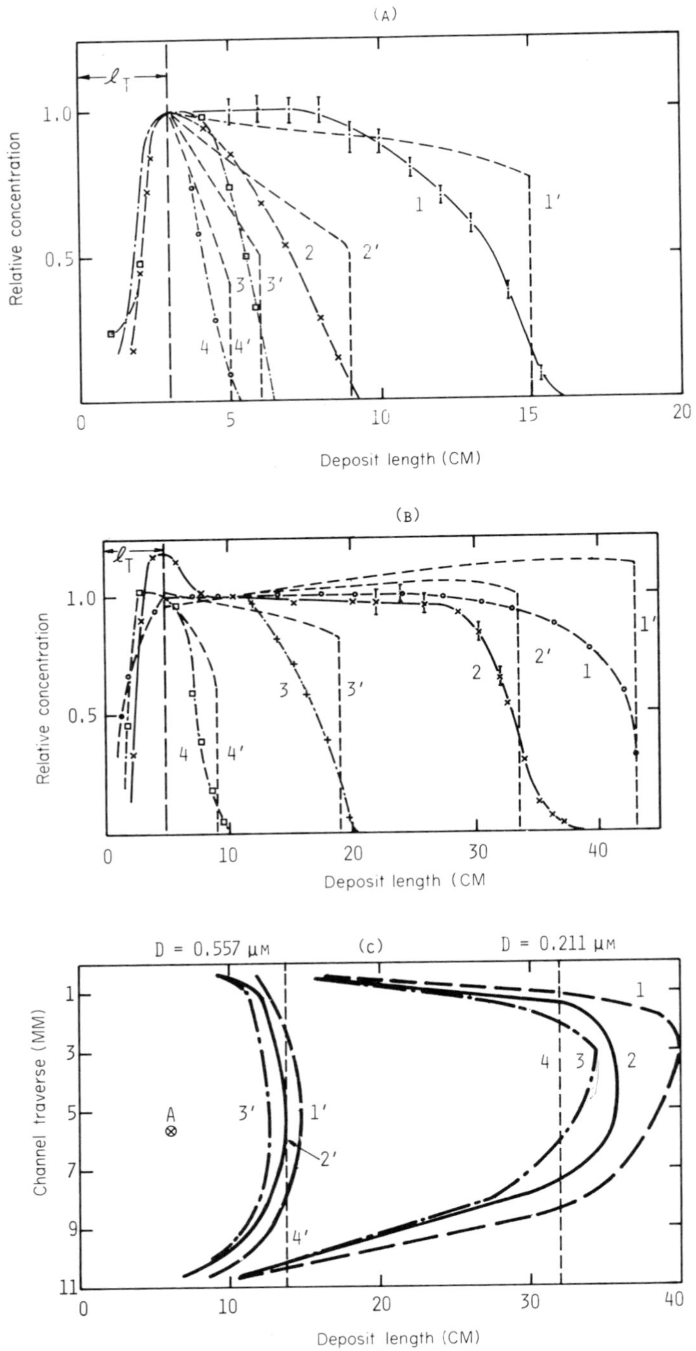
(A)
Relative concentration
Deposit length (CM)
ℓT
1.0
0.5
0
5
10
15
20
1
1′
2
2′
3
3′
4
4′
(B)
Relative concentration
Deposit length (CM
ℓT
1.0
0.5
0
10
20
30
40
1
1′
2
2′
3
3′
4
4′
(C)
D = 0.557 μM
D = 0.211 μM
Channel traverse (MM)
Deposit length (CM)
1
3
5
7
9
11
0
10
20
30
40
A
1′
2′
3′
4′
1
2
3
4

$N \cdot D_j$, where D_j is the diameter of the exhaust jet. They interpreted their measurements in terms of the average linear velocity V_r in the channel, obtaining the relationship

$$V_r = 0.0355 N D_j \qquad \text{cm/sec.}$$

This result is shown in Fig. 6.15, with their experimental data and those of other investigators.

The simplest method for converting the deposition pattern into a size distribution is based on the assumption that particles of a given aerodynamic diameter are uniformly deposited on the collection foil from the channel inlet to their cut-off length, l_D [13]. The distribution function

$$f(D_A) = (1/C_0) \cdot (dC/dD_A),$$

for a polydisperse aerosol at an activity concentration C_0 in the air can be found in a manner similar to that described above for the horizontal elutriator. For an activity concentration C_A per unit area of collection foil, and a total sample volume Ft, the distribution function can be calculated from

$$f(D_A) = \frac{W l_D}{C_0 F t} \cdot \frac{dC_A}{dl_D} \cdot \frac{dl_D}{dD_A},$$

where W is the channel width. This method of evaluation, even when applied only to the center portion of the deposition pattern where the assumption of uniform deposition is most nearly fulfilled, leads to a broadening of the size distribution. An example of this is given in Fig. 6.16, which shows results for

FIG. 6.13. Deposition of monodisperse aerosols in the Goetz spectrometer [15] (22,000 rpm). Courtesy of Staub-Reinhaltung der Luft.

A. Midchannel deposition as a function of distance along channel for 0.5 mm jet. Dashed lines are theoretical curves; l_T is the length of turbulent zone.

Curve:	1	2	3	4
Diameter (μm):	0.163	0.365	0.796	1.305

B. Midchannel deposition as a function of distance along channel for 1.0 mm jet. Dashed lines are theoretical curves; l_T is the length of turbulent zone.

Curve:	1	2	3	4
Diameter (μm):	0.163	0.365	0.796	1.305

C. Profiles of concentrations relative to concentration at A. Curve 1,1′: $C = 0.01\ C_A$; curve 2,2′: $C = 0.1\ C_A$; curve 3,3′: maximum rate of concentration decrease; and curve 4,4′: deposition length as determined visually.

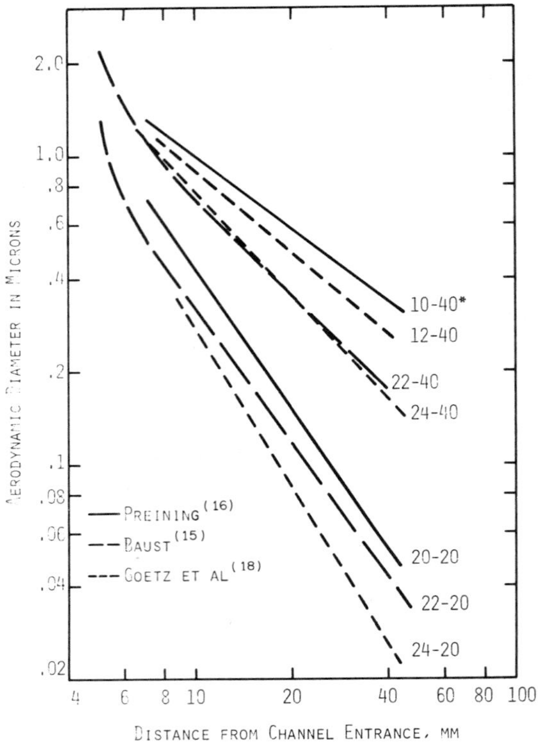

FIG. 6.14. Relationship between aerodynamic diameter and deposition length in Goetz spectrometer at different rotational speeds and jet orifice diameters. First two numbers = rpm/1000; second two = orifice diameter in mils.

a mixture of three monodisperse latex aerosols. Methods which make some allowance for the nonuniformity of the deposition pattern do not eliminate the broadening of the peaks [15]. Moreover, the basic relationships between the amount of aerosol entering the instrument and the amount deposited on the centrifuge itself are upset by the loss of particles at the channel inlet as shown in Fig. 6.17(A). Additional losses occur between the exhaust jets and the outlet as shown in Fig. 6.17(B).

It appears that as much has been written on the calibration of this instrument as on its application to the study of aerosols. Results obtained with it are probably most reliable when they provide comparative measurements of size distributions. The simultaneous measurement of number and radioactivity distributions [19], and the determination of collection efficiency of sampling devices by a comparison of the deposition patterns obtained from centrifuge samples of input and output aerosols [20] are applications of this type.

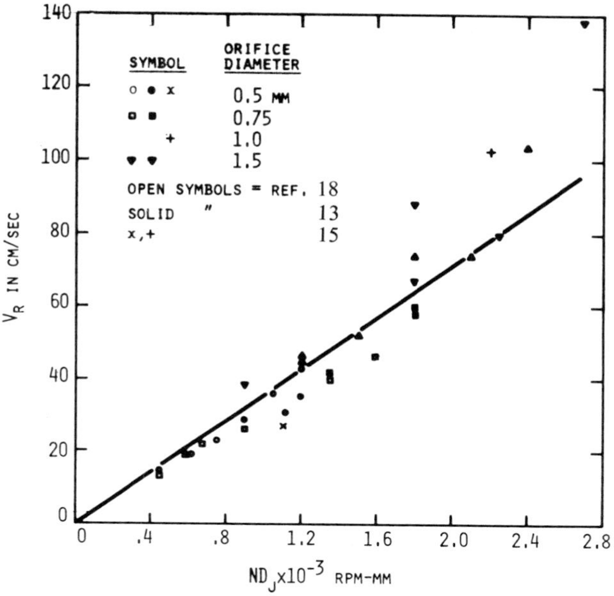

FIG. 6.15. Average air speed in channels as a function of ND_J.

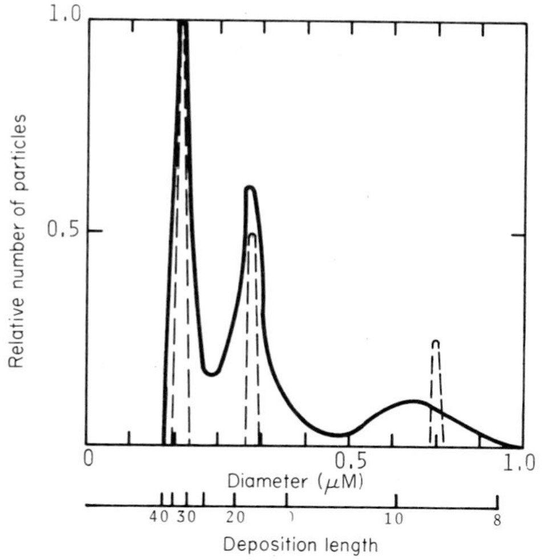

FIG. 6.16. Size distributions of monodisperse aerosols as measured with the Goetz spectrometer [15]. Courtesy of Staub-Reinhaltung der Luft.

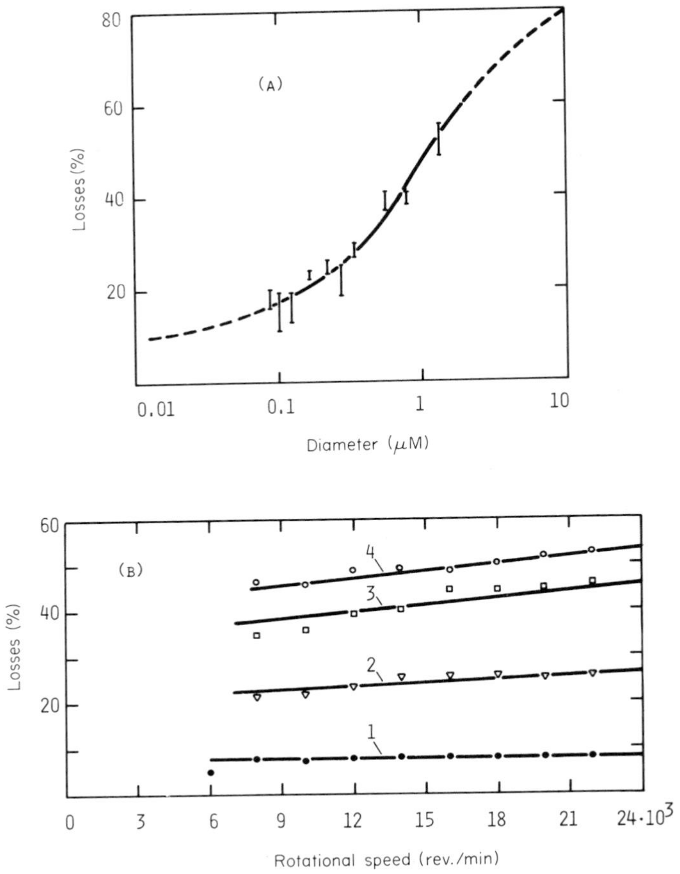

FIG. 6.17. Particle losses in the Goetz spectrometer [15]. (A) In passage through inlet; (B) between exhaust jets and outlet. Courtesy of Staub-Reinhaltung der Luft.

(b) *The Kast Spiral Centrifuge* [11]. Kast developed a centrifuge in the form of a rotating flat spiral (see Fig. 6.18). Like the Goetz centrifuge, the channels are Archimedean spirals. To obtain as large a radial acceleration as possible, the instrument is rotated in the direction of air flow. A blower, applying suction to the spiral outlet, is necessary to obtain the desired air flow. Particles are deposited on a thin aluminum strip, 180 cm long by 2 cm wide, that covers the collecting surface of the spiral. The instrument is usually operated at 3000–4000 rpm and at flow rates up to 125 liter/min, collecting all particles having aerodynamic diameters above about 1 μm. Estimation of particle size distribution is made in the manner described above for the

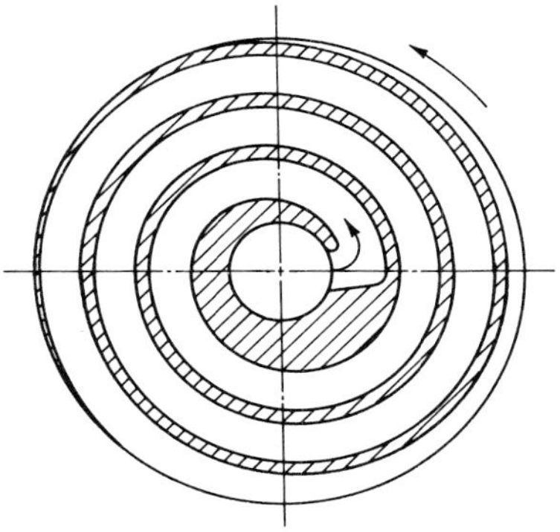

FIG. 6.18. Diagram of the Kast spiral centrifuge [11]. Courtesy of Staub-Reinhaltung der Luft.

horizontal elutriator, under the assumption that particles of a given size are deposited uniformly on the collection surface for a length that can be estimated adequately from theoretical considerations. Experimental calibration has consisted of a comparison between particle size distributions obtained with this instrument and with the Andreasen pipette method.

A stationary precollector, also in the form of an Archimedean spiral, can be fitted over the centrifuge inlet to remove very large particles. In this device, the centrifugal force acting on the particles is due entirely to the flow of air through it and is sufficient, at 125 liter/min, to remove all particles having aerodynamic diameters greater than about 6 μm. Smaller particles are then carried into the rotating spiral where they are subjected to much larger centrifugal forces.

6-2.2 CENTRIFUGAL AEROSOL SPECTROMETERS

(a) *The Conifuge.* The first successful centrifugal aerosol spectrometer was built by Sawyer and Walton [9]. It was designed to collect particles in the aerodynamic size range of about 0.5–30 μm. Modified versions of the conifuge [21, 22] have been designed to collect particles in the range of about 0.05 to 3.0 μm, usually at higher flow rates than those used by Sawyer and Walton. A cross-sectional diagram of an early modification is shown in Fig. 6.19. The instrument consists of a metal cone, mounted directly on the rotor of a high-speed electric motor, and a conical metal cover that can be fastened rigidly to the cone, leaving a conical annular air space between the latter and itself. When the unit is rotated, air is drawn into the opening at the top, pumped through the annulus and exhausted through jet orifices (which control the volumetric flow rate) at the bottom. Since the chamber around the unit is sealed, the air discharged at the base is recirculated, except for a relatively small amount that is removed at a controlled flow rate by suction at T. An equal flow of sample air enters through the inlet tube and spreads in a thin

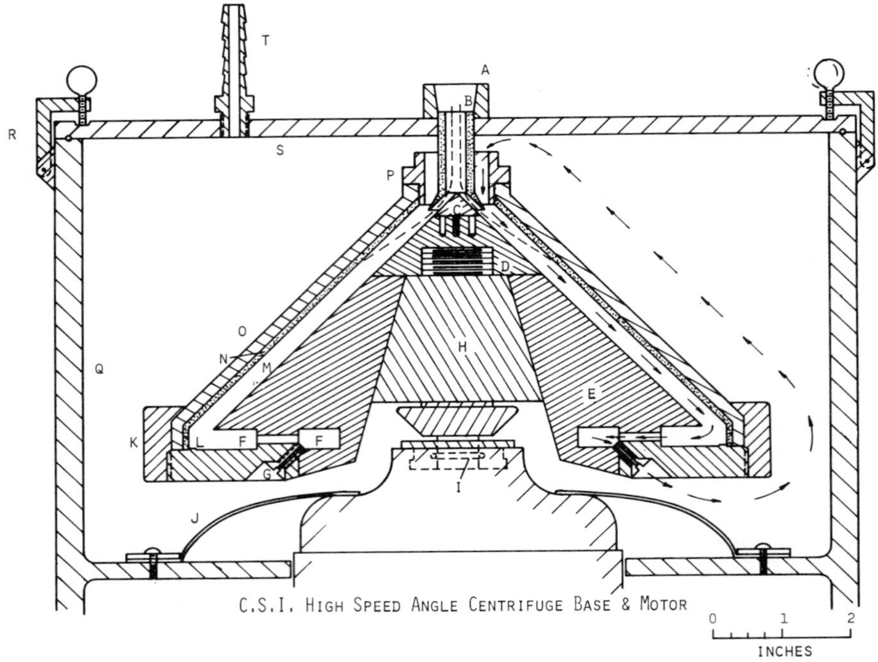

A. SMOKE INLET TUBE
B. TEFLON GUIDE TUBE
C. APEX CAP NUT
D. MOUNTING NUT
E. INNER CONE
F. FLOW EQUILIZING CHAMBERS
G. OUTLET JET
H. C.S.I. ROTOR HUB
I. "O" RING SHAFT SEAL
J. COPPER CHAMBER SEAL
K. CLAMPING RING
L. TEFLON CUSHION FOR BASE OF SLIDE
M. DOVETAILED DARK FIELD SLIDE
N. DOVETAILED NYLON SLIDE HOLDER
O. OUTER CONE
P. AIR ENTRANCE TUBE
Q. CHAMBER WALL
R. CLAMP
S. CHAMBER TOP
T. OUTLET TUBE

FIG. 6.19. Cross-sectional diagram of a modified conifuge [21]. Courtesy of Academic Press, New York.

layer over the inner cone. Particles entering in the sample air are subjected to a centrifugal force which moves them through the much thicker clean layer to the outer cone. As the air moves down the annulus, its linear velocity decreases and the centrifugal force acting on the particles increases, compressing the particle size spectrum so that a wide range of sizes can be collected at a given set of operating conditions. Particles are deposited in narrow bands around the inner surface of the outer cone, the position of the center of the band being characteristic of the aerodynamic diameter of the particles. The outer cone is provided with an appropriate collection surface—microscope slides [9], electron microscope grids [21], or thin foils [23]—which permits direct observation or analysis of the particles.

Stöber and Zessack [24] made a theoretical analysis of the deposition of particles in centrifugal aerosol spectrometers. To make their results generally applicable to conical centrifuges, they used the coordinate system shown in Fig. 6.20, in which l_a represents the slant length from the apex of the center cone to the point at which aerosol enters the annular air space. Assuming a uniform velocity profile between the cones, they obtained the following relationship between the aerodynamic diameter D_A and the slant length l at which the particle would deposit:

$$\rho_0 K_{sA} D_A{}^2 = \frac{27\eta F}{\pi^3 N^2 \cos\theta} \cdot \frac{1}{4(l^3 - l_a{}^3)\sin^2\theta + 3H_0(l^2 - l_a{}^2)\cos^2\theta \sin\theta}, \tag{6.25}$$

where

H_0 = cone separation (cm),
F = total volumetric flow rate (cm^3/sec),
N = angular velocity (revolutions/sec),
η = viscosity of air (g/cm-sec),
K_{sA} = slip factor
ρ_0 = unit density (g/cm^3), and
θ = half angle of cone.

Experimentally, the length $l_D = l - l_a$ (L in Fig. 6.20) is actually measured and

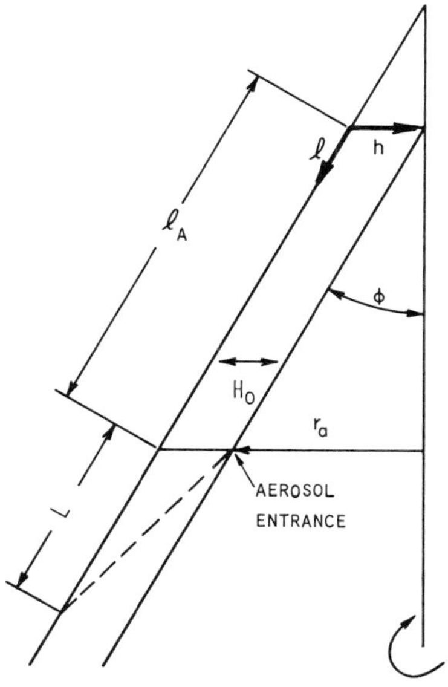

FIG. 6.20. Coordinate system for the conical centrifuge [23]. Courtesy of American Chemical Society.

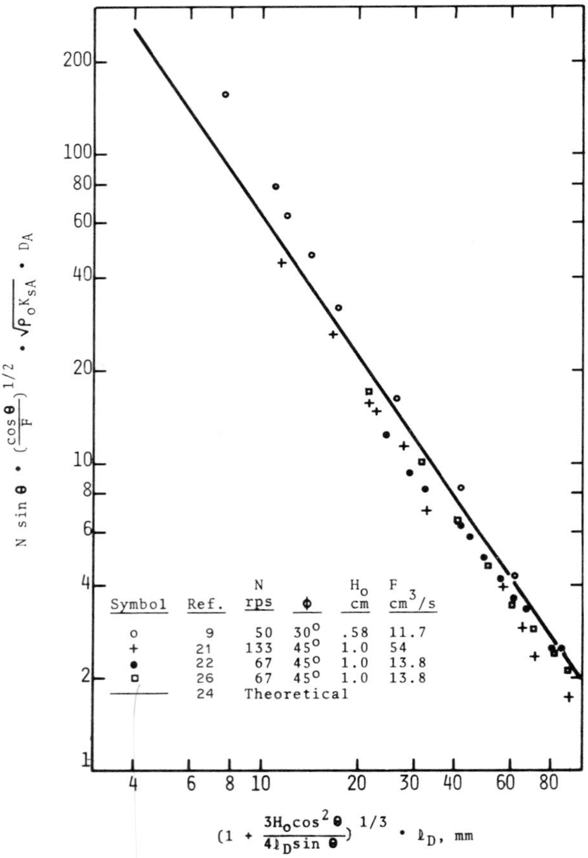

FIG. 6.21. Calibrations of conifuges.

instrument calibrations give D_A as a function of l_D. Figure 6.21 shows several experimental calibrations for instruments in which $l_a = 0$, with the corresponding theoretical relationship according to Eq. (6.25). Agreement between theory and experiment is generally satisfactory. Some of the differences may be related to the fact that the sample air has to be accelerated to fhe rotational speed of the centrifuge after it enters.

Some of the variety of inlet systems that have been developed are shown in Fig. 6.22. Tillery [22] replaced the recirculating clean air system with a controlled air input, the air seal between the rotating and stationary parts being provided by an inverted cup immersed in a trough filled with oil. He showed that the air flow through the instrument increased linearly with rotational velocity above 2000 rpm. Stöber [23] introduced the sample through

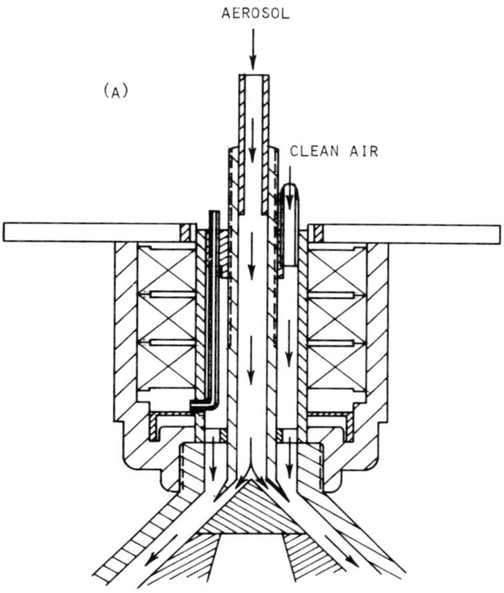

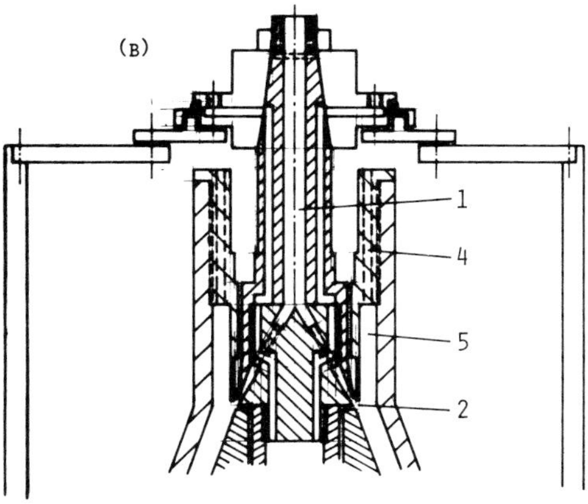

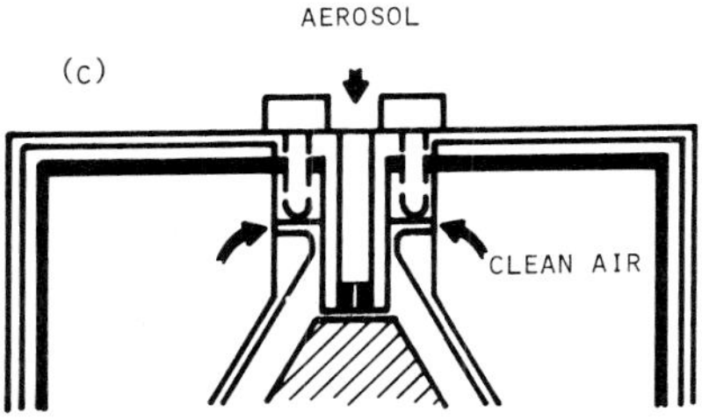

FIG. 6.22. Aerosol inlet systems for conifuges. (A) Tillery [22]. Courtesy of International Atomic Energy Agency. (B) Stöber and Flachsbart [23]. Courtesy of American Chemical Society. (C) Hochrainer and Brown [25].

a ring slit. The necessary clean air sheath was provided by room air that entered the annular airway after having been cleared of particles by centrifugal action during its passage through a system of capillaries rotating with the cones. Hochrainer and Brown [25] used a ring slit arrangement also, but introduced the sample through a fine hole centered on the axis of rotation. They obtained clean air by having room air enter the conifuge after passing through a narrow annular space between the stationary instrument housing and a shell rotating with the conifuge.

The most important application of the conifuge has been in the study of aerodynamic shape factors. Losses at the inlet are not of great significance when the instrument is used for this purpose. They become important, however, when it is applied to the determination of size distributions. According to Sawyer and Walton [9], particles having aerodynamic diameters smaller than 25 μm are sampled without loss at their standard sampling rate of 25 cm^3/min. Losses at inlets of the other types have not been specified. Devices in which the inlet rube rotates with the centrifuge will show a 50% loss of particles at an aerodynamic diameter given by

$$(\rho_0 K_{sA})^{1/2} \cdot D_A = (233/R_i \omega) \cdot (F/l_i)^{1/2} \tag{6.26}$$

where D_A is in micrometers and

R_i = inlet tube radius (cm),
ω = angular velocity (sec^{-1}),
F = sample flow rate (cm^3/sec), and
l_i = inlet tube length (cm).

The use of small diameter inlet ducts, such as that in Fig. 6.22(C), may give rise to severe impaction losses even at aerodynamic diameters below 5 μm. Impaction losses as the sample air leaves the inlet tube will reach 50% at an aerodynamic diameter in micrometers given by

$$(\rho_0 K_{sA})^{1/2} \cdot D_A = 460 \cdot (R_i{}^3/F)^{1/2}. \tag{6.26a}$$

Increasing F or decreasing R_i in an effort to reduce losses during passage through the inlet tube results in increased impaction losses at the exit.

Entrance losses in the ring-slit conifuge (Fig. 6.22B) are due mostly to sedimentation in the conical section of the aerosol inlet system. The aerodynamic diameter in micrometers at 50% losses is given by

$$(\rho_0 K_{sA})^{1/2} \cdot D_A = 7.2 \cdot \sqrt{F}/R_0. \tag{6.26b}$$

R_0 is the distance in cm from the rotor axis to the point at which the aerosol enters the cone.

The resolving power of the conifuge is limited by the finite thickness of the aerosol layer at the inlet, which depends on the ratio of sample flow rate to total flow rate. The dimensions of the inlet itself are less significant, since the aerosol layer will adjust to a thickness corresponding to isokinetic conditions. The ratio $\delta l/l_D$ of a deposit width to its distance from the inlet is related to the ratio $\delta H/H_0$ between the depth of the aerosol layer and the width of the annulus, by

$$\frac{\delta l}{l_D} = -\frac{\delta H}{H_0} \cdot \frac{l_D + 2l_a}{(l_D + l_a)[2 + 4(l_D + l_a)\sin\theta/(H_0 \cos^2\theta]} . \tag{6.27}$$

In general, the resolution $\delta D/D$ is roughly proportional to $-\delta l/l_D$, with a proportionality factor between about 0.6 and 1.5. Also,

$$\delta H/H_0 \simeq (\delta F/3F)^{1/2},$$

where δF is the volumetric aerosol flow rate. These relationships predict that resolution improves with increasing values of l, as observed experimentally, but yield values that are two or three times smaller than experimental resolutions [23]. Part of the difference is, undoubtedly, due to the spread in diameters of even the best of monodisperse aerosols. In addition, disturbances of the flow pattern may have an undesirable effect on resolution. Stöber and Flachsbart [23] have pointed out that Coriolis forces, arising from the interaction between the angular velocity of the centrifuge and the radial component of the air velocity, tend to upset the laminar flow of air through the conifuge. Their quantitative description of the effect is valid only if the relative angular velocity between air and centrifuge is small compared to the angular velocity of the rotor itself. However, the qualitative reality of the effect was demonstrated experimentally.

(b) *The Spinning Spiral Aerosol Spectrometer.* Stöber and Flachsbart [27] developed a flat spiral spectrometer as shown in Fig. 6.23(A). The spiral is formed by six semicircles of increasing radius that are alternately eccentric and concentric with respect to the center of the rotor. Clean air is blown into the system at 1, flows through the laminator section 2, and emerges to form a thick layer of clean air in laminar flow between the aerosol inlet and the outer spiral wall. The sample air is sucked into the center of the inlet section (3) and emerges from it as a much thinner layer of aerosol in laminar flow between the inner wall and the clean air layer. As the air flows along the spiral, the particles are subjected to a centrifugal force that increases in a stepwise fashion with alternate semicircles (see Fig. 6.23B) and are deposited on a strip of metal foil lining the surface of the outer wall. As in the conifuge, particles of a given size are deposited in a narrow band at a distance from the inlet

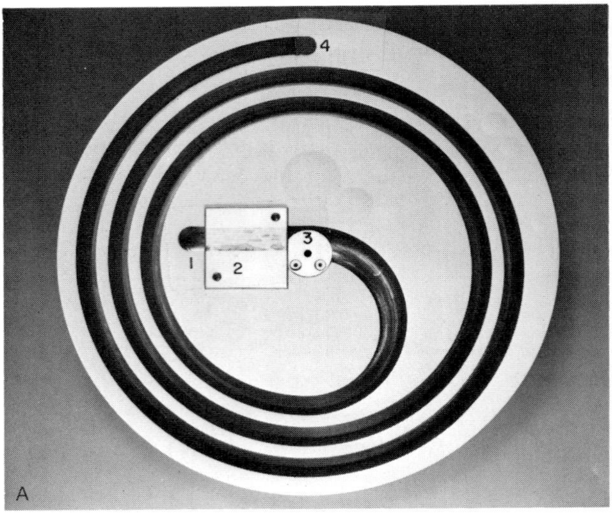

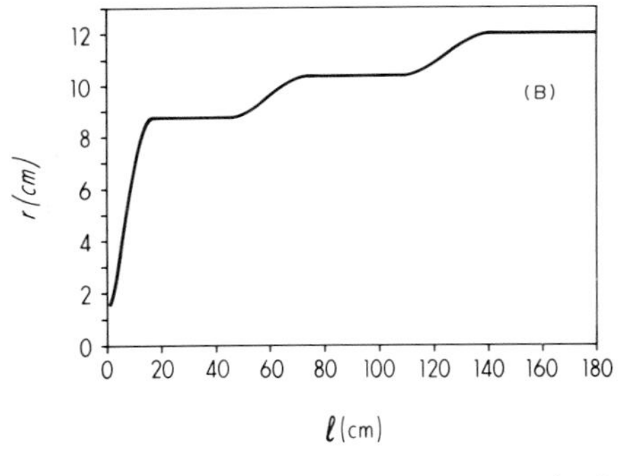

FIG. 6.23. (A) View of the spiral centrifuge spectrometer [27]. Courtesy of American Chemical Society. (B) Radius vector (proportional to centrifugal force) in relation to foil position.

that is a function of the aerodynamic diameter of the particle and the operating conditions of the centrifuge.

The curvature of the flow lines in the presence of the strong centrifugal forces causes a secondary flow in the form of a double vortex (see Fig. 6.24), so that the air streamlines follow counterrotating helical paths around the spiral. For spectrometric purposes, the period of rotation of the vortices must be much longer than the transit time through the spiral duct. Stöber and Flachsbart obtained satisfactory results at 3000 rpm for flow rates up to 19 liter/min. Their calibration results are shown in Fig. 6.25, in which the separation between curves for a given flow rate is a measure of the size resolution. For these measurements, the sample flow was 4% of the total flow, providing size resolution of 4 to 10%. By reducing the sample flow rate to

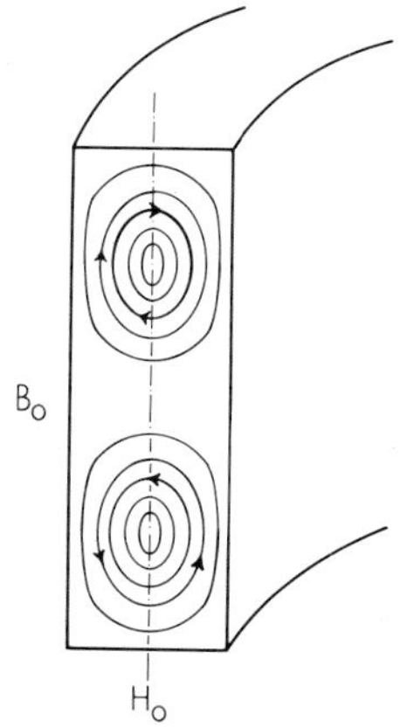

FIG. 6.24. Secondary flow in spiral centrifuge [27]. Courtesy of American Chemical Society.

less than 1% of the total, Stöber and Flachsbart were able to resolve more than 20 deposits representing aggregates of different numbers of monodisperse spherical particles of 0.71 μm diameter.

For their standard aerosol inlet, the authors estimate that losses amount to 50% at about 6 μm.

Kotrappa and Light [28] have developed a small, modified version of the spiral centrifuge in which the cross section of the channel expands by a factor of 3 between the aerosol inlet and the end of the spiral. The length of the collecting foil is 45 cm and particles escaping deposition are caught on a filter set normal to the air flow at the exit. Particles collected on the filter are

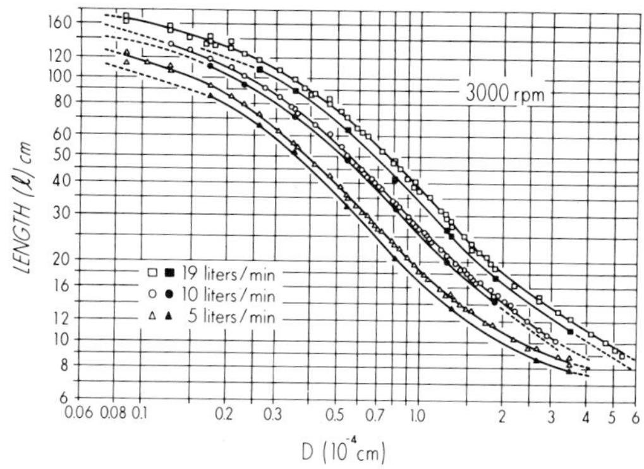

FIG. 6.25. Calibration curves for spiral centrifuge [27]. Courtesy of American Chemical Society.

also separated on the basis of aerodynamic diameter. The instrument is relatively inexpensive to build and shows good resolution of particles between about 0.5 and 5 μm aerodynamic diameter.

6-3 Impaction Methods

Instruments employing the process of impaction have been used for many years to sample toxic airborne particles. Single stage devices, having collection characteristics that were poorly understood but were generally believed to provide an adequate sample of hazardous particles, were designed to collect a sample suitable for analysis in the optical microscope. The Kotze [29] and British konimeters had round jet orifices, as did the Greenburg–Smith [30] and midget [31] impingers; the Owens [32] and Bausch and Lomb [33] dust counters had rectangular jet orifices. All of these instruments were plagued with problems of rebound, re-entrainment, and disaggregation [34]. The principle of cascade impaction, introduced by May [35], made it possible to minimize the significance of these problems and provided a means of segregating particles into several fractions, each of which contained only particles having sizes between two more-or-less sharply defined limits.

Although May was interested primarily in obtaining samples for microscope analysis in which large particles would be fairly represented, he recognized the potential usefulness of the instrument for determining mass distributions. For each stage, he defined an "effective drop size" against which the fraction of the total mass penetrating a given stage could be plotted to obtain an approximate distribution. Laskin [36] adopted the instrument for the study of uranium dioxide particles, making the important modification of adding a final filter stage to capture all particles escaping impaction. Since then the impactor has become one of the most useful and widely used instruments for the study of particle size in relation to inhalation toxicity.

The cascade impactor permits a direct determination of the cumulative distribution of particle activity as a function of aerodynamic size. The accuracy with which such distributions can be estimated from cascade impactor data depends on the degree to which the size distribution of particles passing a given impaction stage overlaps the size distribution of particles retained in the stage. Thus, it is important that the collection efficiency for each impaction stage should increase from zero to unity over a relatively small interval of particle sizes. Moreover, the accuracy of the estimated distributions is affected by certain errors that are inherent in the methods currently used for the interpretation of impactor data. In the following discussion, the theoretical and experimental aspects of collection efficiency, the interpretation of data, and the design of various impactors will be considered.

6-3.1 Impaction from Rectangular Jets

Assume that a stream of particle-laden air emerges from a rectangular orifice of width W and impinges normally upon a plane surface at a distance S from the orifice. The stream will split into two symmetrical branches flowing in opposite directions. The particle trajectories will diverge from the air streamlines and, if the particles have sufficient inertia, may intersect the plane surface. If the jet orifice is assumed to be of very great length, calculating particle trajectories becomes a two-dimensional problem as shown in Fig. 6.26.

If the motion of a particle relative to air is opposed with a purely viscous resistance, and if lengths are measured in units of $W/2$, velocities in units of U_0 (the initial velocity of both air and particles), and time in units of $W/2U_0$, then the equations of motion of a spherical particle in the air stream can be expressed in dimensionless variables as

$$\text{Stk} \cdot (d\tilde{U}_x/d\tilde{t}) = \tilde{V}_x - \tilde{U}_x; \qquad \text{Stk} \cdot (d\tilde{U}_y/d\tilde{t}) = \tilde{V}_y - \tilde{U}_y, \tag{6.28}$$

where Stk is the Stokes number as defined in Chapter 4. The x and y components of the air velocity are, respectively, $\tilde{V}_x$ and $\tilde{V}_y$ and those of the particle are $\tilde{U}_x$ and $\tilde{U}_y$. It is assumed that all particle diameters are negligible compared to the jet width, that a particle is captured and retained if the trajectory of its center intersects the collecting surface, and, for purposes of calculating flow fields, that air is an ideal fluid. Even with these simplifying assumptions, an analytical solution for the particle trajectories cannot be obtained because of the complex nature of the flow fields.

Davies and Aylward [37] studied this process theoretically, making a rigorous development of the flow fields followed by stepwise calculations of the particle trajectories. They assumed that the jet throat was a parallel-sided

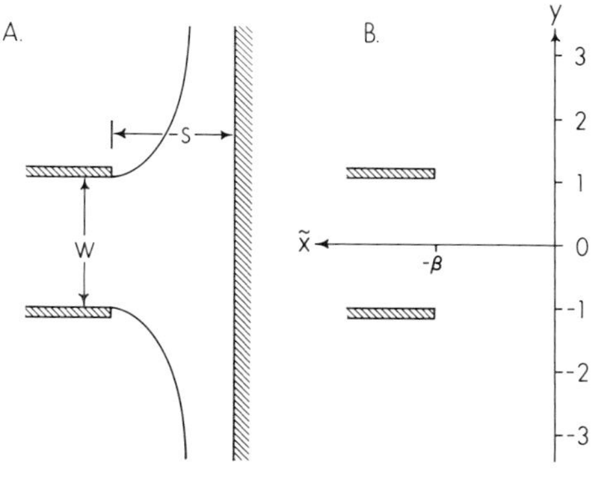

Fig. 6.26. Two dimensional representation of impaction stage [36]. (A) Schematic; (B) coordinate system. Courtesy of Academic Press, New York.

duct of width W and of infinite length along both the x and z axes, that air entered it at a uniform velocity U_0, and that the components of the air velocity were $V_x = 0$, $V_y = \alpha U_0$ at $y = \infty$. They showed that $\alpha \geqslant 1$ and is a function only of S/W, increasing from unity at $S/W = \infty$ to infinity at $S/W = 0$. In their development, the x component of the air velocity as it emerged from the jet orifice was least on the jet axis and increased to αU_0 at the jet wall. They also showed that the overall increase in air velocity from U_0 to αU_0 was accompanied by a pressure drop

$$\Delta P = \tfrac{1}{2}\rho_a U_0^2(\alpha^2 - 1),$$

where ρ_a is the density of air. They calculated curves of efficiency as a function of Stokes number for five different values of α. Their curves, which indicated quite good cut-off characteristics, became steeper and moved to lower values of the Stokes number as the ratio S/W decreased.

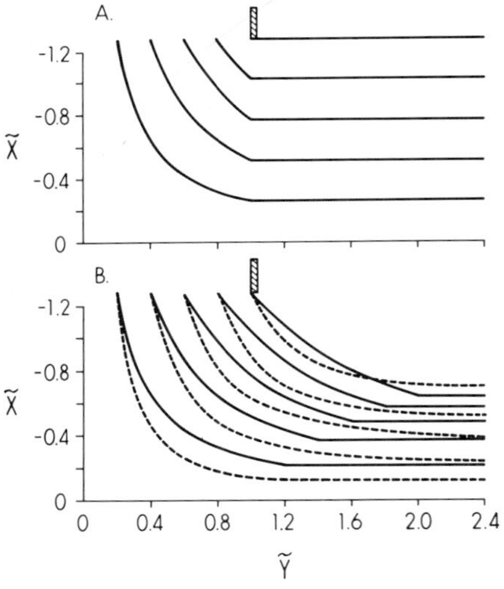

FIG. 6.27. Theoretical flow patterns in impaction stage [41]. (A) Ranz and Wong [38]. (B) Solid lines, Mercer and Chow [41]; broken lines, Davies and Aylward [37]. Courtesy of Academic Press, New York.

Ranz and Wong [38] also treated the problem theoretically, assuming a simplified flow field and certain arbitrary boundary conditions that made it possible to obtain an analytical solution to Eq. (6.28). They did not specify the value of S/W for which their single theoretical curve was derived and it has been assumed to be valid for large separations [39, 40]. On this basis the agreement between the two theoretical developments has been judged to be poor [39] and the experimental results of Ranz and Wong have appeared to be at odds with their theory. In reality, the boundary conditions assumed by

Ranz and Wong were equivalent to setting S/W equal to 0.5; on that basis, their theoretical results are in better agreement with those of Davies and Aylward.

Mercer and Chow [41] modified the theoretical development of Ranz and Wong to take the effect of S/W into account. Setting $\beta = 2S/W$ (see Fig. 6.26), they assumed the following conditions for the flow field:

$$\left.\begin{aligned}\tilde{V}_x &= \tilde{x}/\beta \\ \tilde{V}_y &= \tilde{y}/\beta\end{aligned}\right\} \text{ for } \left\{\begin{aligned}&-(1+\Delta\tilde{y}) < \tilde{y} < (1+\Delta\tilde{y}),\\ &-\beta < \tilde{x} < 0,\end{aligned}\right.$$

$$\tilde{V}_x = 0 \quad \text{for} \left\{\begin{aligned}&-(1+\Delta\tilde{y}) > \tilde{y} > (1+\Delta\tilde{y}),\\ &-\beta < \tilde{x} < 0,\end{aligned}\right. \tag{6.29}$$

$$\tilde{V}_x = \tilde{U}_x = 1 \quad \text{for} \quad \tilde{x} = -\beta$$

$$\tilde{U}_y = 0 \quad \text{for} \quad \tilde{t} = 0.$$

If $\beta = 1$ and $\Delta\tilde{y} = 0$, these assumptions are identical to those made by Ranz and Wong and are equivalent to setting $S/W = 0.5$. To obtain a more realistic flow field, Mercer and Chow set $\Delta\tilde{y}$ equal to $|\tilde{y}_0|$, the ordinate of a given streamline at $x = -\beta$. This assumption permits the various volume elements of air to travel equal distances in the y direction before the x components of their velocities go to zero.

Theoretical flow patterns for the three different methods are shown in Fig. 6.27. Theoretical collection efficiency curves calculated by the three methods

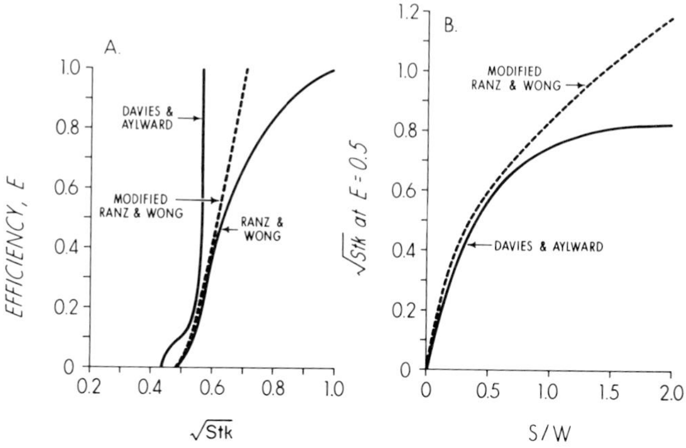

FIG. 6.28. (A) Theoretical collection efficiency curves for $S/W = 0.5$ [41]. (B) Theoretical curves of $\text{Stk}^{1/2}$ at an efficiency of 0.5 as a function of S/W. Courtesy of Academic Press, New York.

are shown in Fig. 6.28(A) for $S/W = 0.5$, and theoretical curves of $\mathrm{Stk}^{1/2}$ at a collection efficiency of 50% are shown as a function of S/W in Fig. 6.28(B). The curve based on the modified form of the Ranz and Wong method diverges rapidly from that of Davies and Aylward as S/W increases above 1. This is due to the modified theory containing the implicit assumption that the air stream begins to diverge immediately upon leaving the jet orifice. This assumption is reasonable when S is small but its validity diminishes rapidly as S/W increases above 1. The air stream actually diverges very little until it is within about one jet width of the plate.

Mercer and Chow also investigated the impaction process experimentally to test the theoretical predictions. They used test jets having geometries as indicated in Fig. 6.29. Curves of efficiency as a function of Stokes number were

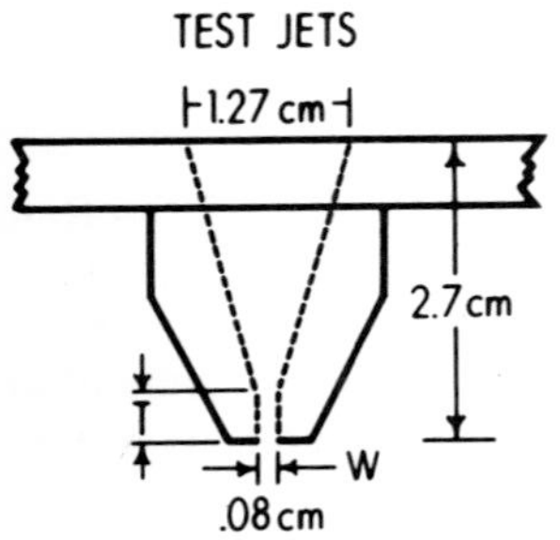

FIG. 6.29. Geometry of test jets.

Stage	Orifice length L (cm)	Orifice width W (cm)	Throat length T (cm)
1	1.27	0.08	0.13
2	1.27	0.08	0.07
2'	1.27	0.088	0.00
3	1.45	0.08	0.80

obtained by varying U_0 at a constant particle diameter. The particles were polyvinyltoluene spheres having diameters of $2.05 \pm 0.018\ \mu$m and a density of 1.026 g/cm^3. Collection efficiency curves for stages 1 and 2 at four different values of S/W are shown in Fig. 6.30. The vertical lines extending from individual data points represent standard deviations calculated on the assumption that the standard deviation on a given count was equal to the square root of the count. The dashed curves are the theoretical results of Davies and Aylward for $S/W = 0.35$ (to the left) and for $S/W = \infty$. Curves for stage 3 at $S/W = 1.0$ and 2.0 and for stage 2′ at $S/W = 0.84$ are shown in Fig. 6.31, together with a theoretical curve for $S/W = 1.0$ (dashed line) estimated from the work of Davies and Aylward. The solid circles on the curve for stage 2′ represent measurements made using 1.305 μm polystyrene particles.

The experimental relationship between α and S/W is compared with theory in Fig. 6.32 for stages 2, 2′, and 3. In each case, there was a range of values of S/W for which α was actually less than 1.

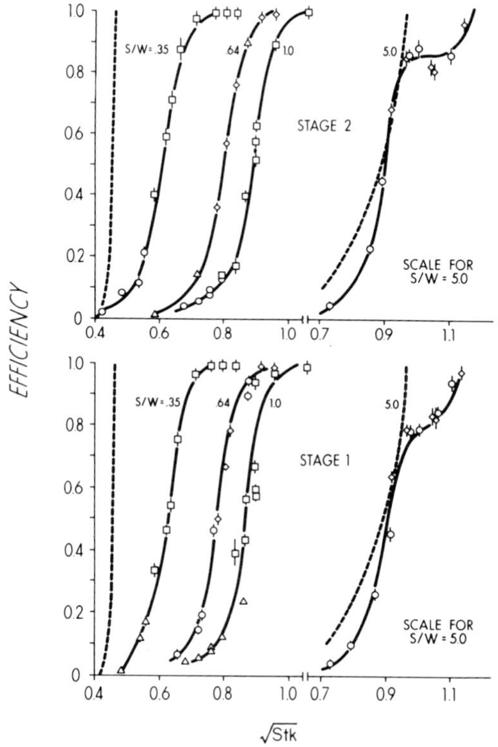

FIG. 6.30. Collection efficiency curves for stages 1 and 2 [41]. Courtesy of Academic Press, New York.

Although the efficiency curves of Figs. 6.30–6.32 generally have greater slopes in the region of $E = 0.5$ than do other published curves, the slopes are not as great as predicted by Davies and Aylward. Moreover, the curves still show a decreasing slope as the efficiency approaches unity. (Although they were extrapolated to $E = 1.0$, the highest efficiency actually observed was 0.9994.) This effect is due in part to a boundary layer of slowly moving air along the jet wall and in part to the fact that the assumption that a purely viscous resistance acts on the particles is not entirely valid, as Hänel [42] has pointed out. In the present case, the effect is much less pronounced than Hänel suggests, but it is still sufficient to move the theoretical curves toward larger Stokes numbers. The displacement is a function of S/W and efficiency,

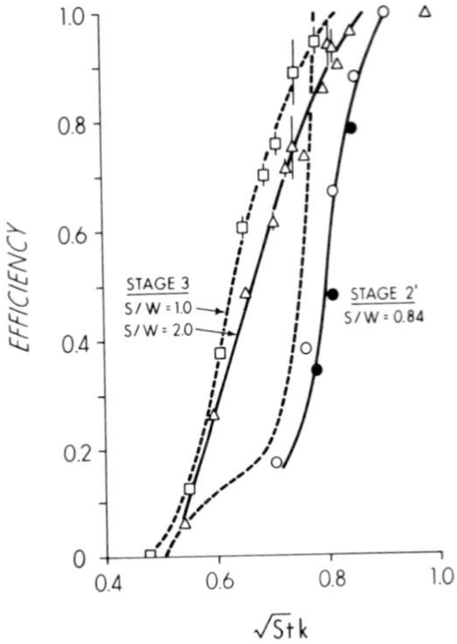

FIG. 6.31. Collection efficiency curves for stages 2′ and 3 [41]. Courtesy of Academic Press, New York.

varying by a factor of about 1.03 at low efficiencies and about 1.15 at efficiencies near unity.

The available data relating $\mathrm{Stk}^{1/2}$ at an efficiency of 0.50 to S/W are shown in Fig. 6.33 with the theoretical curve of Davies and Aylward. Although there is a good deal of scatter about the theoretical line, the data undoubtedly upport the theory.

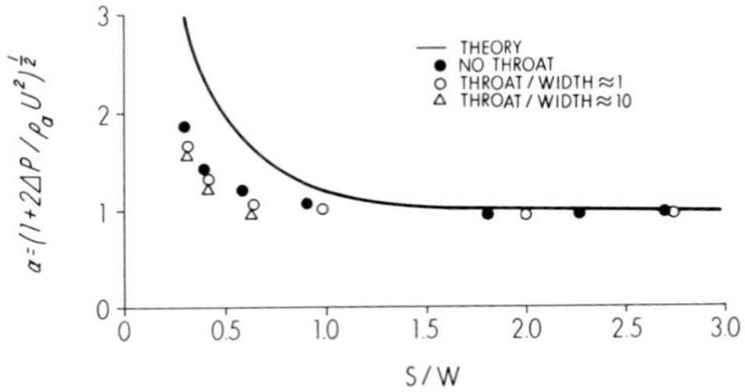

FIG. 6.32. Variation of α with S/W [41]. Courtesy of Academic Press, New York.

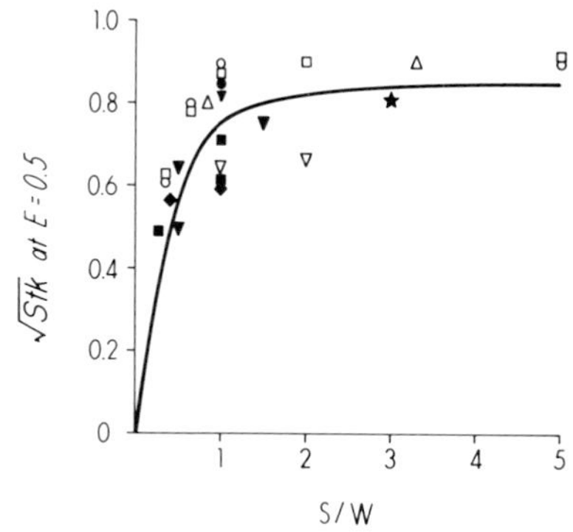

FIG. 6.33. $Stk^{1/2}$ at an efficiency of 0.5 as a function of S/W (rectangular jets) [41]. ▼, Laskin [36]; ●, Lundgren [43]; ■, May [35]; ◆, Stern *et al.* [39]; ★, Ranz and Wong [38]; open symbols, Mercer and Chow [41]. Courtesy of Academic Press, New York.

6-3.2 IMPACTION FROM ROUND JETS

No rigorous theoretical treatment of impaction from round jets is available. Ranz and Wong [38] and Roeber [44] have calculated collection efficiency curves using approximate flow fields and certain arbitrary boundary conditions. Mercer and Stafford [45] modified the theory of Ranz and Wong in a manner similar to that described above for rectangular jets. They obtained more realistic flow patterns and better agreement with experimental curves, but were unable to relate the real values of S/W to the effective values. The effective values seldom exceed unity because the impacting air stream shows little divergence until it is at least within one jet diameter of the impaction surface.

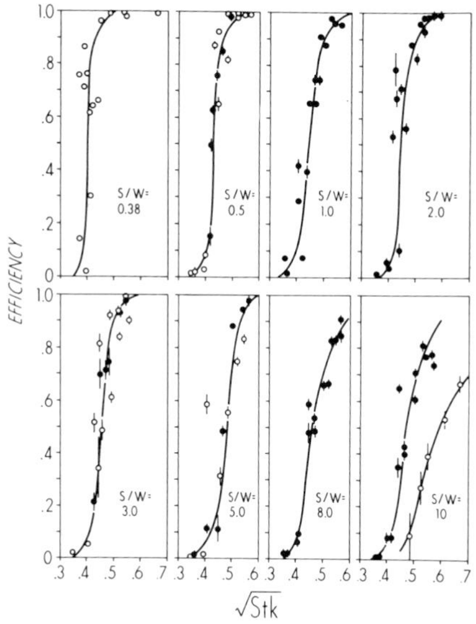

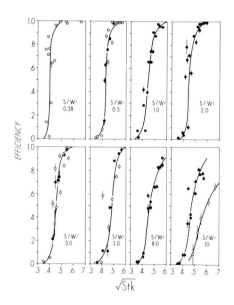

FIG. 6.34. Experimental collection efficiency curves for impaction from round jets [45]. Courtesy of British Occupational Hygiene Society.

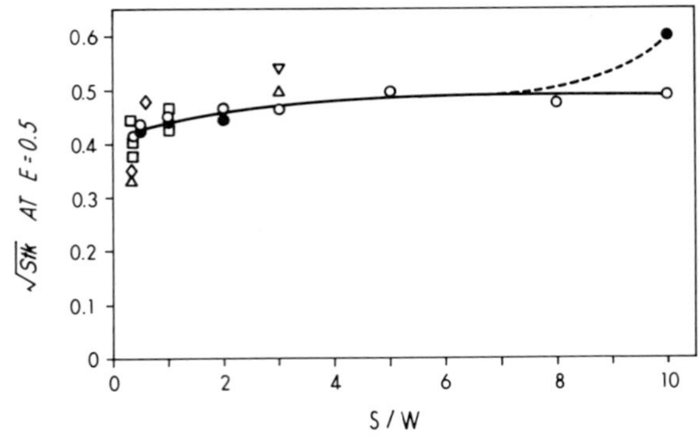

FIG. 6.35. $Stk^{1/2}$ at an efficiency of 0.5 as a function of S/W (round jets) [45]. △, Stern *et al.* [39]; □, Mitchell and Pilcher [46]; ◇, Zeller [47]; ▽, Ranz and Wong [38]; ●○, Mercer and Stafford [45]. Courtesy of British Occupational Hygiene Society.

The impaction system of Fig. 6.26 and the accompanying mathematical description of the process are appropriate for round jets also, if the value of $\tilde{V}_y$ is defined by $\tilde{V}_y = \tilde{y}/2\beta$. Figure 6.34 shows the experimental collection efficiency curves obtained by Mercer and Stafford at several values of S/W.

The values of $Stk^{1/2}$ at which the collection efficiency is 0.5 are shown in Fig. 6.35 as a function of S/W. Data from several authors are included in the figure. The effect of S/W is much less pronounced for round jets than for rectangular jets. For designing cascade impactors, it appears that ratios of S/W up to at least 8 could be tolerated. The values of $S/W < 1$ are accompanied by steeper efficiency curves, but the improvement does not provide a significant advantage for their use.

6-3.3 CASCADE IMPACTORS

The principle of the cascade impactor is apparent from a consideration of the Stokes number corresponding to a collection efficiency of 50% when the linear velocity U_0 is expressed in terms of the total volumetric flow rate F through the jet:

$$\mathrm{Stk}^{1/2}(E = 0.5) = \left(\frac{F}{9\eta L W^2}\right)^{1/2} \cdot (\rho_0 K_{sA})^{1/2} D_A(E = 0.5)$$

(rectangular jets of length L);

(6.30)

$$\mathrm{Stk}^{1/2}(E = 0.5) = \left(\frac{4F}{9\pi\eta W^3}\right)^{1/2} \cdot (\rho_0 K_{sA})^{1/2} D_A(E = 0.5)$$

(round jets).

(If $\mathbf{U}_0 \gtrsim$ one-third the speed of sound, allowance must be made for the rarefaction of air in the jet). Since the term on the left is constant, changes in W at fixed F must be offset by changes in D_A. By arranging several impaction stages in series, each having a jet orifice width W smaller than that of the previous stage, the particles collected at successive stages become progressively smaller. An absolute filter following the last stage collects all particles that escape impaction. If the impactor functioned in an ideal manner, there would be no wall losses and each stage would collect all available particles having diameters greater than a certain characteristic "cut-off" diameter for that stage and no particles of lesser diameter. The interpretation of data obtained with such an impactor would be unequivocal, since all particles collected on a given stage would have diameters between two fixed values. The accuracy with which the cumulative size distribution could be determined would depend only on the number of impaction stages employed and the reliability of the analytical method.

Real impactors, however, have collection efficiency curves such as those described above. The activity deposited in a given stage is a function of the size distribution of the aerosol entering the impactor and of the collection efficiency curves of both the stage under consideration and the one ahead of it. Figure 6.36 gives an example of the segregation of activity among the various stages of a hypothetical four-stage impactor for which the efficiency curves of the different stages are similar to the experimental curve of Ranz and Wong (Fig. 6.28). The aerosol being sampled was assumed to be made up of unit density particles having an activity distribution for which the median

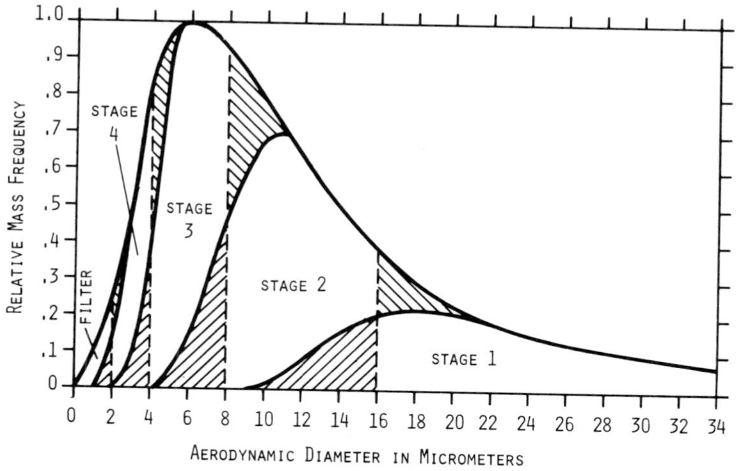

FIG. 6.36. Segregation of a mass distribution among the various stages of a cascade impactor [48]. Courtesy of British Occupational Hygiene Society.

aerodynamic diameter was 10 μm and the geometric standard deviation was 2.0. The values of D_A ($E = 0.5$) for the stages were 16, 8, 4, and 2 μm, respectively. It is apparent from the spread of particles on each stage that the determination of the activity distribution from the gross measurements of activity collected on the individual stages is subject to error.

(a) *Methods of Calibration.* The data obtained from a cascade impactor sample consist of a series of fractions representing the relative amount of activity collected at each impaction stage. When interpreting sampling information, it is generally assumed that the particle diameters of the aerosol from which the sample was drawn were lognormally distributed and the data are suitably arranged for plotting on logarithmic probability paper against the stage constants of the impactor. In the past, it was widely believed that it was necessary to calibrate the instrument for whatever aerosol was to be sampled in order to relate the results to some linear dimension as measured by means of a microscope. This required that calibration be carried out by means of a microscope analysis of the particles collected at each stage after the impactor had been used to sample a test aerosol. The stage constants were then defined as the mass median diameters (MMD) of the size distributions found on the various stages. These MMD's, adjusted for differences in sampling conditions, were used to determine the mass distributions of the aerosols of similar materials. For a given stage, half the mass fraction collected on that stage plus the cumulative mass fraction collected on previous stages was plotted as the mass fraction associated with particles larger than the stage MMD. This method was introduced by Sonkin [49], who attributed it to H. Landahl.

It is recognized now, however, that it is a real advantage to be able to calibrate an impactor using spherical particles of known density to obtain curves of collection efficiency as a function of aerodynamic diameter for each impaction stage. The diameter corresponding to a collection efficiency of 0.5 is taken as an effective cut-off aerodynamic diameter (ECAD). In all subsequent interpretations of impactor data, it is assumed that all particles collected at a given stage, regardless of shape or density, have aerodynamic diameters larger than the cut-off diameter for that stage. Couchman [50] has demonstrated the validity of the relationship between collection efficiency and aerodynamic diameters for spheres having densities up to 19 g/cm^3. The relationship is given added support [51] by Laskin's data for irregularly shaped UO_2 particles. Its validity has yet to be demonstrated for particles of extreme shape.

Figures 6.33 and 6.35 show that for $S/W \gtrsim 1$, the appropriate values of $\mathrm{Stk}^{1/2}$ ($E = 0.5$) to use in Eq. (6.30) are 0.8 and 0.45, respectively, for rectangular and round jets. The corresponding effective cut-off diameters can be calculated from

$$(\rho_0 K_{sA})^{1/2} \cdot D_A(E=0.5) = 324W(L/F)^{1/2} \qquad \text{for rectangular jets,}$$

or

$$= 161W(W/F)^{1/2} \qquad \text{for round jets.}$$

W and L are in centimeters, F is in cubic centimeters/second, and $D_A(E=0.5) =$ ECAD is in micrometers. It is unfortunate that aerodynamic diameter was defined as D_A rather than $(K_{sA})^{1/2} \cdot D_A$, the quantity that is actually calculated from the formulas above. For $D_A \gg 0.05\ \mu m$, the relationship, $(K_{sA})^{1/2} \cdot D_A \simeq 1.01 D_A + 0.072$, simplifies the calculation of aerodynamic diameter.

For linear jet flow rates $\gtrsim$ one-third the speed of sound, these relationships can be used to calculate stage constants for an uncalibrated impactor, instead of making a time-consuming, and perhaps less reliable, calibration. There are two conditions that should be observed, however. The Reynolds number Re at the jet orifice should exceed 100 [52], and the velocity profile transition length

$$L_T = 0.16 \cdot \mathrm{Re} LW/(L+W), \qquad \text{for rectangular jets}$$

or

$$L_T = 0.065W\ \mathrm{Re}, \qquad \text{for round jets,}$$

should be several times the jet length.

(b) *Calibration Errors.* Using the effective cut-off method to estimate an activity distribution implies that the efficiency curves can be approximated by step functions, with the step for a given stage occurring at the ECAD for that stage. The source of error in this method is demonstrated in Fig. 6.36, where the vertical lines bounding shaded areas on one side represent the ECAD's for the different stages. For each stage, the upper shaded area measures the amount assumed to be collected, when actually it is not, and the lower shaded area measures the amount assumed to escape, when actually it is collected. The accuracy of the effective cut-off method depends on the relative magnitudes of these two areas. If they are equal, the ECAD for that stage is also the true effective cut-off diameter for the original distribution; i.e., the fraction of the total activity collected up to and including that stage is equal to the fraction of activity in the original distribution that is associated with particles having diameters larger than the ECAD. In general, the true effective cut-off diameter differs slightly from the ECAD. In the case of the first stage in Fig. 6.36, the true effective cut-off diameter is somewhat smaller than the ECAD. The extent to which the true effective cut-off diameters differ from the ECAD's for stage 2 of the hypothetical impactor of Fig. 6.36 is shown in Fig. 6.37 for a variety of input aerosol distributions. The same figure brings out some of the advantages of ECAD's over MMD's. The "effective drop sizes" defined by May for his original impactor were the true effective cut-off diameters for the particular test aerosol he used. Their use as calibration constants when interpreting data from other size distributions should lead, on the average, to errors comparable to those encountered with the use of ECAD's.

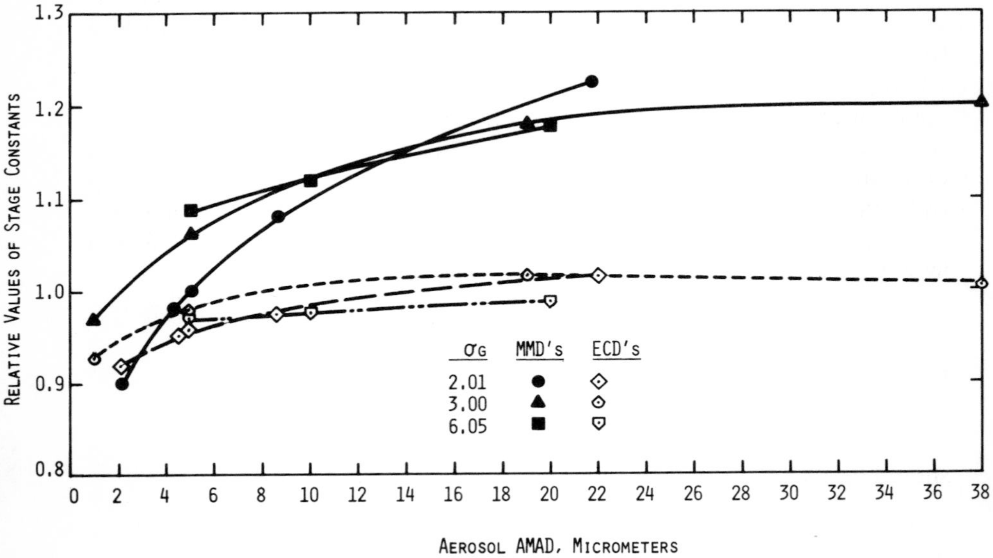

FIG. 6.37. Relative values of true MMD's and true ECAD's as functions of input aerosol size distribution [48]. Courtesy of British Occupational Hygiene Society.

Errors of this sort are generally less with round jet than with rectangular jet impactors, because the former have steeper collection curves. An impactor having round jets with collection characteristics similar to those of Fig. 6.34 has been shown to have ECAD's always within a few percent of the true effective cut-off diameters for input aerosols having geometric standard deviations greater than 1.5 [53]. Methods have been proposed in which the efficiency curves themselves are used to calculate the original size distribution [54], but, while promising, they have not yet received much attention.

(c) *Operating Errors.* The errors discussed above were due to the non-ideal form of the efficiency curves of impactors. In practice, additional errors occur due to changes of the input aerosol distribution as the result of wall losses and disaggregation and due to distortion of the collection efficiency characteristics as a result of rebound and re-entrainment.

When rebound and re-entrainment are negligible, wall losses are due primarily to conditions of nonlaminar flow between stages. The Andersen sampler [55], described below, was found to have negligible wall losses between stages, although the air flow made four 90° changes in direction in going from one jet to the next. For this reason, the Lovelace impactor (see below) was designed to avoid abrupt changes in the cross-sectional area of the air passages. The effectiveness of the design in minimizing wall losses is brought out in Fig. 6.38. The instrument, which was designed to operate at flow rates as low as 50 cm^3/min, has an internal volume of several cubic centimeters, wherein the air remaining between stages at the end of a 1 min sample is contaminated to a degree that depends on the number of stages traversed. This contamination becomes transferred to the walls before the impactor is dismantled. Under these circumstances, the amount of contamination is relatively constant, and the wall losses, relative to the total sample collected, diminish with increasing sample volume, approaching a limit of 2%.

Disaggregation occurs when the air is accelerated upon entering a jet. The more dense particles accelerate less rapidly and a relative velocity develops between particle and air, partially breaking up aggregates. Davies [34] concluded that this was the primary cause of disaggregation of coal particles in dust sampling equipment. He studied the effect experimentally by passing coal particles at various flow rates through the jet of a Bausch and Lomb dust counter. He found the concentration downstream of the jet exceeded the upstream concentration by 50% at a flow velocity of 48.6 m/sec. Disaggregation

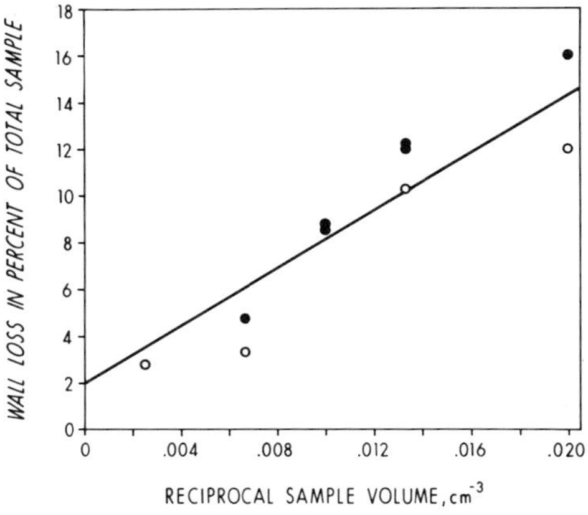

FIG. 6.38. Wall losses in the Lovelace impactor [56]. Courtesy of British Occupational Hygiene Society.

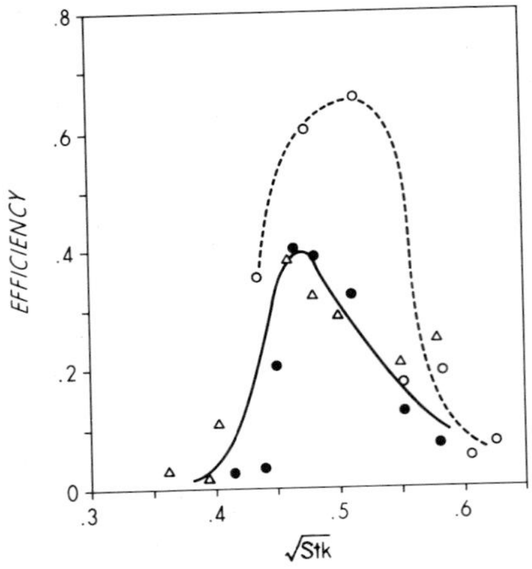

FIG. 6.39. Effect of rebound on efficiency of round jet impactor [45]. ○ ---, $S/W=0.5$; ● ——, $S/W=2.0$; and △ ——, $S/W=10.0$. Courtesy of British Occupational Hygiene Society.

increased with increasing flow rate and was suppressed to some extent when the air was saturated with water vapor. The effect is less significant with cascade impactors, since the early stages operate at much lower flow speeds.

Rebound is a serious problem if the collecting surfaces are not coated with a soft layer to cushion the impact of the particle. Using uncoated slides, collection efficiency goes through a maximum value with increasing Stokes number, as shown in Fig. 6.39. Under the sampling conditions in effect then, polystyrene spheres of 1.17 μm diameter were not re-entrained. Glass beads deposited on glass or metal surfaces are not re-entrained at air velocities of 30 m/sec if their diameters are smaller than 15.9 μm and fly ash of 0.5 μm diameter is not re-entrained by velocities of 150 m/sec [57]. These observations support the conclusion that an adhesive coating on an impactor collection surface is not necessary to prevent re-entrainment of particles that are in direct contact with the surface.

On the other hand, rebound and re-entrainment put an upper limit on the amount of material than can be collected at a given stage. As a deposit builds up on a collecting surface, incoming particles may rebound from deposited particles or, if the surface becomes sufficiently overloaded, parts of the deposit may break loose and be carried deeper into the impactor.

The limits imposed by the need to avoid rebound and re-entrainment can be expressed in terms of "loading capacity" [58]. This is the product of the total cross-sectional area of the jet orifice(s) at a given stage and the average between its effective cut-off diameter and that of the previous stage. For the first stage, the calculation is based on its effective cut-off diameter alone. Thus, the loading capacity is approximately equal to the mass of unit density material contained in a deposit when it covers the jet area with a layer one particle deep. As such, it is a measure of the mass that can be collected before rebound and re-entrainment become a problem. Its absolute value should not be given much credence, but its value relative to other loading capacities permits some reliable comparisons (see Table 6.1).

TABLE 6.1

OPERATING CHARACTERISTICS OF CASCADE IMPACTORS

	May [35]	Andersen [55]	Lovelace [56]	Lundgren [43]
Number of impaction stages	4	6[a]	7	4
Jet geometry	rectangular	round	round	rectangular
Operating flow rate, liters/min	17.5	28.3	0.05–0.15	85
Orifice dimension, W, cm	0.027–0.65	0.025–0.118	0.015–0.081	0.029–0.82
Jet velocities, U_0, m/sec	2.37–77	1.08–23.3	4.85–141[b]	3.4–95
S/W	0.27–1.0	2.1–10.0	1–2	1
ECAD, micrometers	0.5–12.5	0.6–7.1	0.12–2.2[b]	0.3–10.0
Jet Reynolds number	2060–2750	85–400	260–1420	3700
Loading capacity, mg	0.006–1.6	0.016–3.3	$(0.04–1.13) \times 10^{-5}$	4.3–61

[a] 400 holes per stage.
[b] At 0.15 liter/min.

(d) *Impactor Designs.* The diagrams of Fig. 6.40 and the data of Table 6.1 indicate the wide range of operating conditions to which the principle of cascade impaction can be adapted. A commercial version (Casella) of the May impactor was available for many years; its design was modified a few years ago, the dimensions of the last two stages being altered.* The Andersen sampler [55] was designed to collect airborne microorganisms. Andersen was apparently the first to appreciate the fact that collection efficiency is better when a given volume of air is sampled at a given linear velocity through a number of small holes than it is when the same volume of air is sampled at the same linear velocity through one hole. The loading capacities are relatively better also. The Lovelace impactor described in the table was designed to sample atmospheres containing high concentrations of radioactivity. Its loading capacities are so small that it can be used only with aerosols of high specific activity. The Lundgren impactor was designed to study urban particulate matter over day-long periods. The collection surfaces are drums that rotate under the jet, spreading out the deposit in a way that prevents overloading and provides a record of particulate characteristics as a function of time.

* The efficiency curves provided with the instrument were not corrected. however [48, 59].

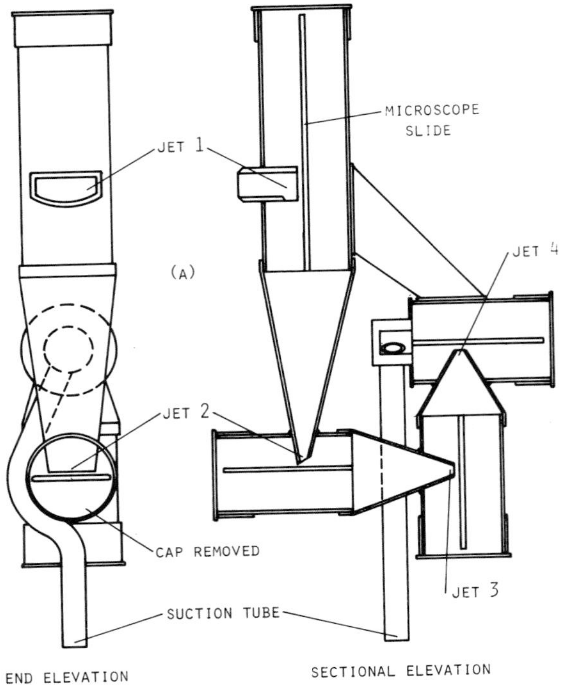

MICROSCOPE SLIDE
JET 1
JET 4
(A)
JET 2
CAP REMOVED
JET 3
SUCTION TUBE
END ELEVATION
SECTIONAL ELEVATION

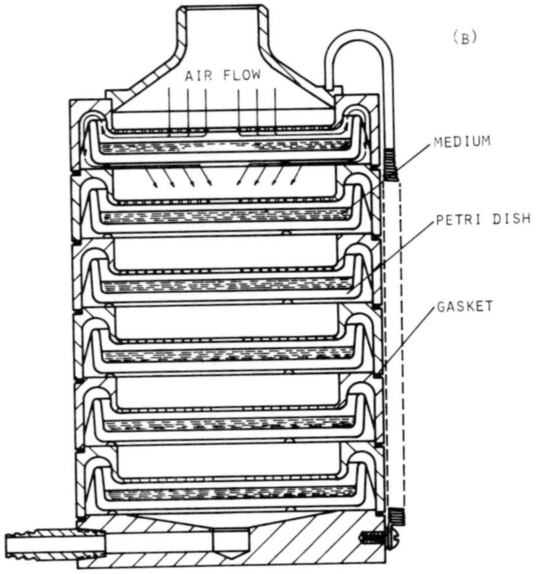

(B)
AIR FLOW
MEDIUM
PETRI DISH
GASKET

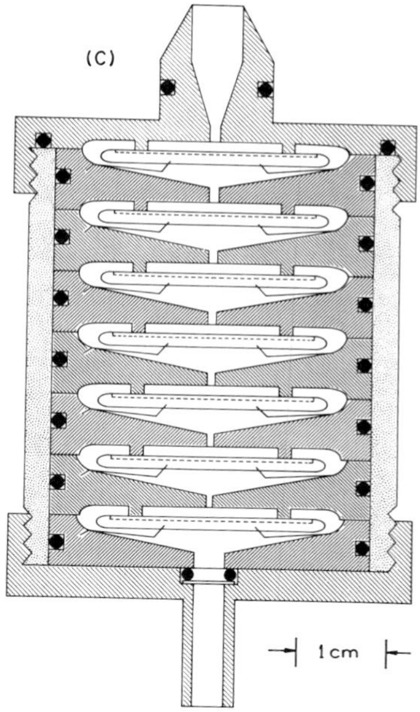

FIG. 6.40. Schematic diagrams of cascade impactors. (A) The May impactor [35]. Courtesy of The Institute of Physics. (B) The Andersen impactor [55]. Courtesy of the American Society for Microbiology. (C) The Lovelace impactor [56]. Courtesy of Pergamon Press Ltd. (D) The Lundgren impactor [43]. Courtesy of the Air Pollution Control Association.

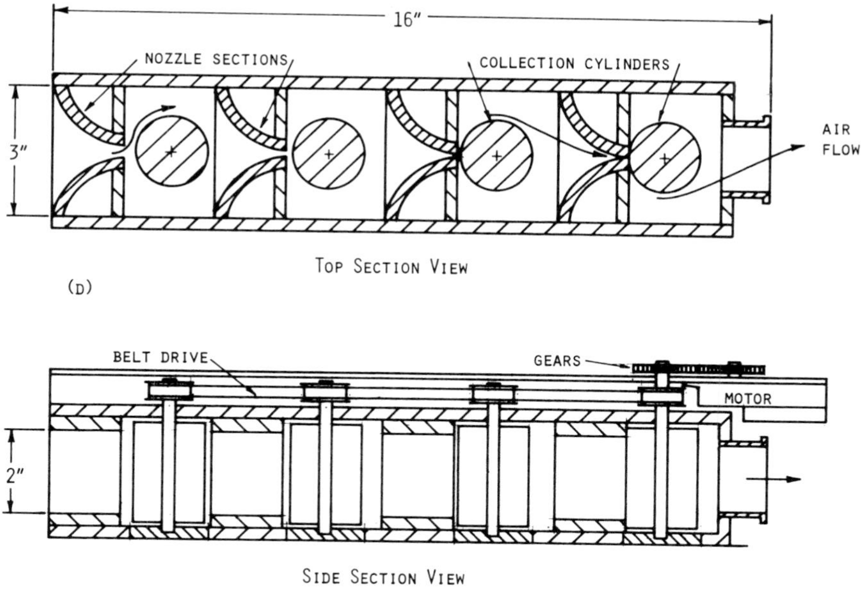

References

1. P. S. Roller, Metal Powder Size Distribution with the Roller Air Analyzer, in *A.S.T.M. Spec. Tech. Publ. No. 140*, 1952.
2. W. H. Walton, Theory of Size Classification of Airborne Dust Clouds by Elutriation, *Brit. J. Appl. Phys., Suppl. 3:* S29–S37 (1952).
3. J. W. Thomas, The Equation for Gravity Settling of Aerosols in Tubes, *J. Colloid Interface Sci., 27:* 2323 (1968).
4. W. Stöber, Zur Bestimmung von Teilchengrössenverteilungen mit einem Horizontal-Elutriator, *Staub, 24:* 221–223 (1964).
5. G. Zebel, The Separation Power of Some Methods to Deposit Aerosol Particles According to Their Size, in T. T. Mercer, P. E. Morrow, and W. Stöber (Eds.), *Assessment of Airborne Particles*, Thomas, Springfield, Illinois, 1972.
6. F. Stein, R. Quinlan, and M. Corn. The Ratio Between Projected Area Diameter and Equivalent Diameter of Particulates in Pittsburgh Air, *Amer. Ind. Hyg. Ass. J., 27:* 39–46 (1966).
7. J. W. Thomas and R. H. Knuth, Settling Velocity and Density of Monodisperse Aerosols, *Amer. Ind. Hyg. Ass. J., 28:* 229–237 (1967).
8. V. Timbrell, An Aerosol Spectrometer and Its Applications, in T. T. Mercer, P. E. Morrow, and W. Stöber (Eds.), *Assessment of Airborne Particles*, Thomas, Springfield, Illinois, 1972.
9. K. F. Sawyer and W. H. Walton, The 'Conifuge'—a Size-Separating Sampling Device for Airborne Particles, *J. Sci. Instrum., 27:* 272–276 (1950).
10. A. Goetz, An Instrument for the Quantitative Separation and Size Classification of Airborne Particulate Matter down to 0.2 Micron, *Geofis. Pura. Appl., 36:* 49–69 (1957).
11. W. Kast, Neues Staubmessgerät zur Schnellbestimmung der Staubkonzentration und der Kornverteilung, *Staub, 21:* 215–223 (1961).
12. O. G. Raabe, Calibration and Use of the Goetz Aerosol Spectrometer, in *Assessment of Airborne Radioactivity*, IAEA, Vienna, 1967.
13. W. Stöber and U. Zessack, Zur Theorie einer konischen Aerosolzentrifuge, *Staub, 24:* 295–305 (1964).
14. T. Kallai and A. Goetz, Instrumentation for Determining Size- and Mass-Distribution of Submicron Aerosols, *APCA J., 12:* 479–487 (1962).
15. E. Baust, Zur Auswertung des Niederschlags Polydisperser Aerosole in einem Goetzchen Aerosolspektrometer nach dem Lichtstreuverfahren, *Staub, 28:* 232–236 (1968).
16. O. Preining, Das Goetzsche Aerosolspektrometer, *Staub, 22:* 129–133 (1962).
17. F. L. Ludwig and E. Robinson, Size Distribution Studies with the Aerosol Spectrometer, *APCA J., 15:* 102–105 (1965).
18. A. Goetz, H. J. R. Stevenson and O. Preining, The Design and Performance of the Aerosol Spectrometer, *APCA J., 10:* 378–383 (1960).
19. R. Reiter and W. Carnuth, Das Partikelspektrum eines mit Radonfolgeprodukten beladenen Aerosols, *Z. Naturwissensch., 54:* 40 (1967).
20. M. Taheri and R. K. Barton, A Technique for Determining Penetration as a Function of Particle Diameter, *Amer. Ind. Hyg. Ass. J., 29:* 252–256 (1968).
21. C. N. Keith and J. C. Derrick, Measurement of the Particle Size Distribution and Concentration of Cigarette Smoke by the 'Conifuge', *J. Colloid Sci., 15:* 340–356 (1960).
22. M. I. Tillery, Design and Calibration of a Modified Conifuge, in *Assessment of Airborne Radioactivity*, IAEA, Vienna, 1967.

23. W. Stöber and H. Flachsbart, Aerosol Size Spectrometry with a Ring Slit Conifuge, *Environ. Sci. Technol.*, *3:* 641–651 (1969).
24. W. Stöber and U. Zessack, Zur Messung von Aerosol-Teilchengrössenspektren mit Hilfe von Zentrifugalabscheidern, *Z. Biol. Aerosolforsch.*, *13:* 263–281 (1966).
25. D. Hochrainer and P. M. Brown, Sizing of Aerosol Particles by Centrifugation, *Environ. Sci. Technol.*, *3:* 830–835 (1969).
26. O. R. Moss, *Shape Factors for Airborne Particles*, Master of Science Thesis, Univ. of Rochester, 1959.
27. W. Stöber and H. Flachsbart, Size-Separating Precipitation of Aerosols in a Spinning Spiral Duct, *Environ. Sci. Technol.*, *3:* 1280–1296 (1969).
28. P. Kotrappa and M. E. Light, *Design and Performance of the Lovelace Aerosol Particle Separator*, USAEC Rep. LF–44, 1971.
29. J. Innes, The Investigation of Injurious Dust in Mine Air by Kotzé Konimeter, *J. Chem. Metal Min. Soc. So. Africa*, *23:* 77 (1922).
30. L. Greenburg and G. W. Smith, *A New Instrument for Sampling Aerial Dust*, U. S. Bureau of Mines, Rep. Invest. 2392, 1922.
31. J. B. Littlefield, F. L. Feicht and H. H. Schrenk, *Bureau of Mines Midget Impinger for Dust Sampling*, U. S. Bureau of Mines, Rep. Invest. 3360, 1937.
32. J. S. Owens, Suspended Impurity in Air, *Proc. Roy. Soc. (London)*, *A101:* 18–37 (1922).
33. C. D. Yaffe, D. H. Byers and A. D. Hosey (Eds.), *Encyclopedia of Instrumentation for Industrial Hygiene*, p. 68, Univ. of Michigan Inst. Ind. Health, Ann Arbor, 1956.
34. C. N. Davies, M. Aylward and D. Leacey, Impingement of Dust from Air Jets, *A.M.A. Arch. Ind. Health Occup. Med.*, *4:* 354–397 (1951).
35. K. R. May, The Cascade Impactor: an Instrument for Sampling Coarse Aerosols, *J. Sci. Instrum.*, *22:* 187–195 (1945).
36. S. Laskin, in C. Voegtlin, and H. C. Hodge (Eds.), *Pharmacology and Toxicology of Uranium Compounds I*, p. 463, McGraw-Hill, New York, 1949.
37. C. N. Davies and M. Aylward, The Trajectories of Heavy, Solid Particles in a Two-Dimensional Jet of Ideal Fluid Impinging Normally upon a Plate, *Proc. Phys. Soc.*, *B64:* 889–911 (1951).
38. W. E. Ranz and J. G. Wong, Impaction of Dust and Smoke Particles, *Ind. Eng. Chem.*, *44:* 1371–1381 (1952).
39. S. C. Stern, H. W. Zeller and A. E. Schekman, Collection Efficiency of Jet Impactors at Reduced Pressures, *Ind. Eng. Chem. Fundamentals*, *1:* 273–277 (1962).
40. H. Green and W. Lane, *Particulate Clouds*, p. 184, Spon, London, 1957.
41. T. T. Mercer and H. Y. Chow, Impaction from Rectangular Jets, *J. Colloid Interface Sci.*, *27:* 75-83 (1968).
42. G. Hänel, Bemerkungen zur Theorie der Düsen-Impactoren, *Atmos. Environ.*, *3:* 69–83 (1969).
43. D. Lundgren, An Aerosol Sampler for Determination of Particle Concentration as a Function of Size and Time, *APCA J.*, *17:* 225–228 (1967).
44. R. Roeber, Untersuchungen zur Konimetrischen Staubmessung, *Staub*, *48:* 41–86 (1957).
45. T. T. Mercer and R. G. Stafford, Impaction from Round Jets, *Ann. Occup. Hyg.*, *12:* 41–48 (1969).
46. R. I. Mitchell and J. M. Pilcher, Design and Calibration of an Improved Cascade Impactor for Size Analysis of Aerosols, in *Proc. A.E.C. Air Cleaning Conf., 5th*, TID–7551, Washington, D.C., 1958.
47. H. W. Zeller, *Balloon-Borne Particulate Fractionator*, Rep. 1922, General Mills, Mech. Div. 1960.

48. T. T. Mercer, The Stage Constants of Cascade Impactors, *Ann. Occup. Hyg.*, *7:* 115–124 (1964).
49. L. S. Sonkin, A Modified Cascade Impactor, *J. Ind. Hyg. Toxicol.*, *28:* 269–272 (1946).
50. J. D. Couchman, Use of Cascade Impactors for Analyzing Airborne Particles of High Specific Gravity, *Conf–650407*, *Vol. 2*, TID–4500, p. 1162, 1965.
51. T. T. Mercer, On the Calibration of Cascade Impactors, *Ann. Occup. Hyg.*, *6:* 1–14 (1963).
52. A. R. McFarland and H. W. Zeller, *Study of a Large-Volume Impactor for High-Altitude Aerosol Collection*, Rep. 2391, Gen. Mills Elect. Div., 1963.
53. T. T. Mercer, M. I. Tillery and C. W. Ballew, *A Cascade Impactor Operating at Low Volumetric Flow Rate*, USAEC Rep. LF–5, 1962.
54. F. L. Ludwig, Behavior of a Numerical Analog to a Cascade Impactor, *Environ. Sci. Technol.*, *2:* 547–550 (1968).
55. A. A. Andersen, New Sampler for the Collection, Sizing, and Enumeration of Viable Airborne Particles, *J. Bacteriol.*, *76:* 471–484 (1958).
56. T. T. Mercer, M. I. Tillery and G. J. Newton, A Multi-Stage, Low Flow Rate Cascade Impactor, *J. Aerosol Sci.*, *1:* 9–15 (1970).
57. M. Corn and F. Stein, Re-entrainment of Particles from a Plane Surface, *Amer. Ind. Hyg. Ass. J.*, *26:* 325–336 (1965).
58. T. T. Mercer, Air Sampling Problems Associated with the Proposed Lung Model, *Proc. Ann. Bioassay Anal. Chem. Meeting*, *12th*, Gatlinburg, Tennessee, 1966.
59. B. W. Soole, Concerning the Calibration Constants of Cascade Impactors, with Special Reference to the Casella Mk 2, *Aerosol Sci.*, *2:* 1–14 (1971).

7

Measurement of Other Diameters Related to Particulate Properties

The size of an irregularly shaped particle may be defined as the diameter of a sphere that is equivalent to the particle with respect to some property that can be measured readily. If the property is itself of interest, then the "equivalent" diameter, defined in this way, is merely a convenient form in which to express the amount of that property associated with the particle. The aerodynamic diameter, which is really a measure of a particle's terminal settling velocity, is an equivalent diameter of this type. On the other hand, a size-related property may be used to define an equivalent diameter merely because the property can be measured conveniently. The best example is the equivalent diameter derived from measuring the intensity of light scattered by a particle. A new set of shape factors is required for each diameter of this latter type. Some methods involving both types of measurement are discussed below.

7-1 Optical Methods of Size Measurement

7-1.1 PARTICLE COUNTERS AND SPECTROMETERS

The desire to make rapid determinations of the size distribution and concentration of particles while they are still airborne stimulated research on techniques for measuring the amount of light scattered from single particles. Early work [1] contributed to the development of successful particle size spectrometers by Gucker and his co-workers [2], O'Konski and Doyle [3], and Fisher *et al.* [4]. Each of these instruments was arranged with the axis of the detector system normal to the axis of the illuminating light beam and intersecting it in the sensitive volume, that is, the illuminated portion of the aerosol stream.

Calculations based on the Mie theory [2] indicated that better results could be obtained by measuring the light scattered in the forward direction, and several recent instruments have been designed to do this. In addition, methods of light collection have been improved and certain advantages have been derived from the use of lasers as light sources. Figures 7.1–7.5 show diagrams of five instruments developed during the past decade. It can be seen from the figures that the solid angles for both illumination and collection of scattered

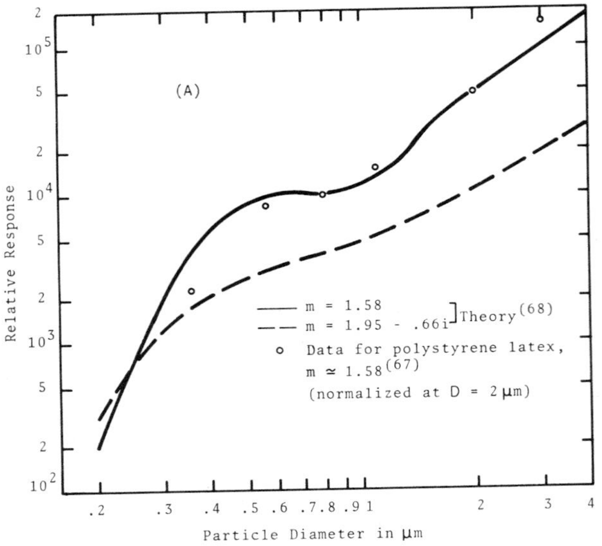

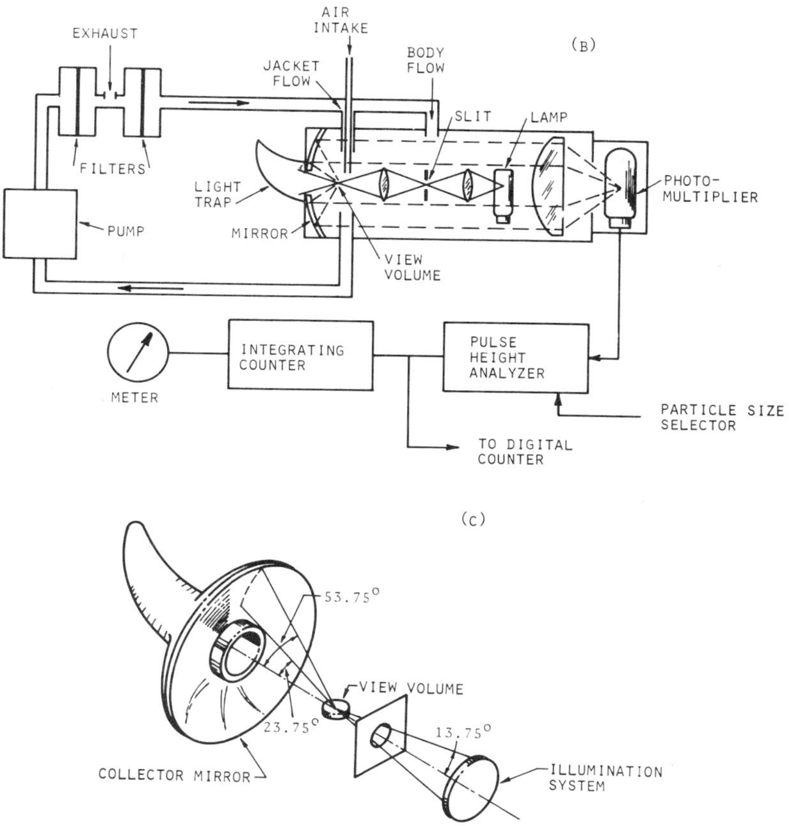

FIG. 7.1. The Bausch and Lomb particle counter [5]. (A) Response curves (for later model in which inner angle of collection cone = 30.75°); (B) Schematic diagram of the counter; (C) light-collecting optical system. *m* is the refractive index. Courtesy of the American Industrial Hygiene Association.

light are sometimes cones and sometimes conical shells. For the former, solid angles of collection are referred to the axis of the cone and for the latter, to a hypothetical conical surface bisecting the angle between the surfaces forming the conical shell. Response curves for spheres of polystyrene and carbon, based on Quenzel's [10] calculations, are included in the first three figures; the other response curves are from the papers describing the instruments. Quenzel took into consideration the distribution of wavelengths in the illuminating light and the spectral response of the photomultiplier tube. Because of the convergence of the incident beam, the limits of the scattering angle over which light is collected exceed those defined by the collecting cones.

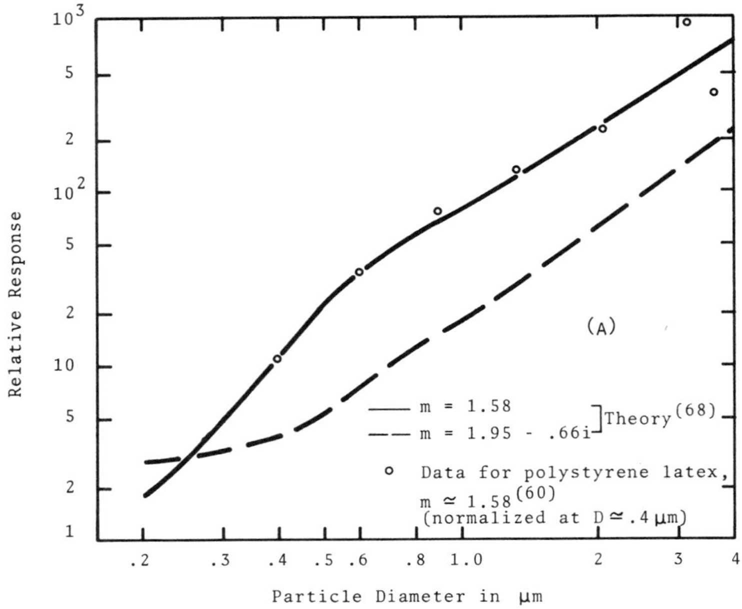

(B)

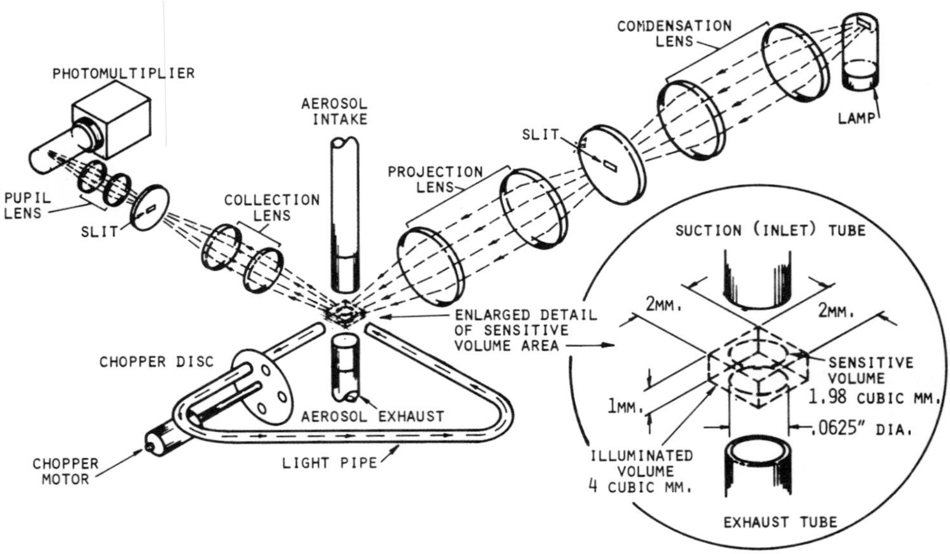

FIG. 7.2. The Royco particle counter [6]. (A) Response curves; (B) schematic diagram of the counting system. Courtesy of the Air Pollution Control Association.

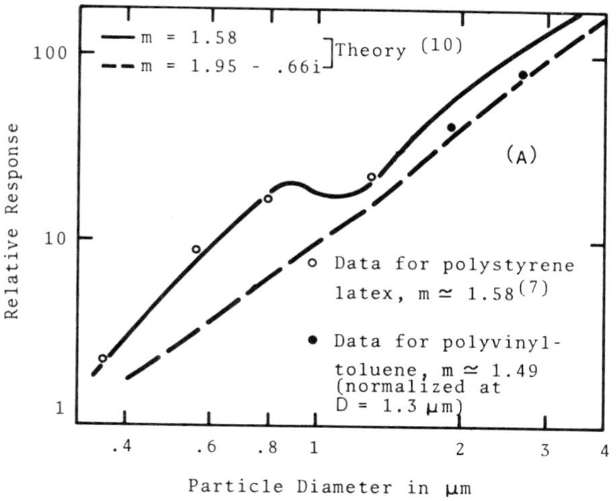

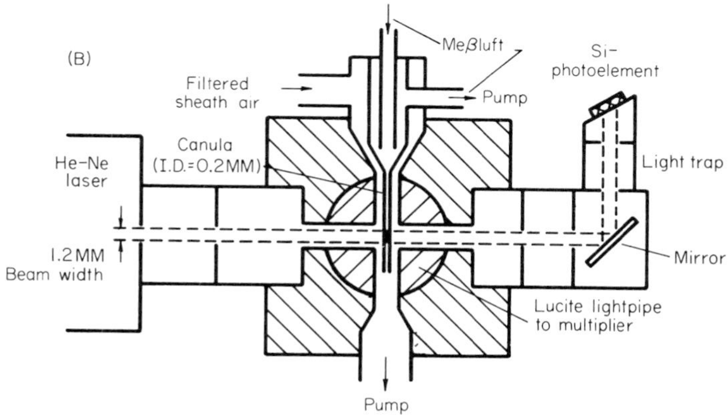

FIG. 7.3. The Jacobi particle counter [7]. (A) Response curves; (B) schematic diagram of the counter system. *m* is the refractive index. Courtesy of Staub-Reinhaltung der Luft.

Desirable characteristics in an aerosol photospectrometer include:

A unique relationship between response and size, regardless of a particle's optical properties;

A large signal-to-noise ratio;

A combination of sampling flow rate and sensitive volume that permits the rapid determination of size distribution with a minimum coincidence error; and

Good size resolution.

Table 7.1 contains pertinent data for the instruments illustrated in Figs. 7.1–7.5.

The first characteristic cannot be realized with single measurements at a

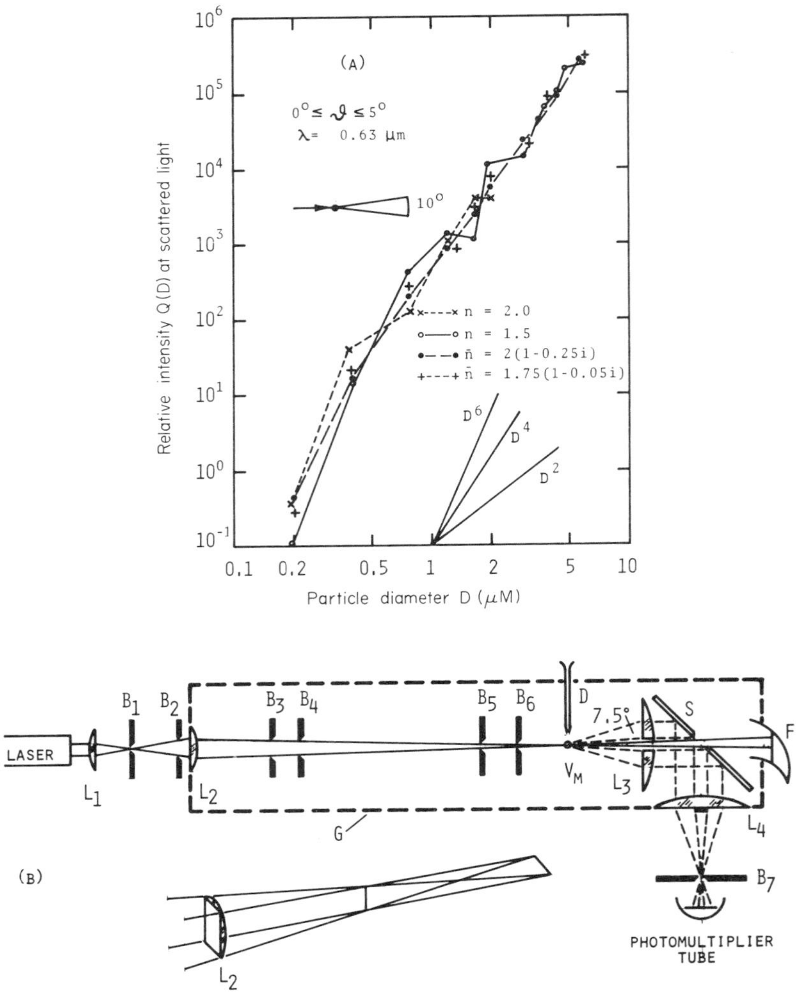

FIG. 7.4. The Gebhart particle counter [8]. (A) Response curves; (B) schematic diagram of the counter system. *m* is the refractive index. Courtesy of Staub-Reinhaltung der Luft.

fixed solid angle. For the instruments illustrated here, the uncertainty in measuring the size of a particle of unknown composition varies from about 30% to 200%, with performance generally improving as the scattering angle is taken more in the forward direction. Some of the instruments show regions of ambiguous response, even for particles of known optical properties.

Aside from the characteristics of the photomultiplier tube, the signal-to-noise

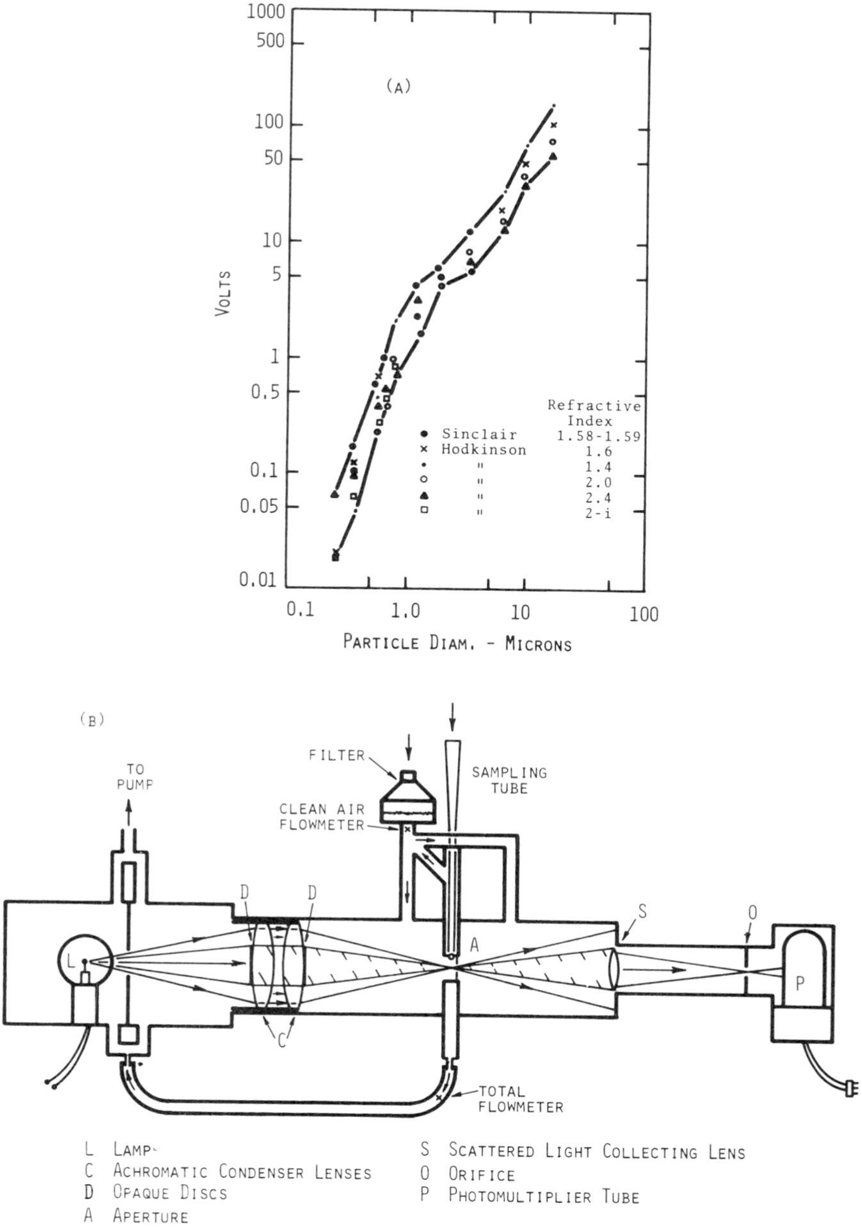

FIG. 7.5. The Sinclair particle counter [9]. (A) Response curves; (B) schematic diagram of the particle counter system. Courtesy of the Air Pollution Control Association.

TABLE 7.1

CHARACTERISTICS OF SOME OPTICAL PARTICLE SIZING INSTRUMENTS

Source	Scattering angle (deg)[d]	Sample flow rate (cm^3/sec)	Sensitive volume (cm^3)	Concentration at 5% coincidence (cm^{-3})	Counts/sec at 5% coincidence	Incident light	Resolution
Sinclair [9][a]	3–20	46.6	8×10^{-4}	64	2830	white	—
Gebhart [8]	1.5–7.5	2.2	1.1×10^{-5}	4660	9760	0.6328 μm	0.11–0.20
Martens [5][b]	10–67.5	2.8	5.0×10^{-4}	105	279	white	0.09–0.50
Jacobi [7]	≃10–170	0.01–0.1	5.0×10^{-5}	1050	10–100	0.6328 μm	0.03–0.15
Zinky [6][c]	75–105	<4.7	2.0×10^{-3}	26	116	white	0.16–0.31

[a] Phoenix Precision Instrument Co.
[b] Bausch and Lomb, resolution calculated from data in Ref. [67].
[c] Royco, resolution calculated from data in Ref. [11] attributed to O. Preining.
[d] Limiting values for light scattered into collecting cone.

ratio depends on the size and position of the solid angle over which light is collected, the size of the scattering volume, and the efficiency with which stray light is suppressed. As Fig. 2.14 shows, the signal from an element of solid angle will be enhanced as the element is moved more to the forward direction. It is advantageous to increase the signal by taking the solid angle over the full 360° around the illumination beam axis, but increasing the width of the solid angle may introduce ambiguities in the response curve. The response curve of Fig. 7.3 for an index of refraction $m = 1.58$ actually exhibits a number of oscillations, but they have been smoothed out in that figure. The scattering volume is the portion of the illuminating beam from which scattered light can reach the detector. It is about twice the size of the sensitive (aerosol) volume, which it includes. Light incident on this volume is subject to Rayleigh scattering by air molecules, which puts a lower limit on the particle diameter that can be measured. For the instrument in Fig. 7.4, the scattering volume of 2×10^{-5} cm^3 corresponds to a particle of 0.12 μm diameter. In combination with the phototube noise, this makes 0.17 μm the minimum diameter that can be measured with this instrument [8].

Coincidence effects due to the simultaneous appearance of two or more particles in the sensitive volume distort size distributions and number concentrations by registering multiple particles as a single large particle. The probability that the sensitive volume V_s will contain more than one particle at any given instant usually has been assumed to be given by the Poisson distribution:

$$P(n) = (NV_s)^n \cdot e^{-NV_s}/n!, \tag{7.1}$$

where N is the number of particles per unit volume. The fraction of all measurements that involve multiple particles is

$$\frac{P(n>1)}{P(n>0)} = \frac{1-P(0)-P(1)}{1-P(0)} = 1 - \frac{(NV_s)\,e^{-NV_s}}{1-e^{-NV_s}}. \tag{7.2}$$

This equation, or approximations to it, has been used to describe coincidence losses in the Coulter counter [61, 64] also, where a similar problem is encountered. While it is correct for systems in which V_s is a "grab" sample from a very much larger volume, it is inaccurate when the fluid making up the element V_s is continually changing. Particles that are randomly distributed with respect to the volume being sampled will be randomly distributed with respect to the time at which they enter V_s during a given counting period. In this case, the time interval between successive intrusions of particles into the sensitive volume is a continuous random variable having an exponential distribution [71]:

$$f(t) = NF \cdot e^{-NFt}, \tag{7.3}$$

where F is the volumetric flow rate through the sensitive volume. If τ is the transit time for a particle passing through V_s, then one count will be lost for each interval for which $t < \tau$, and the observed number per unit volume will be

$$N_0 = N[1 - \int_0^\tau f(t)\,dt] = N \cdot e^{-NF\tau}. \tag{7.4}$$

This equation predicts that the observed concentration goes through a maximum as N increases, so that N_0 is an ambiguous number and care must be taken to work at concentrations that are dilute enough to ensure that N_0 is below the maximum value. At small values of N,

$$N_0 \simeq N - N^2 F\tau. \tag{7.5}$$

The exponential relationship given above was used by Derjaguin *et al.* [72] to describe the observed concentration of particles in the flow-ultramicroscope. It was derived also by Pisani and Thomson [73], who verified it experimentally using the Royco counter. Princen and Kwolek [62] adopted a procedure that should have yielded the approximate equation for small values of N, but their derivation contains a factor-of-2 error.

Since the transit time is equal to V_s/F,

$$N_0 = N \cdot e^{-NV_s}. \tag{7.6}$$

Coincidence losses amount to 5% when $NV_s \simeq 0.05$. It is apparent from Table 7.1, in which the values of N corresponding to 5% coincidence are tabulated, that very small sensitive volumes have marked advantages with respect to coincidence losses. The product of this value of N_0 and the sampling flow rate, which is also included in Table 7.1, is the maximum number of particles that can be counted per unit time before coincidences in the sensitive volume reach 5% of the total count. Bol and others [66] recently described an instrument in which the sensitive volume had been reduced to 2×10^{-8} cm^3. This was accomplished by focusing the aerosol beam aerodynamically before it entered a finely focused laser beam.

When a number of identical particles are measured with a photospectrometer, the phototube output is a pulse-height distribution having a finite width that limits the instrument's ability to distinguish between particles of comparable diameter. This ability is described quantitatively by the resolution of the instrument, defined here, as in gamma-ray spectrometry, as the ratio between the width ΔD of the size distribution, measured at one-half the peak value, and the value of D at the peak,

$$R = \Delta D / D_{max}. \tag{7.7}$$

For a normal distribution of standard deviation σ

$$\frac{\Delta D}{D_{max}} = \frac{2.36\sigma}{D_{max}}. \tag{7.8}$$

For a lognormal distribution of standard deviation, $\sigma = \ln \sigma_g$,

$$\Delta D/D_{\max} = e^{1.178\sigma} - e^{-1.178\sigma} = \sigma_g^{1.178} - \sigma_g^{-1.178}. \tag{7.9}$$

For small values of σ, $\Delta D/D_{\max} \simeq 2.36\sigma$. If the particles are not truly monodisperse, but have a lognormal size distribution of standard deviation σ_a and the measured size distribution is lognormal, with a standard deviation σ_0, then the value of σ to be used in calculating resolution is

$$\sigma = (\sigma_0{}^2 - \sigma_a{}^2)^{1/2} \tag{7.10}$$

provided σ_a is not too large. Some of the data on resolution are given in Table 7.1.

The tabulated resolutions refer to measurements of monodisperse particles made with multichannel analyzers in which the channel widths were equivalent to small fractions of a micrometer. The instruments available commercially have wider, and fewer, channels. The Bausch and Lomb instrument reads out the number of particles having equivalent optical diameters greater than a threshold value that can be set at 0.3, 0.5, 1.0, 2.0, 3.0, 5.0 or 10.0 μm. The Royco can be operated in a similar manner or it can be used to sort pulses into size intervals. The instrument described by Zinky [6] provided five size ranges. Whitby and Vomela [11] tested a later model having 15 size ranges. Their work indicates that the resolution of the latter instrument, when used to measure the diameters of perfect spheres, does not differ significantly from that obtained when the pulse heights are sorted with a multichannel analyzer. The resolution appeared much poorer when the particles had the same geometric diameter but were not perfectly spherical. This showed that the optical equivalent diameter depended on the orientation of the particle as it moved through the sensitive volume, an effect that was more pronounced for instruments having a 90° scattering angle.

Whitby and Vomela included aerosols of india ink, a strongly absorbing material, in their tests. Polystyrene particles had been used to calibrate the instruments under study, so that the optical equivalent diameters of the india ink particles were smaller than their geometric diameters. Their data for the Royco and Bausch and Lomb instruments are shown in Fig. 7.6, together with theoretical curves based on Quenzel's work. The agreement between theory and experiment is only passable for the Bausch and Lomb instrument, and the Royco measurements for particle diameter $\gtrsim 2$ μm are significantly lower than predicted theoretically, an effect that Whitby and Vomela also attributed to irregularities of shape.

The photospectrometers register the same diameter for all particles from which they collect the same amount of scattered light, regardless of the physical and optical characteristics of the particle. This optical "equivalent" diameter is referred to a calibration particle, which is usually one of the polystyrene

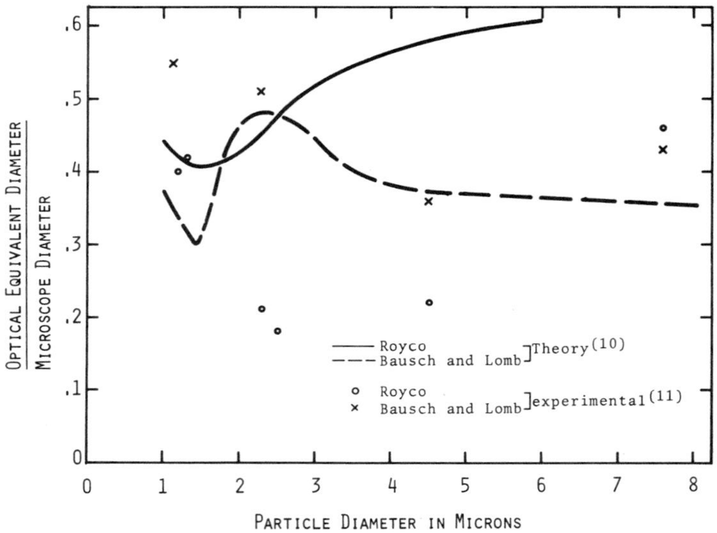

FIG. 7.6. Comparison of optical equivalent diameter and microscope diameter for particles of india ink.

latexes. If the particles being measured are spheres of known refractive index, their real diameters can be calculated with reasonable accuracy from their optical equivalent diameters. If they have irregular shapes, however, size distribution parameters measured in this way may be seriously in error, particularly with respect to the dispersity of the aerosol. On the other hand, an instrument of this type can be particularly useful as a means of comparing two aerosols of similar materials.

Except for the Royco instrument, all of those illustrated here surround the aerosol with a sheath of clean air, which is intended to minimize contamination of the scattering system by airborne particles. The instrument designed by Jacobi *et al.* [7] integrates the scattered light reaching the detector as a particle passes through the sensitive volume, whereas the other instruments measure the peak intensity of the scattered light. Since particles remain in the sensitive volume for different time intervals, depending on their position with respect to the axis of the aerosol stream, Jacobi's method introduces an additional source of error into the measurement. However, with the aerosol sheath that he uses, the error is less than 3%.

Laser illumination has several advantages: the use of monochromatic light simplifies the calculation of scattering functions; lenses are not necessary (see Fig. 7.3) but can be used to minimize the sensitive volume; and strong source intensities are possible.

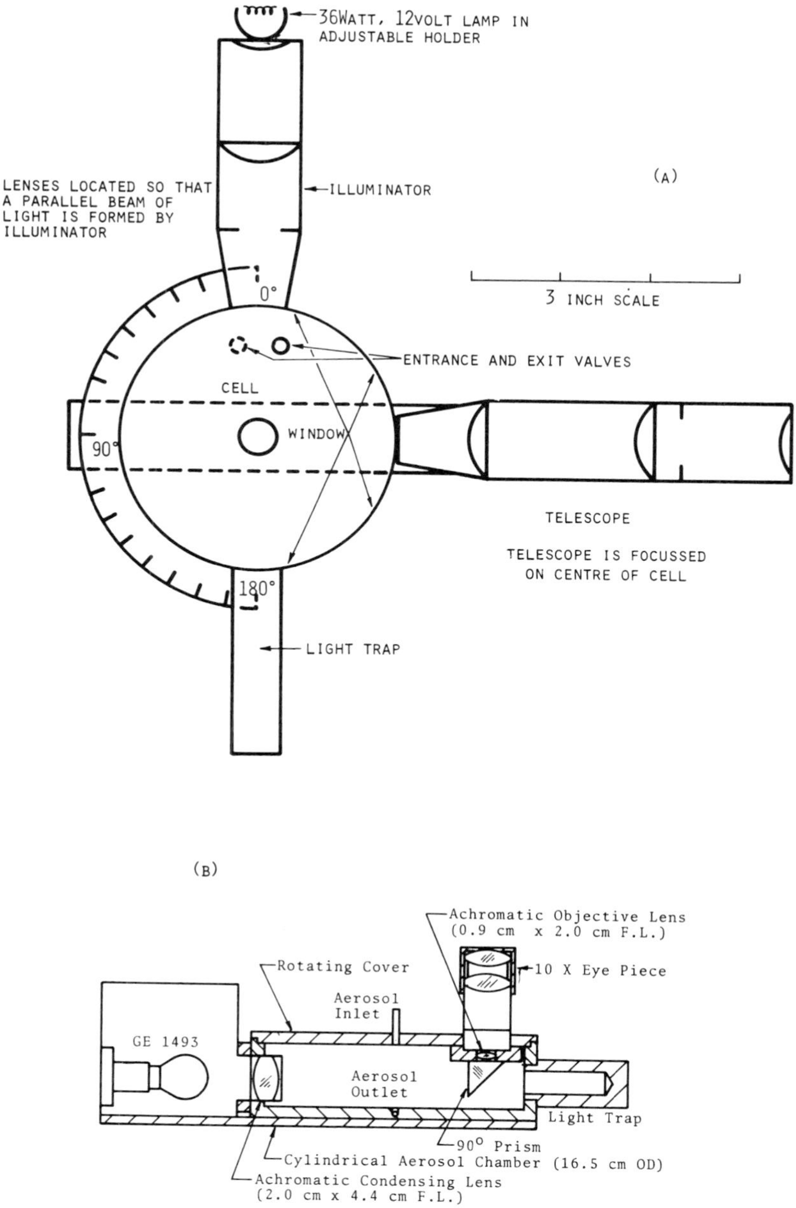

FIG. 7.7 Schematic diagrams of the "Owl." (A) Original instrument described by Sinclair and LaMer [13]; (B) modified instrument described by Liu *et al.* [12]. Courtesy of American Chemical Society.

7-1.2 THE "OWL"

This instrument, two forms of which are shown schematically in Fig. 7.7, is useful for the size measurement of monodisperse aerosols. Its principle of operation is described in Chapter 2 in the discussion of the higher order Tyndall spectra (HOTS) produced by the scattering of white light. Particle size is related to the position and number of red bands occurring between 0° and 180°. Sinclair and LaMer [1] estimated particle size by using the following relationship:

$$\bar{D} = 0.2n,$$

where n is the observed number of reds and $\bar{D}$ is the mean particle diameter in micrometers. For refractive indices between 1.33 and 1.17, Kitani [14] obtained a theoretical relationship between $\bar{D}$ and θ_1, the angle in degrees at which the first red appears, that can be expressed as [12]

$$D = 10\theta^{-0.7}.$$

The clarity of the spectra depends on the monodispersity of the aerosol. Kitani found that this method did not work if the standard deviation of the aerosol exceeded 0.12 μm. The limit of applicability varies somewhat with refractive index, but does not differ much from that reported by Kitani, according to work cited by Fuchs and Sutugin [15].

7-2 Electrical Methods of Size Measurement

The mobility of a charged spherical particle in an electric field is

$$Z_E = \frac{U_E}{E} = \frac{qK_s}{3\pi\eta D} = qZ, \tag{7.11}$$

where $q = n_E\varepsilon$ is the charge carried by a particle of diameter D, E is the strength of the electric field, U_E is the particle's velocity, and Z is its mechanical mobility. Rohmann's theoretical work [16] indicated that the number n_E of elementary charges ε should be proportional to D^2 when the particles are charged in the presence of a strong electric field and unipolar ions, so that

$$Z_E \propto D \cdot K_s.$$

A number of investigators have attempted to design instruments that would separate particles on the basis of this relationship [16–20]. In general, they could not achieve an unequivocal separation of sizes because particle charge does not remain proportional to D^2 at sizes below a few micrometers (see Chapter 4). Langer *et al.* [21] were able to separate particles down to 0.006 μm by using electric fields of 30 kV/cm in a unique charging arrangement, but the method has not yet been developed for routine use.

Whitby and his associates [22, 23] developed a spectrometer in which diffusion charging was used, so that $n_E \not\propto D$ and

$$Z_E \not\propto K_s,$$

permitting the separation of spherical particles on the basis of slip factor, a quantity that decreases monotonically with increasing particle size. Figure 7.8 is a diagram of their spectrometer with the various accessories used in its operation. The aerosol first passes through the charger, where particles are charged by diffusion in a unipolar ion field produced by the sonic jet ionizer.

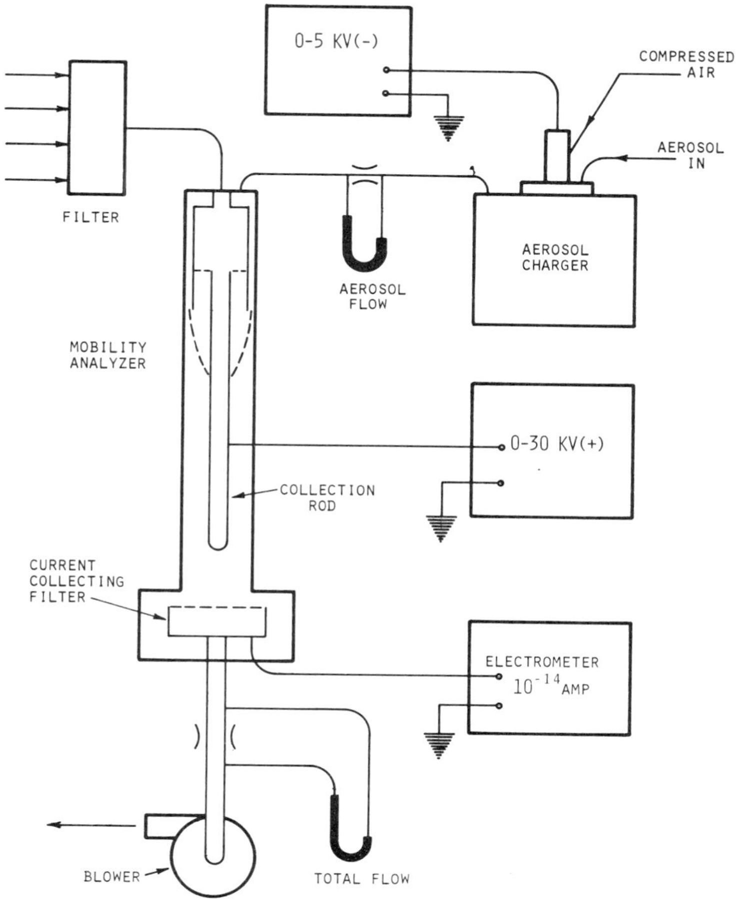

FIG. 7.8. Schematic diagram of the electric particle classifier and accessory equipment [23]. Courtesy of Svenska Geofysika Föreningen.

The charged aerosol is then introduced into the concentric cylinder condenser as an annulus of contaminated air about an inner core of clean air, all in laminar flow. The electric field imparts a radial velocity to the charged particles and they move inward to be deposited on the center electrode at a distance from the aerosol entrance that depends on their mobility. Particles that do not have sufficient mobility to reach the inner electrode are collected on a glass fiber filter supported by a porous stainless steel disk. The latter is grounded through a vibrating reed electrometer, permitting measurement of the current carried by particles reaching the filter.

Whitby and Clark [23] used spherical methylene blue particles to calibrate the instrument in terms of deposition length as a function of size. Flesch [24] calibrated the instrument with monodisperse polystyrene particles and with polydisperse aerosols of zinc ammonium sulfate and sodium chloride. All of their calibration results are plotted in Fig. 7.9. The difference between the zinc ammonium sulfate curve and the methylene blue curve suggests that the nature of the particle affects the charging characteristics. The calibrations were under different operating conditions, but those effects cancel out.

After charging, their mutual repulsion causes some of the particles to be deposited on surfaces within the charger. In addition, some particles are "lost" analytically because they do not pick up a charge. Theoretical values for these losses, at an aerosol flow rate of 28 liters/min, were provided by Whitby and Clark and are also included in Fig. 7.9.

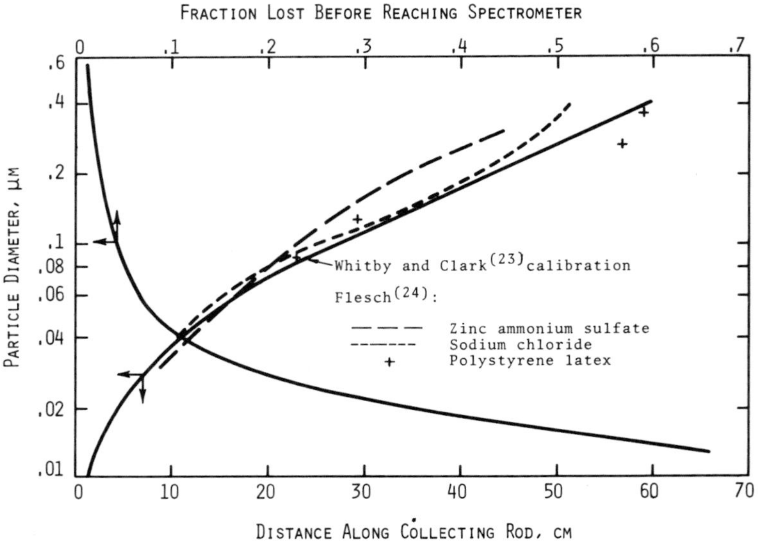

FIG. 7.9. Calibration data for electric particle classifier.

Two methods of determining size distribution are possible. The voltage difference between the electrodes can be varied from 0 to 30 kV and the current measured at each voltage. With each incremental increase in voltage, the mobility of particles that just escape deposition on the inner electrode decreases, causing an incremental decrease in the number of particles reaching the filter and a corresponding reduction in current. For a given voltage change, the concentration of particles per unit size interval is given by

$$\frac{\Delta N}{\Delta D} = \frac{\Delta I}{n\varepsilon F_{\mathrm{A}}\,\Delta D}, \tag{7.12}$$

where ΔI is the observed change in current, ε is the elementary charge, F_{A} is the aerosol flow rate, ΔD is the change in diameter of the largest particle that is collected on the inner electrode, and n is the average number of charges carried by particles having diameters in the increment ΔD. $\Delta N/\Delta D$ is centered on the diameter at the midpoint of that increment; it must be corrected for losses according to the curve of Fig. 7-9.

The total flow rate F_{T}, the voltage V, and the limiting mobility Z_{EM}, for deposition on the inner electrode, are related by

$$VZ_{\mathrm{EM}}/F_{\mathrm{T}} \approx \text{constant} \approx 0.0035 \qquad \mathrm{cm}^{-1},$$

for VZ_{EM} in cm²/sec and F_{T} in cm³/sec. Under normal operating conditions, particle charge is not affected by flow rate through the charger and particles of a given diameter always have the same mobility. The relationships between diameter and mobility and between diameter and charge, which are necessary for the above calculations, are shown in Fig. 7.10.

The second method of determining size distributions depends on the fact that the center electrode is segmented, permitting measurement of the total mass or activity collected between fixed deposition lengths. The distances from the inlet at which the different segments terminate and the corresponding values of $VZ_{\mathrm{EM}}/F_{\mathrm{T}}$ are given in Table 7.2.

The diameters whose mobilities, Z_{EM}, correspond to the segment endpoints in the second column can be treated as effective cut-off diameters [24] and the data analyzed in a manner similar to that used with cascade impactors. The accuracy of such an analysis depends on the resolution of the spectrometer. Theoretically, the resolution should be about 0.1 [23], but Flesch's measurements indicate that it is approximately 0.3 for the range of sizes he studied. He also found that there was little classification of particles above about 0.2 μm. Efficiency curves for segments 4–7, analogous to those for a cascade impactor, have been estimated on the basis of a resolution of 0.3 and are shown in Fig. 7.11. With such good cut-off characteristics, the use of effective cut-off diameters should provide an accurate method of data interpretation for the range of sizes in which separation is achieved.

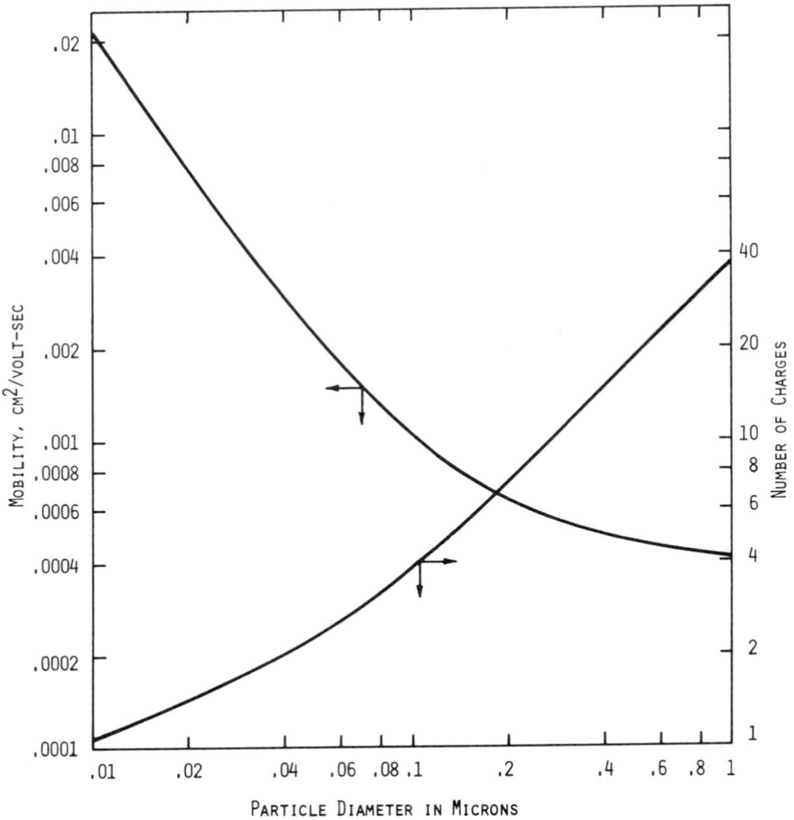

FIG. 7.10. Particle mobility and charge as a function of diameter.

TABLE 7.2

SEGMENTED ELECTRODE METHOD OF DETERMINING SIZE DISTRIBUTIONS

Segment	Distance from inlet at which segment terminates (cm)	VZ_{EM}/F_T (cm^{-1})
1	1.27	0.212
2	3.175	0.085
3	6.147	0.044
4	10.617	0.025
5	17.27	0.016
6	27.43	0.0096
7	42.67	0.0063
8	59.44	0.0045
9	76.2	0.0035

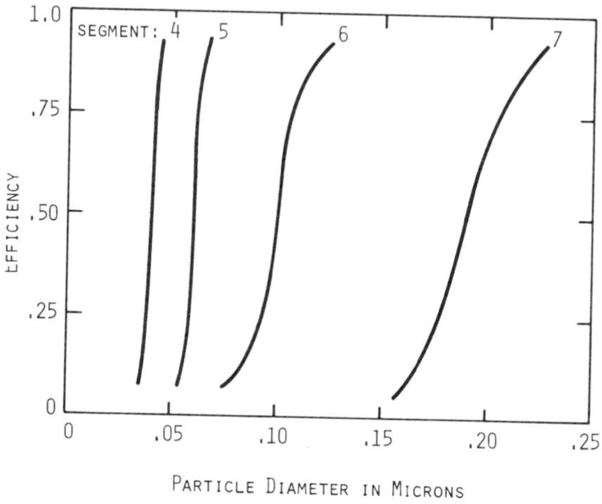

FIG. 7.11. Efficiency curves for segments 4–7 of electric particle counter.

It is difficult to define the particulate property actually measured with this instrument. A particle is assigned a diameter on the basis of its site of deposition, which depends on both its charging and drag characteristics. It is not clear to what extent different values of these characteristics can combine to yield a common site of deposition. Flesch's data suggest that particles deposited at a given site have similar projected areas, whether they are spherical or of irregular shape.

7-3 Diffusion Measurements

As particle size decreases below about 0.5 μm, aerodynamic diameter becomes increasingly difficult to measure and its role as the significant parameter in lung deposition is taken over by particle diffusivity. This particulate property is measured by means of diffusion batteries, which comprise a number of cylindrical tubes or rectangular ducts in a parallel arrangement. For monodisperse particles, the fraction P penetrating the battery is measured at a given flow condition. The equivalent diffusion diameter of the particles under observation is then defined as the diameter of spherical particles that would penetrate the battery to the same extent. P can relate to any measured property of the particle, i.e., $P = A/A_0$, where A_0 is the activity entering the battery and A is the activity leaving it. The most common property measured is the number concentration. If the particles are monodisperse, P is the same whatever A means; otherwise, P varies with A in a manner depending on the relationship between A and particle diameter.

The value of P can be predicted quite accurately for spherical particles. If air flows under laminar conditions through a cylindrical tube, carrying particles of diffusion coefficient Δ, then the relative concentration of particles at a distance L from the tube entrance is [25, 26]:

$$P = 0.819 \cdot e^{-1.828\mu} + 0.0975 \cdot e^{-11.15\mu} + 0.032 \cdot e^{-28.48\mu} + 0.0157 \cdot e^{-53.8\mu} + \cdots, \tag{7.13}$$

where

$$\mu = 2\pi\Delta L/F, \qquad \text{and} \qquad F = \text{volumetric flow rate.}$$

For $\mu < 0.063$,

$$P = 1 - 1.615\mu^{2/3} + 0.6\mu + 0.070\mu^{4/3}. \tag{7.14}$$

Thomas [69] has reported that the equation

$$P = 0.819e^{-1.828\mu} + 0.097e^{-11.15\mu} + 0.032e^{-28.5\mu} + 0.027e^{-61.5\mu} + 0.025e^{-375\mu} \tag{7.15}$$

is accurate to within 0.001 for all values of μ.

For a given volumetric flow rate, the tube diameter does not affect the rate of loss of particles. The increased residence time brought about by an increase in tube diameter, for instance, is exactly offset by the increased time needed for particles to diffuse to the more distant wall. Moreover, the transition length L_T (the distance from the tube inlet at which the axial flow rate is within 1% of its value for parabolic flow) is unaffected by tube diameter [27]:

$$L_T = 0.083F/v,$$

where v is the kinematic viscosity.

The corresponding equation for flow through a rectangular duct of width B and height $2H$ is [28–30]:

$$P = 0.9104 \cdot e^{-1.885\mu} + 0.0531 \cdot e^{-21.43\mu} + 0.0153 \cdot e^{-62.32\mu} + 0.0068 \cdot e^{-124.5\mu} + \cdots, \tag{7.16}$$

where

$$\mu = 2\Delta BL/FH. \tag{7.16a}$$

According to Sparrow *et al.* [31], the transition length for fully developed parabolic flow in a rectangular duct is

$$L_T = 0.08H' \,\mathrm{Re} = 0.64HF/Bv,$$

where $H' = 4BH/(B+2H) \simeq 4H$ is the hydraulic diameter of the duct. For most diffusion batteries, L_T will be a fraction of a centimeter.

Equations (7.13) and (7.16) refer to a large population of particles, all of which have the same diffusion coefficient. For a given battery and flow rate, a measurement of P can be used to determine Δ, from which an estimate of particle diameter can be made. It is customary to prepare a graph of P as a function of μ to facilitate the determination of μ from measured values of P. It is also useful to have a graph of the theoretical relationship between diffusion coefficient and diameter.

If the particles have a range of diffusion coefficients, measurement of P provides a means of calculating an effective diffusion coefficient and an effective equivalent diffusion diameter. Estimates of this sort are not very satisfactory because the effective diffusion coefficient of a polydisperse aerosol is a function of the flow rate at which P is measured and the length of the battery. According to Pollak [32], a 25% change in air flow alters the diffusion coefficient by about 12%. If this method is used, it is advantageous to adjust the flow rate to make $P \simeq 0.4$. The effective equivalent diffusion diameter is then a reasonably good estimate of the corresponding geometric mean diameter (see below).

More accurate particle size data can be obtained by measuring P at several values of L or of F. Penetration can be measured as a function of L by lining a cylindrical diffusion tube with a thin foil or filter paper [33] that can be removed after sampling and analyzed in segments. Alternatively, the tube itself can be cut into segments and each analyzed for its content of activity [34]. An example of this method is given in Fig. 7.12, which shows how three radioisotopes, released by the melt-down of neutron-irradiated uranium, were deposited in a cylindrical diffusion tube. The data for iodine were analyzed by fitting the last four points to a straight line, using a method of least squares. It was assumed that this line represented the first term of Eq. (7.13) for particles having the lowest diffusivity. The slope of the line is equal to -1.828μ and its intercept equals 0.819 times the fraction of the activity associated with these particles. With these data, Eq. (7.13) can be used to subtract from the curve all of the activity due to the most slowly diffusing particles, and the process can be repeated for the next component. In this case, the curve could be described adequately by two groups of particles. Sinclair and Hinchliffe [70] have used this method by measuring penetration through tubular diffusion batteries of a wide range of lengths.

In general, batteries of rectangular ducts are more versatile* than those of cylindrical tubes, but they are not readily adapted to the method just described. Pollak and Metnieks [35] have described a similar method in which the relative amount of activity penetrating the battery is measured at several flow rates. They assumed that activity was lost to the battery according to the

* In a properly designed battery, alternate plates can be charged to permit measurement of the fraction of particles that are charged [74].

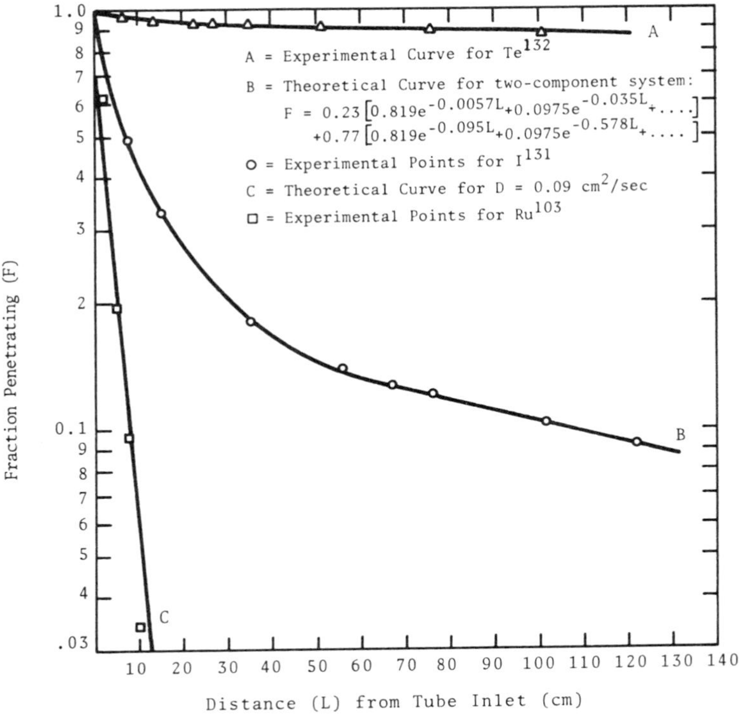

FIG. 7.12. Diffusional deposition of radioactive particulates.

approximate relationship

$$A = \sum_{i=1}^{j} A_i \cdot \exp[(-k/F) \cdot \Delta_i],$$

where $k = 3.770BL/H$ and A_i is the activity associated with particles of diffusion coefficient Δ_i. A plot of A versus $1/F$, with $A(1/F=0)$ equal to the input activity, is similar to the curve of Fig. 7.12. The A_i and Δ_i for the different components can be calculated in the manner described above. Metnieks [36] has shown that this method accurately determines the arithmetic mean diameter and standard deviation and the geometric mean diameter and standard deviation.

Fuchs and others [37] have described a method for estimating the parameters of a lognormal distribution from measurements of penetration through a rectangular duct battery at three values of F. They plotted eleven families of theoretical curves showing penetration as a function of $\log(3.77\Delta BL/FH)$. Each set of curves was calculated for five lognormal distributions having different standard deviations, but the same median diameters. Data points are plotted on tracing paper as penetration versus log $(1/F)$, using coordinate scales identical to those used in plotting the theoretical curves. The tracing paper is then laid over the theoretical plot and adjusted until the points fall on a single curve, taking care that the two sets of axes are properly aligned. Fuchs recommends the use of three data points having penetration values in the ranges 0.05–0.2, 0.40–0.45, and 0.7–0.9. Metnieks [36] found that this method provided reasonably good estimates of the geometric mean diameter even for distributions that were not lognormal, but the estimates of geometric standard deviation were consistently high by 5 to 15%.

The method of Fuchs *et al.* can be generalized to a single family of curves by assuming that activity is lognormally distributed with respect to diffusion coefficient and, consequently, with respect to μ, i.e.,

$$dA = \frac{1}{\sigma'(2\pi)^{1/2}} \cdot \exp\left[\frac{-(\ln \mu - \ln \mu_g)^2}{2\sigma'^2}\right] d \ln \mu.$$

Using Eq. (7.16) penetration can be calculated as a function of μ_g for different values of σ', giving the family of curves shown in Fig. 7.13. Data points are

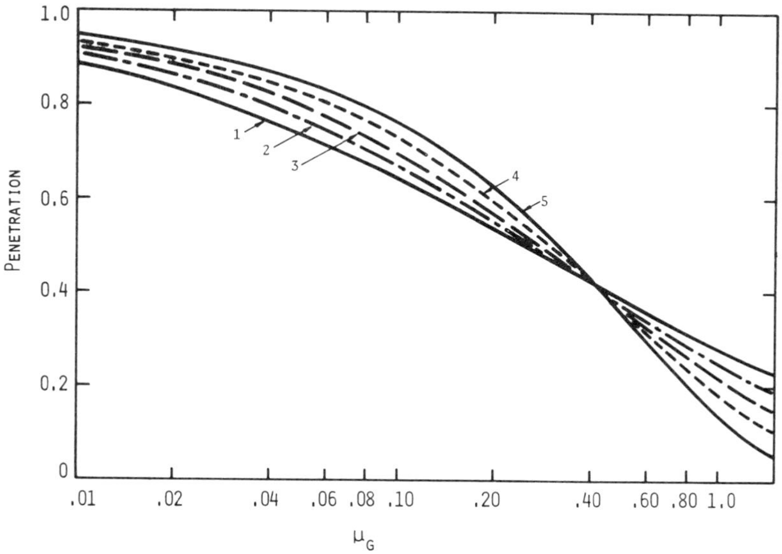

FIG. 7.13. Penetration of polydisperse aerosols through a diffusion battery of rectangular cross section as a function of the median diffusion parameter.

Curve:	1	2	3	4	5
σ'	1.93	1.54	1.16	0.77	0

handled in the manner described in the preceding paragraph to obtain estimates of μ_g and σ'. The value of Δ_g can be calculated from μ_g and the experimental conditions, and D_g can be calculated from Δ_g on the assumption that in the size range of interest

$$\Delta = aD^b.$$

Once the data are properly aligned with the graph, each point provides an independent estimate of Δ_g when the appropriate values of μ_g and F are introduced into Eq. (7.16a). The geometric standard deviation σ_g of the particle size distribution is related to σ' by

$$\ln \sigma_g = \sigma'/b.$$

Some representative values of a and b are:

Diameter range (μm)	a	b
0.001–0.02	5.35×10^{-8}	−2.0
0.008–0.10	7.65×10^{-8}	−1.93

7-4 Surface Area Measurements

One of the most important parameters of an aerosol is its specific surface, that is, its surface area per unit mass or unit volume. A knowledge of specific surface provides sufficient information about the particles for an estimation of speeds of chemical reactions [38]. Moreover, some believe the concentration of airborne particles in terms of their surface area is often a better index of their hazardous nature than are concentrations by number or mass [39]. For a spherical particle of diameter D the surface per unit volume is

$$S_V = 6/D.$$

If D is in micrometers, S_V is in square meters per cubic centimeter. For a population of particles of any shape

$$S_V = 6/D_{SV},$$

where D_{SV} is the equivalent specific surface diameter. It is also referred to as the volume–surface mean diameter and as Sauter's mean diameter.

The measured specific surface of a population of particles depends on the experimental method used. Photometric methods, based on the extinction of a light beam by particles in its path, yield an estimate of the total cross-sectional area of the particles from which the surface area, essentially measured over a particle envelope that smooths out concavities, can be calculated. Adsorption methods, in which one measures the amount of material required to cover the particles with a single layer of molecules (accomplished either by low temperature adsorption of nitrogen or an inert gas from the vapor phase, or by the adsorption of solutes from the liquid phase), provide an estimate of the total surface area—including that of pores and crevices— accessible to the adsorbate molecules. Permeability methods, based on measurement of the pressure difference necessary to move a fluid at a given flow rate through a packed bed of particles, yield an estimate of surface area that is likely to be higher than that measured by photometry and smaller than that measured by adsorption.

Photometric determination of specific surface has been described in Chapter 4 in the section on optical measurement of mass concentration and will not be considered further. Adsorption and permeability measurements are considered below.

7-4.1 Low Temperature Adsorption from the Vapor Phase

This method of surface area measurement depends on the physical, or Van der Waal's, adsorption of molecules of vapors or gases near their boiling temperatures. Molecules striking a solid surface may not rebound instantaneously, so that at any time a portion of the solid surface is covered with molecules. At a given temperature, the amount of adsorbate q attached to the solid depends on the pressure of adsorbate molecules in the volume around the solid relative to their saturation vapor pressure P_0 at that temperature. A graph of q as a function of the relative pressure P/P_0 is called an adsorption isotherm. There are several types of adsorption isotherms, depending on the nature of the substances involved [40]. This method of surface measurement should be limited to systems exhibiting isotherms of the types shown in Fig. 7.14.

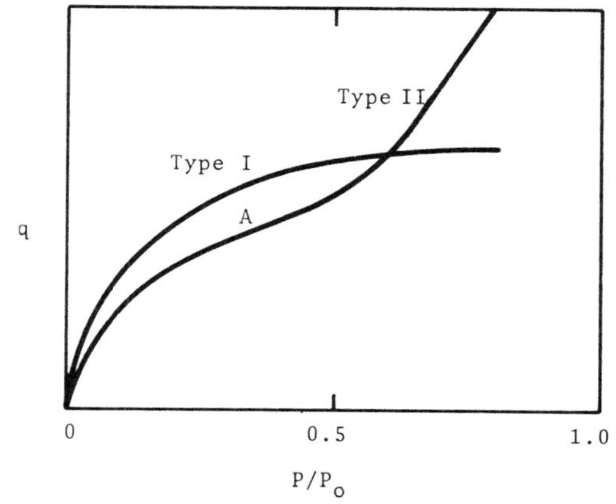

FIG. 7.14. Types of adsorption isotherms.

Langmuir [41] interpreted the Type I isotherm as being due to a process in which only a single layer of molecules could build up on the solid surface. Assuming that the rate at which molecules were deposited on the solid was proportional to the gas pressure, the free surface area, and a factor related to the activation energy for adsorption, while the rate at which they left the solid was proportional to the covered surface area and a factor related to the activation energy for desorption, he obtained the equation

$$q = aq_0 P/(1+aP),$$

where q_0 is the amount of adsorbate required to completely cover the solid with a monomolecular layer and a is a constant related to the two activation energies. The equation can be put in the following form suitable for graphical analysis:

$$\frac{P}{q} = \frac{1}{aq_0} + \frac{P}{q_0}. \tag{7.17}$$

If q is measured as a function of P, a graph of P/q versus P will be linear, with slope equal to $1/q_0$ and intercept equal to $1/aq_0$.

Many materials exhibit adsorption isotherms of Type II, which are interpreted as representing the formation of several layers of adsorbate molecules on the solid surface. Early attempts [42] to estimate surface area from these isotherms were based on the assumption that a monolayer had been achieved at the pressure corresponding to the point (A) where the isotherm first becomes a straight line. The uncertainty associated with this method of data interpretation was eliminated when Brunauer *et al.* [43] obtained a theoretical expression for Type II isotherms:

$$\frac{P}{q(P_0-P)} = \frac{1}{q_0 C} + \frac{(C-1)}{q_0 C} \cdot \frac{P}{P_0}. \tag{7.18}$$

The constant C is related to the heats of adsorption and liquefaction. By plotting $P/q(P_0-P)$ against P/P_0, the quantities q_0 and C can be determined. When $C \gg 1$ and $P \ll P_0$, this reduces to the equation for the Langmuir isotherm. This procedure for determining the mass or volume of gas that will just provide a monomolecular layer over the surface of a solid is called the BET (Brunauer–Emmett–Teller) method.

Determining the adsorption isotherm requires the measurement of q at different values of P for a known temperature at which the saturation vapor pressure is P_0. Figure 7.15 is a diagram of an apparatus with which q is measured gravimetrically as a function of P [44]. Identical glass envelopes (L) surround matched weighing pans suspended from a beam consisting of two quartz rods fused to a quartz ring that is supported on torsion wires. The ring carries a

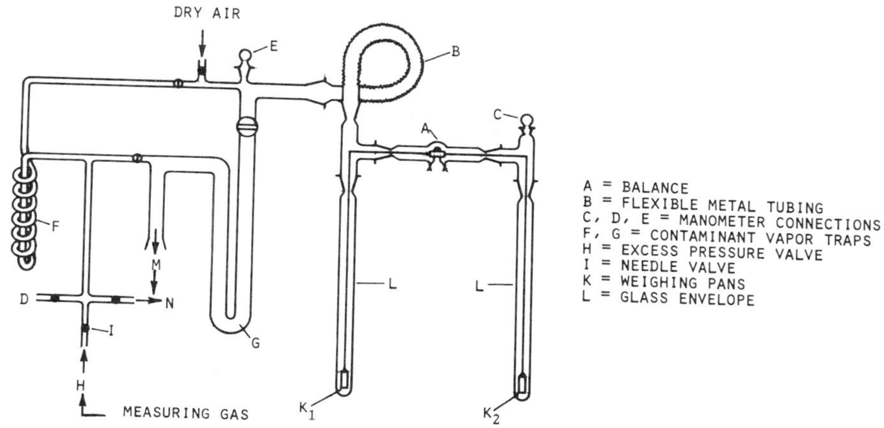

FIG. 7.15. Apparatus for gravimetric determination of adsorbed gases [44]. Courtesy Verlag Chemie.

field coil and lies in an externally applied high-frequency field. A deflection of the beam induces an HF current in the field coil that is amplified and rectified to provide a d.c. current directly proportional to the weight difference causing the deflection.

To measure an isotherm, a weighed amount of sample is placed in the weighing pan and tared with an equal weight of solids. Buoyancy corrections can be minimized or avoided by combining two solids of different densities so that the tare is approximately the same volume as the sample. The glass envelopes are surrounded with heating coils and the system degassed under vacuum. After degassing, the coils are replaced with temperature baths, usually Dewar flasks containing liquid nitrogen, and the adsorbate, usually nitrogen, introduced into the system. Simultaneous measurements of pressure and weight of adsorbed gas provide the data with which to plot the curves described above. For an adsorbed mass q_0 at one monolayer, the surface area of the sample is

$$S = (A/M) \cdot q_0 \cdot S_1, \tag{7.19}$$

where A is Avogadro's number, M is the molecular weight of the adsorbate, and S_1, the area covered by a single molecule, is given by [45]

$$S_1 = 1.091\,(M/A\rho)^{2/3}. \tag{7.20}$$

ρ is the density of the liquid adsorbate. This yields a value of 16.2 Å^2 for nitrogen gas, which is the most commonly used adsorbate. The molecular area varies somewhat with temperature, but the effect is negligible for surface area measurements.

Many gases other than N_2 have been used as adsorbates, including the inert gases argon, krypton, and xenon, and a variety of organic compounds. McClellan and Harnsberger [45] recently compared data from published accounts of adsorption studies and recommended values for molecular cross sections of many adsorbate materials.

Early methods of measuring surface area by gas adsorption relied on volumetric determinations of the quantity of adsorbed gas. Apparatus for this type of measurement, used by Lauterbach [46] to determine the specific surface of uranium dusts, is shown schematically in Fig. 7.16. Basically, the method consists of discharging a measured volume of gas from the McLeod gauge into the evacuated manifold and sample tube, which is immersed in a coolant and which contains a weighed, degassed sample, and measuring the pressure in the system after the adsorption process has come to equilibrium. The method is complicated by the need to convert pressure readings to gas volumes, which makes it necessary to determine the volumes of the manifold to cut-off A, the sample line to cut-off A, and the sample tube (with sample in place). This is accomplished by expanding known amounts of helium into the volumes of interest and measuring the pressure of the system. The vapor pressure of helium is so high, even at the temperature of liquid nitrogen, that its adsorption is always negligible.

A consideration of Eq. (7.18) shows that the adsorbed quantity is the same for a given value of P/P_0, regardless of the particular value of P_0 at which the measurements are made. If an adsorbate gas is used that has a very low value of P_0 at the temperature of the coolant bath, then the amount adsorbed at a given value of P/P_0 represents a large fraction of the amount introduced into the system, and the sensitivity of the method is much enhanced. The use of organic adsorbates with the volumetric method makes it possible to measure surface areas as small as 25 cm^2 [47]. The apparatus of Fig. 7.16 was used with ethane as the adsorbate. An example of an ethane adsorption isotherm, plotted to yield a straight line, is shown in Fig. 7.16(B). Only the experimental points for values of P/P_0 between about 0.05 and 0.35 lie satisfactorily in a straight line. This is generally true of adsorption data, so that points used to measure specific surface should be limited to values of P/P_0 within that range. The same graph brings out the fact that the intercept $1/Cq_0$ is approximately zero. This also is frequently observed in adsorption isotherms, especially when nitrogen is the adsorbate. When measuring the specific surface of a number of samples of the same material, the work load can be minimized by measuring q at a single value of $P/P_0 \simeq 0.3$. The slope of a line through the origin and the experimental point is equal to $1/q_0$.

The need for vacuum techniques can be eliminated by using a mixture of nitrogen and helium in a continuous flow system [48]. Metered flows of the two gases, adjusted to provide a partial pressure of nitrogen equal to the

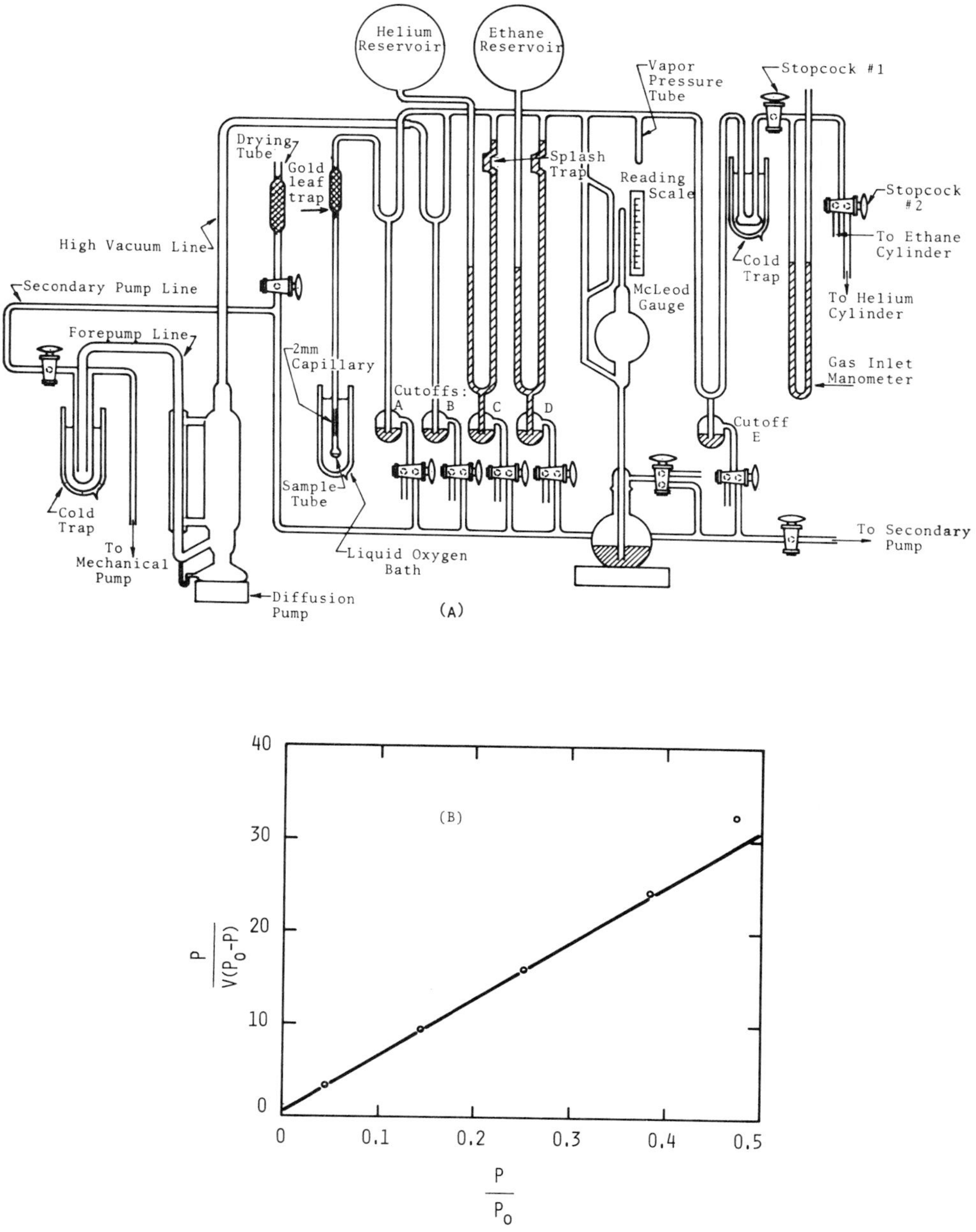

FIG. 7.16. (A) Apparatus for the volumetric determination of adsorbed gases [46]. (B) Adsorption of ethane on uranium dioxide. Courtesy of The Franklin Institute.

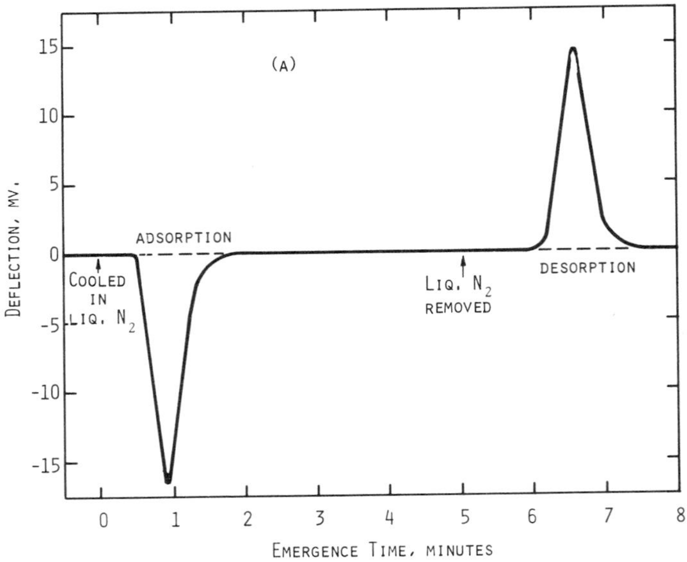

USED CRACKING CATALYST, 0.21 GRAM; SURFACE AREA, 101 SQUARE METERS PER GRAM; P/P_0, 0.30

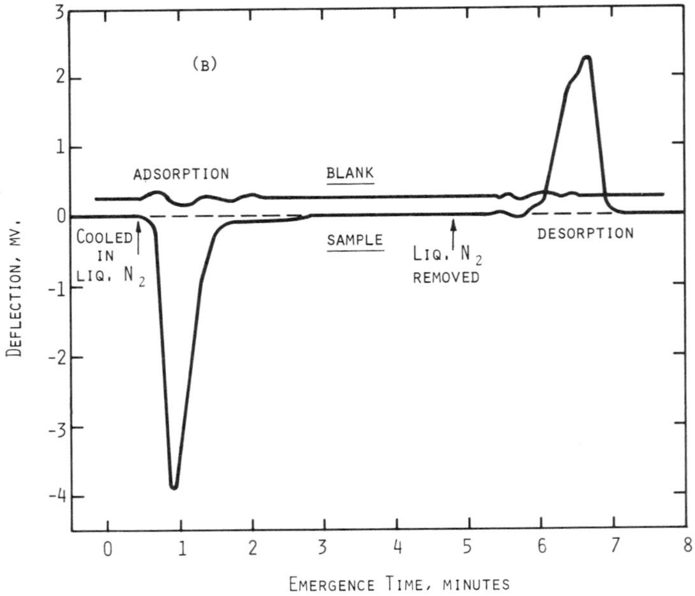

CRUSHED FIREBRICK, 0.75 GRAM; SURFACE AREA, 3 SQUARE METERS PER GRAM; P/P_0, 0.25.

FIG. 7.17. Nitrogen adsorption and desorption curves as determined with the continuous flow system [48]. Courtesy of American Chemical Society.

desired adsorption pressure, are mixed and passed in sequence through two reference thermal conductivity cells, the sample tube, and two sensor thermal conductivity cells. The conductivity cells form a Wheatstone bridge network with the output going to a recording potentiometer. A base line is established with the sample tube at room temperature. When the tube is immersed in liquid nitrogen, the sample adsorbs nitrogen, changing the composition of the gas passing through the sensor cells. The resulting unbalance of the bridge circuit persists until adsorption is complete and is recorded as shown in Fig. 7.17. After the bridge returns to the balanced condition, the coolant is removed and the resulting desorption of nitrogen causes a second deflection of the recorder. The area under either of these curves is proportional to the amount of adsorbed nitrogen. The method is rapid but less sensitive than the other methods.

Although there is doubt about the accuracy of specific surfaces measured by adsorption, the BET method has become a standard. It has been used to good advantage both in inhalation studies and in investigations of urban aerosols [49].

7-4.2 Permeability Measurements

For viscous flow of fluid through a capillary, the average linear velocity is given by the Poiseuille–Hagen equation:

$$V_c = d^2\,\Delta P/16 \cdot k_0 \cdot \eta \cdot L_c, \tag{7.21}$$

where d is the diameter and L_c the length of the capillary, η is the fluid viscosity, k_0 is a shape factor (equal to 2, for cylindrical tubes), and ΔP is the pressure drop producing the flow. Kozeny [50] applied this equation to flow through a porous bed of length L, cross-sectional area A, and porosity p, by treating its void volume as the lumen of a capillary and its particulate surface as the capillary wall, i.e.,

$$\frac{p}{(1-p)\,S_v} = \frac{d}{4}.$$

With the additional assumption that the void volume is uniformly distributed along the length of the bed,

$$V_c = \frac{FL_c}{pAL} = \frac{\Delta P}{k_0\,\eta L_c} \cdot \frac{p^2}{(1-p)^2 S_v^{\,2}}.$$

The volumetric flow rate F is then given by

$$F = \frac{A \cdot \Delta P}{k \cdot \eta \cdot L} \cdot \frac{p^3}{(1-p)^2 S_v^{\,2}}, \tag{7.22}$$

where $k = k_0 (L_c/L)^2$. $S_v^{\,2}$ can be determined from measured values of F and

ΔP for a particulate column of known porosity and dimensions. Carman [51] demonstrated that the available experimental data for spheres of known size verified Kozeny's equation and that the constant k was very nearly equal to 5.0. The fluids used included oils, water, and air, for which the range of viscosities had limits in the ratio of 23:1. Moreover, some of the experiments were made with air at so high a pressure that it expanded to about 30 times its initial volume in passing through the bed. Carman also showed that the Kozeny equation (7.22), with $k = 5.0$, was valid for porous beds of irregularly shaped polydisperse particles and concluded that k was independent of particle shape.

Lea and Nurse [52] measured the specific surface of the same sample using both air and water as the permeating fluids. For values of S_v up to about 0.25 m^2/cm^3, the two methods were in good agreement with each other and with specific surface based on the particle size distribution of the sample. As S_v increased, the ratio $S_v(\text{water})/S_v(\text{air})$ increased to 1.4 for $S_v(\text{air}) \simeq 0.75$ m^2/cm^3, with the particle size distribution yielding specific surfaces in close agreement with $S_v(\text{air})$. Later [38], they reported that S_v depended on the porosity at which it was measured and recommended that a standard porosity be used for a given type of material, with S_v calculated from Eq. (7.22) and taking $k = 5.0$. From their results, they concluded that the method gave specific surfaces for spherical and near-spherical particles that were consistent with the results of other methods.

Figure 7.18 is a diagram of a simple apparatus for measuring specific surface by the permeability method [53]. A sample of mass M and density ρ is held in place by the porous plug. It is compacted by rapping on or vibrating the sample tube until the upper level of the sample becomes constant. If V_T is the volume of the tube within which the sample is confined, the porosity is

$$p = 1 - (M/\rho V_T).$$

Porosities should be in the range of 0.4 to 0.8. The flow rate F is determined from the pressure drop across the resistance, which is a column of compacted sand. The input pressure is regulated by a standpipe filled with water to a fixed level. The combined resistance of drier and porous plug is negligible, so that the pressure drop across the sample is equal to the difference between the pressure drops across the regulator and the sand column. S_v can then be calculated from a rearranged form of Eq. (7.22):

$$S_v = \frac{1}{1-p}\left(\frac{g\rho_f p^3}{kK_p\eta}\right)^{1/2} = \frac{14p}{1-p}\left(\frac{\rho_f p}{K_p\eta}\right)^{1/2}, \tag{7.23}$$

where $K_p = FL/HA$ is the permeability of the sample, g is the acceleration due to gravity, $H = \Delta P/g\rho_f$ is the pressure drop in centimeters of the manometer fluid of density ρ_f, and $k = 5$. All dimensions are in cgs units.

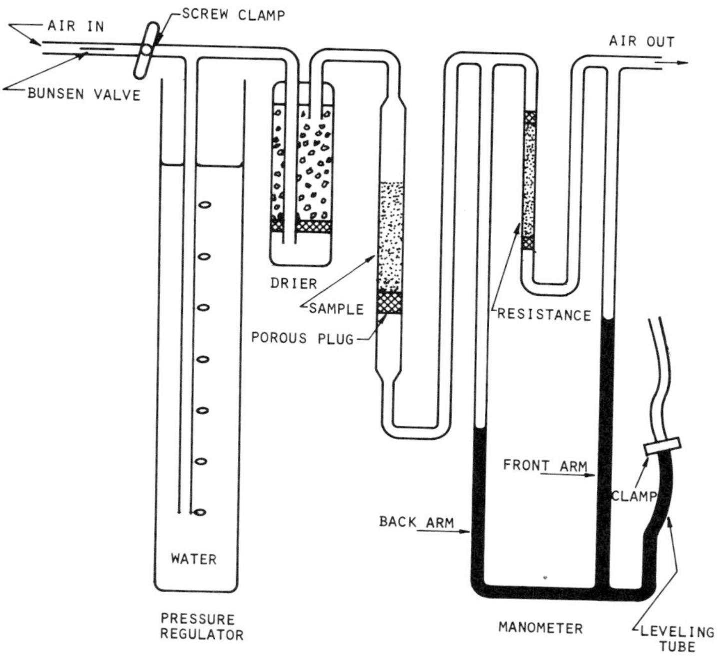

FIG. 7.18. Apparatus for measuring surface area by the permeability method [53]. Courtesy of the American Chemical Society.

Another common apparatus for the measurement of permeability is the Blaine fineness tester (Fig. 7.19) [54]. A machined rod (V) is used to compress the sample (P) to a fixed volume between two filter disks (U) in the sample container (R). The container is fitted tightly into the ground-glass, tapered joint (E) at the top of one arm of the U-tube manometer. The open end of the sample container is stoppered, the stopcock (G) is opened, and the suction bulb (H) is used to bring the manometer fluid up to line A. G is closed, the stopper is removed from R, and the time required for the manometer fluid to drop from line B to line C is measured. In this case, the pressure drop ΔP changes continuously as the fluid level falls. At a given value of ΔP, the element of volume passing through the sample in time dt is

$$dV = F\,dt = \beta \cdot \Delta P \cdot dt = \beta(\Delta P_0 - \alpha V)\,dt$$

where $\beta \cdot \Delta P$ equals the right-hand term of Eq. (7.22), and α is the pressure drop per unit volume of manometer tube. Rearranging and integrating this equation gives

$$\beta t = \ln[1-(\alpha V_{BC}/\Delta P_0)]^{-1/\alpha}, \qquad (7.24)$$

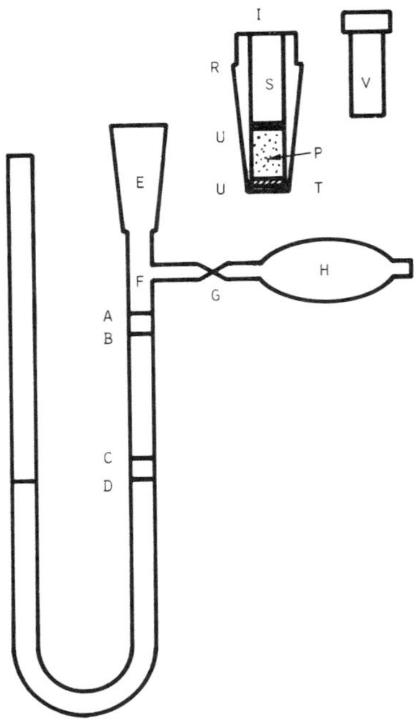

FIG. 7.19. The Blaine fineness tester [54]. Courtesy of the American Society for Testing Materials.

from which the specific surface is

$$S_v = K'[p^{3/2}/(1-p)] \cdot t^{1/2}, \tag{7.25}$$

where

$$K' = \left(\frac{-\alpha A}{kL\eta \cdot \ln[1-(\alpha V_{BC}/\Delta P_0)]}\right)^{1/2}$$

is a constant for the instrument for a given manometer fluid, and V_{BC} is the volume included between lines B and C.

For a given powder, the porosity should not affect the determination of S_v, whichever apparatus is used. At values of $S_v \simeq 5\ \mathrm{m^2/cm^3}$, however, marked effects of porosity are observed, with S_v decreasing as p increases. This is apparently the result of a complex variety of effects related to particle size and shape and to the transition from viscous flow to molecular flow as pore dimensions become comparable to the mean free path of the permeating gas. Theoretical corrections to the basic equation have been proposed, but they tend to make data interpretation difficult or to require more elaborate experimentation. Empirical corrections have been tried also. Ober and Frederick [54] replaced the term $(1-p)$ with $(1-xp)$ and determined x experimentally for the Blaine fineness tester. They were able to make accurate measurements of S_v up to $\sim 6.4\ m^2/cm^3$ ($D_{SV} \simeq 1\ \mu m$). Carman and Arnell [55], applying a slip correction factor, reported specific surfaces up to $303\ m^2/cm^3$ for carbon black, with the data in reasonable agreement with the results of electron microscope measurements up to $S_v \simeq 150\ m^2/cm^3$ ($D_{SV} \simeq 0.04\ \mu m$).

7-5 Measurement of Particulate Volume

The Coulter counter has been widely used in hematology and powder technology for the measurement of particles larger than several micrometers in diameter. Schrag and Corn [56] used the method to determine shape factors of particles in the respirable size range. The instrument is unique in that it measures the volumes of individual particles, for a certain range of operating conditions, making it possible to determine directly the cumulative number distribution with respect to equivalent volume diameter.

Figure 7.20 is a diagram of the Coulter counter system [57]. The particles to be measured are suspended in an electrolyte solution surrounding a glass tube filled with a similar solution. A small aperture in the tube wall permits flow of current between electrodes submerged in the solution on either side of the aperture. A particle passing through the aperture displaces an amount of solution equal to its volume, temporarily raising the resistance of the fluid in the aperture and creating a voltage pulse that is proportional to the product

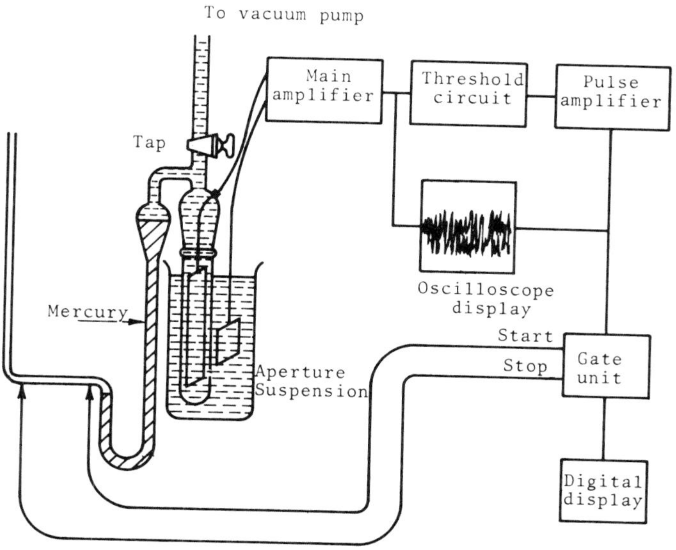

FIG. 7.20. Schematic diagram of the Coulter counter system [57]. Courtesy of the Society for Analytical Chemistry, London.

of the aperture current and the resistance change. The latter is given by [58]

$$\Delta R = \frac{\rho_0 V}{A^2} \cdot \frac{A(1-\rho_0/\rho)}{A-a(1-\rho_0/\rho)} \simeq \frac{\rho_0 V}{A^2} \Big/ \left(1 - \frac{a}{A}\right), \tag{7.26}$$

where ρ_0 and ρ are the electrical resistivities of the electrolyte and particle, respectively, A is the cross-sectional area of the aperture, and a and V are respectively the cross-sectional area and volume of the particle. Strictly speaking, the equation refers to a cylinder with its axis parallel to that of the aperture. However, it can be applied quite accurately to spheres of diameter D by setting $a = 0.2\pi D^2$. The assumption that $\rho_0/\rho \simeq 0$ is supported by experimental evidence showing that in electrolyte solutions even copper particles act like insulators [59] at the field strengths occurring in the Coulter counter.

Equation (7.26) indicates that the resistance change increases with increased electrolyte resistivity. Values of ρ_0 up to 10^3 ohm-cm have been used to improve response [58]. However, Allen [57] found that the advantages gained by increasing ΔR to improve sensitivity are largely offset by an increased noise level. An aqueous sodium chloride solution comparable to physiological saline (0.9% NaCl) is widely used. The value of ρ_0 has little effect on the aperture current, which is determined by a much larger resistance in the external circuit.

Equation (7.26) also shows that ΔR can be considered proportional to V as long as $a/A \ll 1$. For nearly spherical particles having diameters less than about one-fourth the aperture diameter, the deviation from proportionality is less than 6%. The volume measured is apparently that of the particle envelope rather than the true particle volume, the fluid in re-entrant spaces acting like part of the particle [59].

The sensitivity of the method increases with decreasing aperture diameter, the minimum detectable particle size being limited by amplifier noise for the larger apertures and by thermal effects due to electric heating of the fluid in the aperture. According to Berg [58], the limiting size in the former case is about 2% of the aperture diameter, and, in the latter case, it is about 0.3 μm. Apertures having diameters as small as 10 μm are available. Table 7.3 lists the dimensions of some of the more commonly used apertures.

To measure a particle size distribution, the sample is thoroughly dispersed in the electrolyte solution. The stopcock leading to vacuum (Fig. 7.20) is opened, starting a flow of suspension through the aperture and drawing the mercury into the unbalanced position indicated in the figure. When the stopcock is closed, the mercury falls back, maintaining the flow of suspension through the aperture. A known volume of suspension is analyzed as the mercury flows between the start and stop contacts. A range of contact positions permits a choice of sample volumes of 0.05, 0.5, and 2.0 ml. All particles above a set threshold level are counted. The threshold level is then set to another value and the analytical process is repeated. Ten or twelve threshold levels

TABLE 7.3

DIMENSIONS OF COULTER COUNTER APERTURES[a]

Nominal diameter (μm)	Actual diameter (μm)	Approximate thickness (μm)
30	32.7	50
50	52.7	50
70	67.3	50
100	89.2	70
140	138.2	100
200	199.9	140
280	287.2	200

[a] Data from Coulter Electronics Ltd., cited in discussion of Ref. 57.

are available, providing data for a cumulative frequency distribution of particle volumes.

The probability of coincidence counting with this instrument is described by Poisson's law just as it is for optical counters. It appears that coincidence losses are usually calculated on the assumption that the instrument looks successively at discrete volumes of liquid [61, 64, 65], although the equation of Princen and Kwolek has been used also [63]. Equation (7.6) should be used in the form

$$N_0 \cdot V_s = N \cdot V_s e^{-NV_s}, \tag{7.27}$$

where V_s is the sensitive volume of the counter and N is the true concentration of particles in the liquid. A graph of the term on the right as a function of NV_s makes it possible to determine N for a measured value of N_0 and the known volume V_s. At small values of NV_s, $N_0 \simeq N - N^2 V_s$ and

$$N \simeq [1 - (1 - 4N_0 V_s)^{1/2}]/2V_s. \tag{7.28}$$

The threshold level affects the number of particles counted, but does not alter the frequency with which multiple particles occur in the sensitive volume. The latter can be kept below 5% by using dilutions for which $NV_s \gtrsim 0.05$. It is not easy to put an accurate value on V_s, which is larger than the aperture volume, because the lines of electrical force extend beyond the aperture boundary [64].

The most straightforward treatment of Coulter Counter data consists of plotting the cumulative relative frequency of particles observed at a given threshold level against the volume corresponding to the pulse height at that particular threshold level. The resulting curve can be converted to other distribution forms the conversion being particularly simple if the volume distribution is lognormal. Since the mass or volume distribution is usually of more interest than is the number distribution, the data are often converted to relative mass (volume) by multiplying the number of particles observed between two threshold levels by an average mass (volume). This is most simply defined as one-half the mass (volume) of the sum of the masses (volumes) corresponding to the two threshold levels. Although more complex definitions have been advanced [65], they offer little advantage compared to this.

References

1. D. Sinclair and V. K. LaMer, Light Scattering as a Measure of Particle Size in Aerosols, *Chem. Rev., 44:* 245–267 (1949).
2. F. T. Gucker and D. G. Rose, A Photoelectric Instrument for Counting and Sizing Aerosol Particles, *Brit. J. Appl. Phys., Suppl., 3:* S138–S143 (1954).
3. C. T. O'Konski and G. J. Doyle, Light-Scattering Studies in Aerosols with a new Counter-Photometer, *Anal. Chem., 27:* 694–701 (1955).
4. M. A. Fisher, S. Katz, A. Lieberman, and N. E. Alexander, The Aerosoloscope: an Instrument for the Automatic Counting and Sizing of Aerosol Particles, *Proc. Nat. Air Pollut. Symp.*, p. 112, Los Angeles, 1955.
5. A. E. Martens and J. D. Keller, An Instrument for Sizing and Counting Airborne Particles, *Amer. Ind. Hyg. Ass. J., 29:* 257–267 (1968).
6. W. R. Zinky, A New Tool for Air Pollution Control: the Aerosol Particle Counter, *APCA J., 12:* 578–583 (1962).
7. W. Jacobi, J. Eichler and N. Stolterfoht, Teilchen-grössenspektrometrie von Aerosolen durch Lichtstreuung in einem Laserstrahl, *Staub, 28:* 314–319 (1968).
8. J. Gebhart, J. Bol, W. Heinze and W. Letschert, Ein Teilchengrössenspektrometer für Aerosole unter Ausnutzung der Kleinwinkelstreuung der Teilchen in einem Laserstrahl, *Staub, 30:* 238–245 (1970).
9. D. Sinclair, A New Photometer for Aerosol Particle Size Analysis, *APCA J. 17:* 105–108 (1967).
10. H. Quenzel, Influence of Refractive Index on the Accuracy of Size Determination of Aerosol Particles with Light-Scattering Aerosol Counters, *Appl. Opt., 8:* 165–169 (1969).
11. K. T. Whitby and R. A. Vomela, Response of Single Particle Counters to Non-Ideal Particles, *Environ. Sci. Technol., 1:* 801–814 (1967).
12. B.Y.H. Liu, V. A. Marple, and H. Yazdani, Comparative Size Measurements of Monodisperse Liquid Aerosols by Electrical and Optical Methods, *Environ. Sci. Technol., 3:* 381–386 (1969).
13. D. Sinclair, Optical Properties of Aerosols, *Handbook on Aerosols*, Chapter 7, USAEC, Washington, D.C., 1950.
14. S. Kitani, Measurement of Particle Sizes by Higher Order Tyndall Spectra (θ_1 Method), *J. Colloid Sci., 15:* 287–293 (1960).
15. N. A. Fuchs and A. G. Sutugin, Generation and Use of Monodisperse Aerosols, in C. N. Davies, (Ed.), *Aerosol Science*, Academic Press, London and New York, 1966.
16. H. Rohmann, Methode zur Messung der Grösse von Schwebeteilchen, *Z. Phys., 17:* 253–265 (1923).
17. B. G. Saunders, *Electrostatic Precipitator for Measuring Particle Size Distribution in Aerosols*, USAEC Rep. ORNL–1655, 1954.

18. T. T. Mercer, *Charging and Precipitation Characteristics of Submicron Particles in the Rohmann Electrostatic Precipitator*, USAEC Rep. UR–475, 1956.
19. W. N. Lipscomb, T. R. Rubin, and J. H. Sturdivant, An Investigation of a Method for the Analysis of Smokes According to Particle Size, *J. Appl. Phys.*, *18:* 72–79 (1947).
20. G. Langer and J. L. Radnik, Development and Preliminary Testing of a Device for Electrostatic Classification of Submicron Airborne Particles, *J. Appl. Phys.*, *32:* 955–957 (1961).
21. G. Langer, J. Pierrard and G. Yamate, Further Development of an Electrostatic Classifier for Submicron Airborne Particles, *Int. J. Air. Water Pollut. 8:* 167–172 (1964).
22. K. T. Whitby, R. C. Jordan, and C. M. Peterson, *Development of a Particle Counter System and Development of a Technique for Studying the Charge on an Evaporating Drop*, Particle Lab. Rep. No. 80, Mech. Eng. Dept. Univ. of Minnesota, to USPHS, 1964.
23. K. T. Whitby and W. E. Clark, Electric Aerosol Particle Counting and Size Distribution Measuring System for the 0.015 to 1μ Size Range, *Tellus*, *18:* 573–586 (1966).
24. J. P. Flesch, Calibration Studies of a New Submicron Aerosol Size Classifier, *J. Colloid Interface Sci.*, *29:* 502–509 (1969).
25. P. G. Gormley and M. Kennedy, Diffusion from a Stream Flowing through a Cylindrical Tube, *Proc. Roy. Irish Acad.*, *A52:* 163–169 (1949).
26. S. Twomey, *Bull Observatoire du Puy de Dome*, p. 173, 1963, cited by K. Spurny and J. Pich, in Zur Frage der Filtrationsmechanismen bei Membranfiltern, *Staub*, *24:* 250 (1964).
27. O. G. Tietjens, *Applied Hydro–and Aerodynamics*, p. 22, Dover, New York, 1957.
28. J. J. Nolan, P. J. Nolan, and P. G. Gormley, Diffusion and Fall of Atmospheric Condensation Nuclei, *Proc. Roy. Irish Acad.*, *A45:* 47–63 (1938).
29. S. Twomey, The Determination of Aerosol Size Distributions from Diffusional Decay Measurements, *J. Franklin Inst.*, *275:* 121–138 (1963).
30. T. T. Mercer and R. L. Mercer, Diffusional Deposition from a Fluid Flowing Radially between Concentric, Parallel, Circular Plates, *Aerosol Sci.*, *1:* 279–286 (1970).
31. E. M. Sparrow, C. W. Hixon and G. Shavit, Experiments on Laminar Flow Development in Rectangular Ducts, *J. Basic Eng.*, *89:* 116–124 (1966).
32. L. W. Pollack, Counting of Aitken Nuclei and Applications of the Counting Results, *Int. J. Air. Pollut.*, *1:* 293–306 (1959).
33. A. C. Chamberlain, W. J. Megaw and R. D. Wiffen, Role of Condensation Nuclei as Carriers of Radioactive Particles, *Geofis. Pura e Appl.*, *36:* 233–242, (1957).
34. J. C. Gallimore and T. T. Mercer, *The Particulate State of Fission Products Released from Irradiated Uranium when Heated in Air*, USAEC Rep. LF–9, 1963.
35. L. W. Pollack and A. L. Metnieks, On the Determination of the Diffusion Coefficient of Heterogeneous Aerosols by the Dynamic Method, *Geofis. Pure Appl.*, *41:* 211–217 (1957).
36. A. L. Metnieks, On the Various Methods of Deducing Aerosol Size Distribution from Diffusional Decay Measurements, *Pure Appl. Geophys.*, *61:* 183–190 (1965).
37. N. A. Fuchs, I. B. Stechkina and V. I. Starosselskii, On the Determination of Particle Size Distribution in Polydisperse Aerosols by the Diffusion Method, *Brit. J. Appl. Phys.*, *13:* 280–282 (1962).
38. F. M. Lea and R. W. Nurse, Permeability Methods of Fineness Measurement, Symposium on Particle Size Analysis, *Trans. Inst. Chem. Eng.*, *25:* 47–56 (1947).
39. J. R. Hodkinson, The Optical Measurement of Aerosols, in C. N. Davies (Ed.), *Aerosol Science*, p. 307, Academic Press, London and New York, 1966.

40. S. J. Gregg, Adsorption and Heat of Wetting Methods of Measuring Surface Area, *Trans. Inst. Chem. Eng., 25:* 40–46 (1947).
41. I. Langmuir, The Adsorption of Gases on Plane Surfaces of Glass, Mica, and Platinum, *J. Amer. Chem. Soc., 40:* 1361–1403 (1918).
42. S. Brunauer and P. H. Emmett, The Use of van der Waals Adsorption Isotherm in Determining the Surface Area of Iron Synthetic Ammonia Catalysts, *J. Amer. Chem. Soc., 57:* 1754–1755 (1935).
43. S. Brunauer, P. H. Emmett, and E. Teller, Adsorption of Gases in Multimolecular Layers, *J. Amer. Chem. Soc., 60:* 309–319 (1938).
44. G. Sandstede and E. Robens, Gravimetric Determination of Gas Adsorption with an Electronic Microbalance, *Chemie–Ingen. Technol., 32:* 413–417 (1960).
45. A. L. McClellan and H. F. Harnsberger, Cross-Sectional Area of Molecules Adsorbed on Solid Surfaces, *J. Colloid Interface Sci., 23:* 577–599 (1967).
46. K. E. Lauterbach, Specific-Surface Determinations of Uranium Dusts by Low-Temperature Adsorption of Ethane, *J. Franklin Inst., 250:* 13–24 (1950).
47. L. A. Wooten and C. Brown, Surface Area of Oxide Coated Cathodes by Adsorption of Gas at Low Pressures, *J. Amer. Chem. Soc., 65:* 113–118 (1943).
48. F. G. Nelsen and F. T. Eggertsen, Determination of Surface Area. Adsorption Measurements by a Continuous Flow Method. *Anal. Chem., 30:* 1387–1390 (1958).
49. M. Corn, T. L. Montgomery and N. A. Esmen, Suspended Particulate Matter: Seasonal Variation in Specific Surface Areas and Densities, *Environ. Sci. Technol., 5:* 155–158 (1971).
50. J. Kozeny, Über kapillare Leitung des Wassers in Boden, Sitz. Ber. Akad. Wiss. Wien, *Math-Naturwiss., 136:* 271–306 (1927).
51. P. C. Carman, Fluid Flow through Granular Beds, *Trans. Inst. Chem. Eng., 15:* 150–166 (1937).
52. F. M. Lea and R. W. Nurse, The Specific Surface of Fine Powders, *J. Soc. Chem. Ind. 58:* 277–283 (1939).
53. E. L. Gooden and C. M. Smith, Measuring Average Particle Diameter of Powders, *Ind. Eng. Chem. (Anal. ed.), 12:* 479–482 (1940).
54. S. S. Ober and K. J. Frederick, A Study of the Blaine Fineness Tester and a Determination of Surface Area from Air Permeability Data, *ASTM Spec. Publ. No. 234*, p. 279–285, 1958.
55. P. C. Carman and J. C. Arnell, Surface Area Measurements of Fine Powders Using Modified Permeability Equations, *Can. J. Res., A26:* 128–136 (1948).
56. K. R. Schrag and M. Corn, Comparison of Particle Size Determined with the Coulter Counter and by Optical Microscopy, *Amer. Ind. Hyg. Ass. J., 31:* 446–453 (1970).
57. T. A. Allen, A Critical Evaluation of the Coulter Counter, in *Particle Size Analysis*, Soc. Anal. Chem. London, 1967.
58. R. H. Berg, Electronic Size Analysis of Subsieve Particles by Flowing through a Small Liquid Resistor, *ASTM Spec. Publ. No. 234*, p. 245–255, 1958.
59. R. K. Eckoff, A Static Investigation of the Coulter Principle of Particle Sizing, *J. Sci. Instrum. Ser. 2, 2:* 973–977 (1969).
60. F. G. Jacobus, cited in Ref. 39.
61. M. Wales and J. N. Wilson, Theory of Coincidence in Coulter Particle Counters, *Rev. Sci. Instrum., 32:* 1132–1136 (1961).
62. L. H. Princen and W. F. Kwolek, Coincidence Corrections for Particle Size Determinations with the Coulter Counter, *Rev. Sci. Instrum., 36:* 646–653 (1965).
63. I. C. Edmundson, Coincidence Error in Coulter Counter Particle Size Analysis, *Nature (London), 212:* 1450–1452 (1966).

64. C. F. T. Mattern, F. S. Brackett and B. J. Olson, Determination of Number and Size of Particles by Electrical Gating: Blood Cells, *J. Appl. Physiol., 10:* 56–70 (1957).
65. S. Barnes, D. C.-H. Chen and H. R. Yarde, The Analysis of Coulter Counter Data, *Brit. J. Appl. Phys., 17:* 1501–1506 (1966).
66. J. Bol, J. Gebhart, W. Heinze, W.-D. Petersen, and G. Wurzbacher, Ein Streulicht-Teilchengrössenspektrometer fur submicroskopische Aerosole hoher Konzentration, *Staub, 30:* 475–479 (1970).
67. A. E. Martens and D. D. Doonan, Comments on: Influence of Refractive Index on the Accuracy of Size Determination of Aerosol Particles with Light-Scattering Aerosol Counters, *Appl. Opt., 9:* 1930–1931 (1970).
68. H. Quenzel, Reply to Comments by Martens and Doonan, *Appl. Opt., 9:* 1931 (1970).
69. J. W. Thomas, Particle Loss in Sampling Conduits, in *Assessment of Airborne Radioactivity*, IAEA, Vienna, 1967.
70. D. Sinclair and L. Hinchliffe, Production and Measurement of Submicron Aerosols. II., in T. T. Mercer, P. E. Morrow, and W. Stöber (Eds.), *Assessment of Airborne Particles*, Thomas, Springfield, Illinois, 1972.
71. A. B. Clarke and R. L. Disney, *Probability and Random Processes for Engineers and Scientists*, Wiley, New York, 1970.
72. B. V. Derjaguin and G. Ja. Vlasenko, Flow-Ultramicroscopic Method of Determining the Number Concentration and Particle Size Analysis of Aerosols and Hydrosols, *J. Colloid Sci., 17:* 605–627 (1962).
73. J. F. Pisani and G. H. Thomson, Coincidence Errors in Automatic Particle Counters, *J. Phys. E. Sci. Instrum., 4:* 359–361 (1971).
74. T. A. Rich, Apparatus and Method for Measuring the Size of Aerosols, *J. Rech. Atmos., 2:* 79–85 (1966).

8

Respirable Activity Samplers

Introduction

It has been known for many years that certain airborne particles, by virtue of their size, are not a threat to health, although other particles of the same composition may present a serious hazard. For this reason, a continuing effort has been made to develop methods with which particles that are relevant to hazard could be assessed separately from those that are not. At first, separation could be achieved only by examining samples with the optical microscope, limiting the analysis to particles smaller than some specified size. The early studies of inhalation hazards in the dusty trades, from which the remarkable correlations between exposure and occurrence of silicosis shown in Fig. 1.5 were derived, reported levels of dustiness in terms of the number concentration of particles smaller than 10 μm in diameter. Following a study of airborne particles in British anthracite coal mines, Bedford and Warner [1] recommended that standards be based on the mass concentration of particles $\leqslant 5$ μm in diameter. Recognizing, however, that such concentrations could not be measured gravimetrically, they further recommended that the criterion of hazard be the number concentration of particles having diameters $\geqslant 1$ μm. Their studies had shown that an adequate estimate of the former could be made from a measured value of the latter. Total mass concentration, however, was poorly correlated with the size-limited mass concentration and could not be used as a criterion of hazard. Thus, early definitions of hazard were made in terms of the number concentration of "respirable" particles in air, even though their mass concentration was believed to be the proper criterion of hazard.

The theoretical studies of Findeisen [2] and Landahl [3] and the experimental studies of Van Wijk and Patterson [4], Landahl and his associates [5, 6], and Brown *et al.* [7] made it possible to define respirable particles in quantitative terms. Davies [8] used the results of these and other studies to get a curve of alveolar retention as a function of aerodynamic diameter. He then described the performance characteristics of a horizontal elutriator that would permit the collection of a sample in which particles could be represented approximately in proportion to their probability of deposition in the alveolar regions of the lung. The penetration curve of the elutriator described by Davies was adopted by the British Medical Research Council [9] as the definition of the respirable fraction of airborne dust. They recommended that only the respirable fraction be used for compositional analysis or for the estimation of concentrations by mass or by surface area. The Pneumoconiosis Conference in Johannesburg in 1959 adopted the same definition of respirable activity. They concluded that the best single parameter to describe health risk was the mass concentration of respirable dust in the case of coal and the surface area concentration of respirable dust in the case of quartz [10].

In the U.S., the impetus to develop respirable activity samplers came from those concerned with inhalation hazards in the atomic energy industry, where standards of air contamination were expressed in terms of the gross concentration of radioactivity or radionuclide mass in the air. Harris and Eisenbud [11] pointed out that in deriving these MPC's it was assumed that radiation dose from insoluble particles was distributed uniformly throughout the lung. Radiation of the upper respiratory tract by the large, highly active particles being cleared from the lung with a half-time of less than a few days was ignored. Following a recommendation by Burnett [12], they suggested the use of a two-stage sampler, the first stage collecting particles with an efficiency comparable to that of the upper respiratory tract, which they considered to be made up of those regions that are cleared by ciliary action. Their interpretation of the available experimental data contained significant errors that, fortuitously, enabled them to derive three curves, in good agreement with each other, which showed deposition in the upper respiratory tract as a function of diameter for particles of density 2.6 g/cm^3. Their observation that these curves were closely approximated by efficiency curves for the Design 2 Aerotec Tube, a high-volume cyclone sampler calibrated by Dennis *et al.* [13] for particles of the same density, introduced the cyclone as a respirable activity sampler.

Although concern about the irradiation of the upper respiratory tract waned, interest in respirable activity samplers quickened as it became increasingly evident that the fixed respirable fraction built into the definitions of maximum permissible concentrations was often grossly in error for radioactive dusts encountered in practice [14]. The development of cyclone samplers was advanced, particularly by Hyatt [15] and by Lippmann and Harris [16], with emphasis on the production of a family of instruments capable of operating at a variety of flow rates and under a variety of field conditions. Of special significance was the development of cyclones small enough to be used as personal samplers. The collection efficiency characteristics of all of these samplers can be made to conform closely to the criteria adopted at a meeting at Los Alamos Scientific Laboratory sponsored by the AEC Office of Health and Safety, January 18–19, 1961. These criteria can barely be distinguished from those adopted subsequently by the ACGIH [17] and the U.S. Department of Labor [18].

A respirable activity sampler should function in such a way that the probability $P(D)$ that an incoming particle of diameter D will be included in the respirable fraction is proportional to the probability $P'(D)$ that an inhaled particle of the same diameter would be deposited in the alveolar regions of the lung:

$$P(D) = \text{constant} \cdot P'(D). \tag{8.1}$$

Since P' is a function of various physiological and anatomical parameters, the definition of respirable activity must be, to some extent, arbitrary. Usually, a standard breathing pattern is assumed and the alveolar regions are assumed to include the respiratory bronchioles. No single-stage device can provide the desired sample because it must collect particles according to the same physical characteristics that bring about their deposition in the lung, which is, itself, a multistage sampler. The simplest instruments that approximate the desired sampling conditions are two-stage samplers in which the collection efficiency of the first stage is defined by

$$E(D) \simeq 1 - P(D),$$

and the second stage, which contains the respirable activity, collects everything that penetrates the first stage.

The basic advantage derived from measuring respirable activity (which is not necessarily mass) as opposed to measuring number concentration is that the former permits the direct determination of the concentration of toxic agent that is subject to alveolar deposition. Other significant advantages that have been cited for it, however, include [19]: the ability to sample over a full shift; the ability to use personal samplers and obtain truer "breathing zone" exposures; the greater possibility for standardizing weight determinations as contrasted with optical count; and a much lower cost per determination of weighted exposure to dust. It appears to be generally accepted that when "activity" means mass, gravimetric analysis is a much simpler operation than particle counting and requires little time or training [20]. In this connection, it is of some interest that Bloomfield and DallaValle [21], when pointing out the high correlation between total mass concentration and number concentration encountered in the surveys of the dusty trades, stated that the count method was used because general experience showed that it was less time-consuming than was the determination of mass concentrations. Morse [22] recently made a similar point in a discussion of some of the shortcomings of respirable activity sampling.

It is important to keep in mind that sampling for respirable activity is applicable only to those particles that, when deposited in the upper respiratory tract, are cleared before they undergo any significant dissolution. If this is not the case, or if the particles tend to dissolve in the gastrointestinal tract, then the total activity concentration must be measured.

8-1 Definitions of Respirable Activity

Since the collection characteristics of the two samplers most widely used to estimate respirable activity were based on the experimental studies of Brown *et al.* [7], some attention must be given first to their results. Their subjects were placed in a Drinker respirator, which controlled their rate of breathing and maintained a fairly constant tidal volume of about 700 cm^3. They inhaled china clay particles (density $= 2.6$ g/cm^3) that had been dispersed from suspension in water by a high-pressure atomizer. The size distributions of the suspended particles were lognormal, having geometric standard deviations in the range 1.20–1.25. The measured size was almost certainly Feret's diameter. The subject's exhalation was serially separated into seven fractions, each of which was analyzed for CO_2 and particle concentration. The amounts of lung air and upper respiratory tract air in each fraction were calculated from the CO_2 concentrations relative to the alveolar CO_2 concentration, which was measured at the beginning and end of each experiment. With these data and the measured number of particles in each fraction, they calculated the average

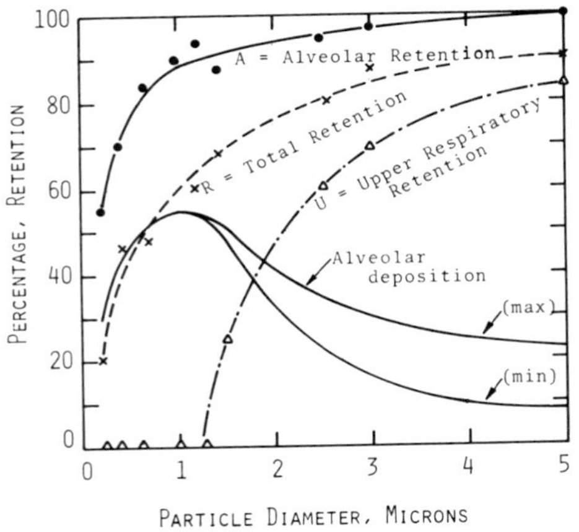

FIG. 8.1. Experimental deposition curves of Brown *et al.* [7].

particle concentration in alveolar and upper respiratory tract air, from which they estimated the total deposition R, deposition U from upper respiratory tract air (i.e., air confined to the anatomical dead space), and deposition A in the alveolar region of particles reaching that region. The calculations contained the implicit assumption that CO_2 and particles are partitioned identically between dead space air and lung air. Their results are shown in Fig. 8.1. The abscissa is probably the median Feret's diameter of the test dust. They estimated alveolar deposition, that is, the number of particles deposited in the alveolar region relative to the number inhaled, as

$$\text{alveolar deposition} = AK(1-U)^{1/2},$$

where $K = (CO_2$ output per breath)/(alveolar CO_2 concentration $\times$ tidal volume). They found $K = 0.6$ on the average. Their alveolar deposition curves are included in Fig. 8.1. The curve labeled "minimum" is derived when deposition U is assumed to occur entirely during inspiration; the other curve is derived when upper respiratory tract deposition is equal in both directions.

The authors interpreted $(1-K) \times$ tidal volume as the anatomical dead space, so that the alveolar regions were intended to include the respiratory bronchioles. For $K = 0.6$, the anatomic dead space was > 250 cm^3. However, the same data were used later by Hatch *et al.* [23] to demonstrate that the anatomical dead space of the subjects was, on the average, 130 cm^3 and was independent of tidal volume. If the definition of alveolar region were maintained, this value of anatomical dead space would alter the calculated alveolar deposition by increasing K and decreasing A, which would have the effect of increasing alveolar deposition of larger particles relative to the peak deposition observed for particles of about 1 μm diameter.

8-1.1 "Respirable Fraction"—BMRC

In the paper referred to above, Davies accepted the "maximum" curve of Fig. 8.1 as defining alveolar deposition of particles having a density of 2.6 g/cm^3. He adjusted the curve for density, but not for shape factors, to obtain the curve shown in Fig. 8.2 in which deposition is put in terms of particles of unit density. He also included the respiratory bronchioles in the alveolar region. Arguing that there was little point in trying to match the alveolar deposition curve at small particle sizes, since their contribution to the respirable mass would be small, he recommended that the first stage of a respirable activity sampler be a horizontal elutriator having the characteristics shown by the solid curve in Fig. 8.2. Davies felt that an elutriator would not cut off quite so sharply in practice so that the match to the relative alveolar deposition curve would be even better than it appears in the figure. On this basis, the proportionality factor between the respirable fraction collected by the sampler and alveolar deposition would be approximately constant at about 1.8.

Shortly after the BMRC Panels [9] had recommended that the respirable fraction be defined by the equation proposed by Davies for an elutriator sampler, Watson [24] showed that the aerodynamic diameters of china clay particles were actually smaller than Feret's diameter by the factor 0.87 when

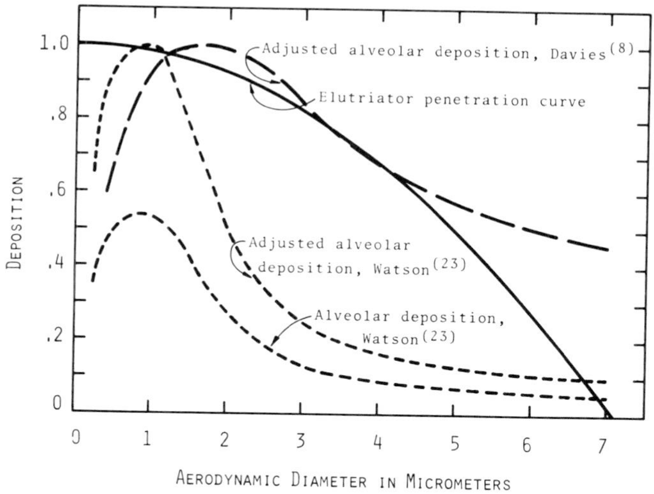

Fig. 8.2. Alveolar deposition curves based on Fig. 8.1.

due allowance was made for both particle density and shape. His curve for alveolar deposition according to the data of Brown *et al.* is included in Fig. 8.2 with its normalized counterpart. The latter does not compare favorably with the BMRC curve.

8-1.2 "RESPIRABLE" DUST—AEC OHS

At the meeting at Los Alamos, respirable dust was defined as "that portion of the inhaled dust which is *deposited* in the non-ciliated portions of the lungs" [16], a definition in keeping with those of Davies and of Brown *et al.* It was described quantitatively as tabulated below.

Aerodynamic diameter (μm)	Percent respirable
10	0
5	25
3.5	50
2.5	75
2.0	100

The collection efficiency curve of a device which would pass particles in this proportion is shown in Fig. 8.3. The curve was based principally on the work

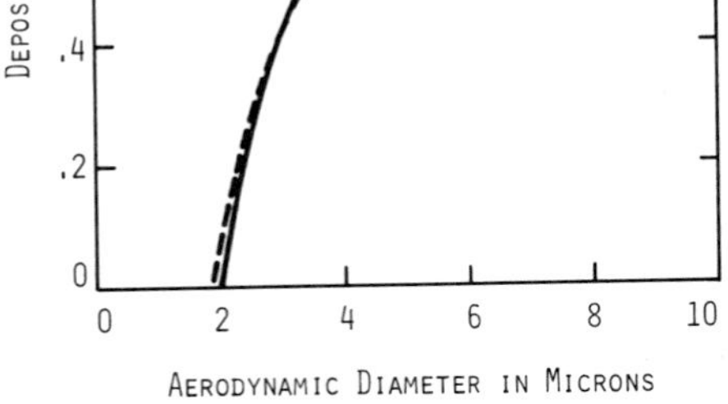

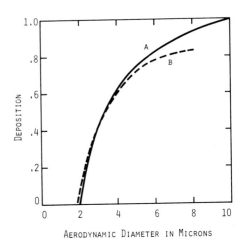

FIG. 8.3. Deposition as a function of aerodynamic diameter. (A) AEC efficiency curve for a device that would pass only respirable dust [16]. (B) Upper respiratory tract deposition. Data of Brown *et al.* [7] adjusted for density but not for shape.

of Brown *et al.*, the dotted line in the figure being their curve of upper respiratory tract deposition corrected for density but not for shape. Since *U* refers to air confined to the upper respiratory tract, the particles collected behind such a sampler do not fit the definition of respirable dust given above, but are better described as "that portion of the inhaled dust which penetrates to the non-ciliated portions of the lung" [25].

8-1.3 "Respirable Activity"—ICRP Task Group on Lung Dynamics

The Task Group on Lung Dynamics [26] divided the lung into three compartments: nasopharyngeal, tracheobronchial, and pulmonary. The pulmonary compartment corresponded to the previously defined alveolar regions in that it included the respiratory bronchioles. Deposition in the nasopharyngeal compartment was calculated from the empirical equation due to Pattle [27]; deposition beyond the entrance to the trachea was calculated using Findeisen's model and method. Calculations were made for 15 respirations per minute and for tidal volumes of 750, 1450, and 2150 cm^3. These curves were intended to be used with samplers that are capable of estimating activity median aerodynamic diameters; however, it will be shown below that they are applicable to respirable activity sampling. The Task Group's curve for pulmonary deposition (respirable activity) at 1450 cm^3 is shown in Fig. 8.4, with an adjusted form of it obtained by multiplying each abscissa value by 3.33. The BMRC and AEC curves are shown in the same figure, also.

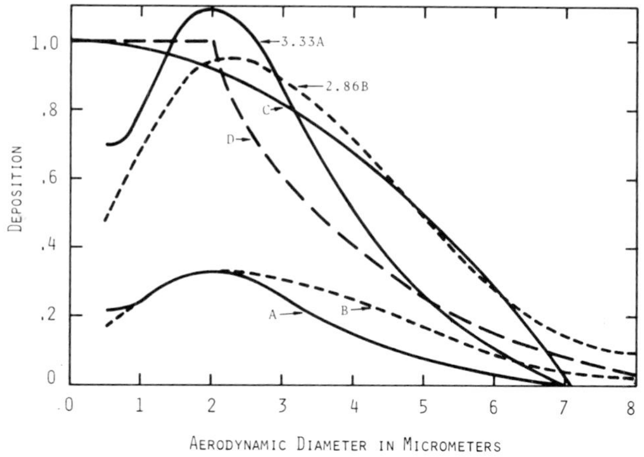

FIG. 8.4. Deposition as a function of aerodynamic diameter. (A) Pulmonary deposition at 15 respirations/min, 1450 cm^3 tidal volume, nasal breathing, Task Group on Lung Dynamics [26]. (B) Alveolar deposition at 20 respirations/min, 1000 cm^3 tidal volume, mouth breathing, Beeckmans [28]. (C) Elutriator penetration curve. (D) Cyclone penetration curve.

8-1.4 "Alveolar Deposition"—Beeckmans

Beeckmans' [28] definition of respirable activity lacks whatever virtue one acquires when endorsed by a quasiofficial organization. It is included here for two reasons:

(1) Beeckmans calculated alveolar deposition for aerosols of several different size distributions and compared the results with the respirable fraction of the same aerosols according to the BMRC and AEC samplers; and

(2) His results permit a comparison between deposition curves based on different assumed lung structures and breathing patterns.

Beeckmans used Weibel's model A [29], defining his standard breathing pattern as 20 respirations per minute at a tidal volume of 1000 cm^3 with 0.3 sec pauses at the end of inspiration and expiration. His model, unlike the others, was based on mouth breathing and did not include the respiratory bronchioles as part of the alveolar region. His alveolar deposition curve is shown in Fig. 8.4, with an adjusted deposition curve obtained by multiplying each point by 2.86. Had Beeckmans included the respiratory bronchioles in his alveolar region, his deposition values would have been higher at all points on the curve; the inclusion of nasal breathing, on the other hand, would have reduced deposition for diameters above about 1.5 μm, with the effect being greater at larger sizes.

8-1.5 "Respirable Fine Dust"—Breuer

Breuer [30] adopted the curve for lung dust based on Cartwright's [31] analysis of the residual particles in the lungs of deceased miners. In this case, the definition is related not to the probability that particles of a given size will deposit in the alveolar regions, but to the combined probability that they will be deposited and will be retained for a long period. The curve is shown

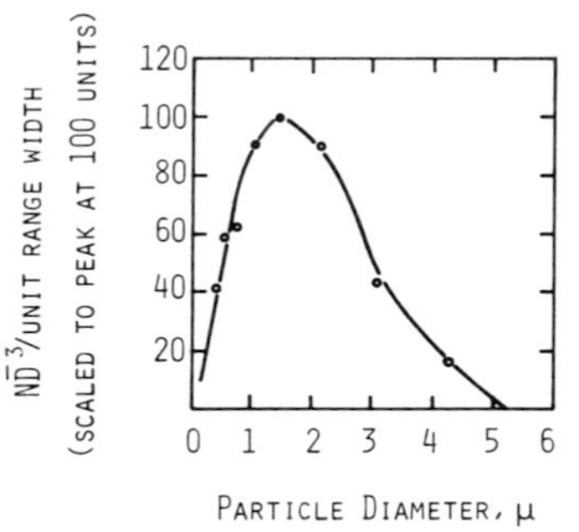

FIG. 8.5. Retention of quartz particles as a function of Stokes diameter [31].

in Fig. 8.5. Breuer has also developed high flow rate (200 liters/min) cyclone samplers which permit collection of fine particles in proportion to their respirability as he has defined it.

8-2 Operating Characteristics of Respirable Activity Samplers

8-2.1 Horizontal Elutriators

It was shown in Chapter 6 that particles having a terminal settling velocity U_G penetrate a horizontal elutriator according to

$$P = 1 - U_G \cdot A/F, \tag{8.2}$$

where A is the floor area of the duct and F is the volumetric flow rate through it. The critical settling velocity, i.e., the lowest velocity at which no penetration will occur, is

$$U_G = F/A.$$

To satisfy the BMRC definition of respirable fraction, the value of this critical velocity must be 0.15 cm/sec, for which the corresponding cut-off diameter is 7.1 μm. The conditions assumed when the equation for penetration was derived require that flow be laminar and that the elutriator be made up of similar flow tubes. The latter requirement cannot be rigorously satisfied because the air velocity goes to zero at the duct sides as well as at the floor and ceiling. The error is negligible, however, if the duct width is very much greater than its height. It is also desirable to have a large length-to-width ratio [32] and a relatively large pressure drop through the duct. Both of these conditions tend to minimize any effects of ambient air conditions on the flow pattern within the elutriator. On the other hand, in the effort to achieve a satisfactory pressure drop, it is not wise to make the channels too narrow because re-entrainment of deposited particles may occur. Hamilton and Walton [32] consider that this becomes a problem when the ratio of mean air velocity to duct height reaches a critical value that they put in the range 240–650 sec^{-1}.

(a) *The Hexhlet.* The horizontal elutriator designed by Wright [33] is shown schematically in Fig. 8.6(A). It was the first developed for the purpose of measuring respirable fraction as defined by the BMRC. It comprises 118 rectangular ducts formed of 0.5 cm thick aluminum plate and stacked in two banks. Each duct is 0.08 cm high by 3.55 cm wide by 25.1 cm long, providing a total floor area of 10,477 cm^2. An additional area between the end of the ducts and the critical orifice opening into the back-up soxhlet filter brings the total area for deposition to 10,903 cm^2. The instrument was designed to operate at 100 liters/min, giving a value of $F/A = 0.153$ cm/sec.

Suction is provided by a compressed air ejector, and flow rate is regulated by the critical orifice. The latter was designed with a tapered outlet, permitting a recovery of pressure head sufficient for sonic speeds to be achieved at an applied pressure drop of 100 mm Hg.

Wright stressed the importance of maintaining close tolerances on the spacing between plates. At a given pressure drop across the ducts, flow is proportional to the cube of the duct height, and small differences in spacing can cause disproportionately large effects on the sampling characteristics. Hamilton and Walton [32] demonstrated that such differences would bring about a tailing effect in the penetration curve of the elutriator.

For field applications, the Hexhlet serves primarily as a general area sampler, but even for that purpose it is inconvenient in many cases because it requires a source of compressed air. Its ratio of mean air speed to duct height is about 625 sec^{-1}, which is near the upper limit of the range within which redispersion becomes a problem. This problem has been overcome in a modified version, available commercially, that operates at 50 liters/min. It has 26 ducts, approximately 0.22 cm high and 7.6 cm wide, giving a velocity-to-height ratio of about 90 sec^{-1}, well below the range of critical values cited above. The pressure drop through the elutriator is very much less than that through the 100 liters/min instrument.

(b) *The MRE Elutriator* [35]. A diagram of this instrument is shown in Fig. 8.6(B). It has four ducts 0.238 cm high by 4 cm wide by 17.19 cm long and operates at 2.5 liters/min. Flow is provided by a reciprocating diaphragm pump driven by a battery-powered motor, which makes its use independent of local utilities. The respirable activity is collected on a 5.5 cm diameter glass fiber filter and, under normal operating conditons, should be readily detected by a balance sensitive to 0.01 mg.

The instrument has been calibrated in a wind tunnel under a variety of conditions with respect to wind speed and direction relative to the position of the sampling inlet. It was found that restriction of the duct heights at the inlet, as shown in the figure, permitted sampling either into the wind or cross-wind without detectable error at air speeds up to 1000 ft/min (~11 mph). At 2000 ft/min, the respirable fraction is overestimated by 20% when sampling into the wind and underestimated by 15% when sampling cross-wind.

(c) *The Long-Running Thermal Precipitator* [37]. This is the standard thermal precipitator fitted with a small elutriator at the inlet as shown in Fig. 8.6(C). It was intended not only to limit the precipitator sample to the respirable fraction, but also to lower the density of the deposit to reduce the effect of overlap.

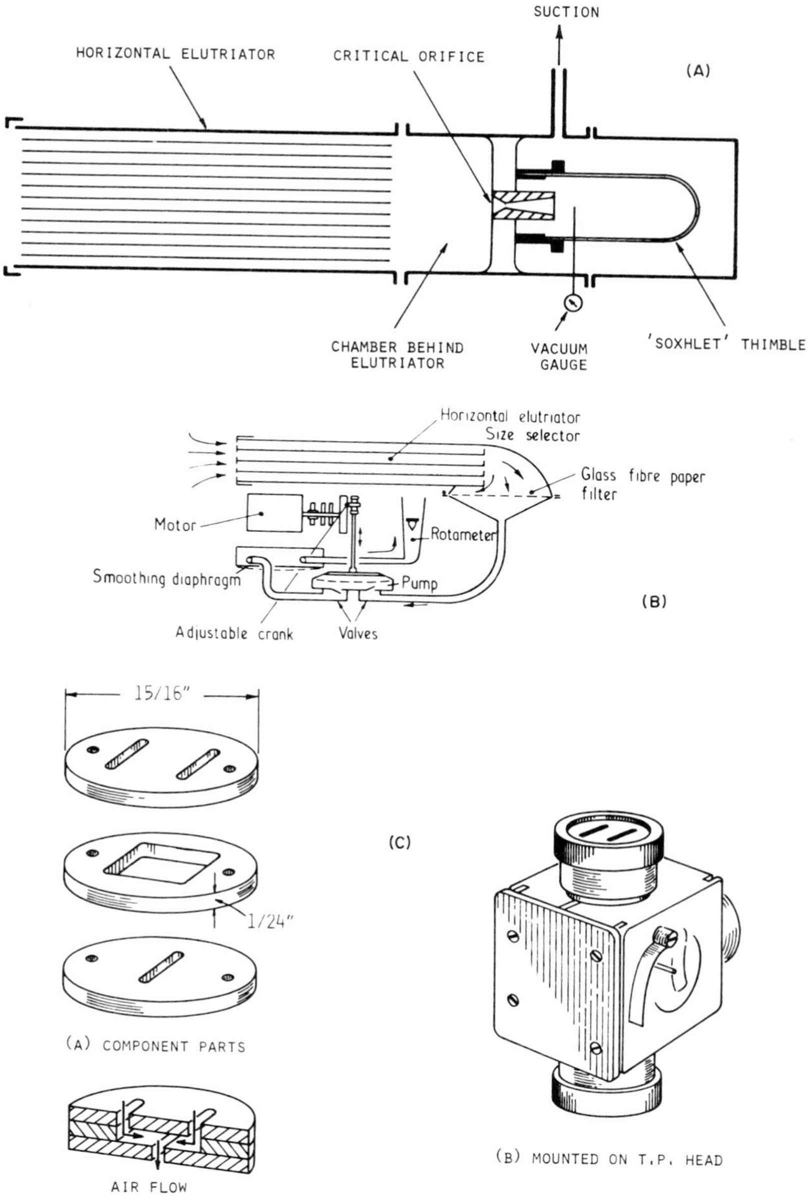

FIG. 8.6. Schematic diagrams of some horizontal elutriators. (A) The Hexhlet [33, 34]; courtesy of American Industrial Hygiene Association. (B) The MRE [35]; courtesy of The Institute of Physics. (C) The long-running thermal precipitator [36]; courtesy of The Institute of Physics and The Physical Society.

(d) *Experimental Penetration Curves of Horizontal Elutriators.* The instruments described above should all estimate a common theoretical curve when penetration is plotted as a function of the ratio D_A/D_{AC}, where D_{AC} represents the aerodynamic diameter of the particle corresponding to the cutoff settling velocity. A plot of this type, prepared from published data, is given in Fig. 8.7. All of the data refer to spherical particles of known density. In some cases the particles were monodisperse. When data were presented as histograms, the penetration was plotted against the relative aerodynamic diameter at the midpoint of the corresponding bar. The results bear out Davies' contention [8] that a real elutriator would not show the sharp cutoff that is predicted throretically. However, the effect is not enough to cause any marked improvement in the agreement between respirable fraction as defined by the elutriator and the adjusted alveolar deposition curve (see Fig. 8.2). Equally significant is the fact that the data generally fall below the expected values for penetrations greater than about 50%.

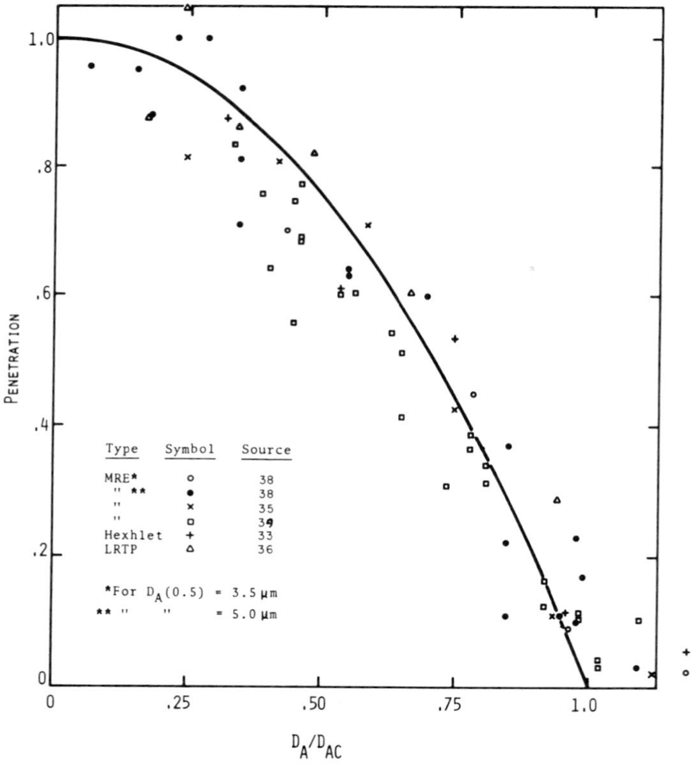

FIG. 8.7. Experimental data on penetration of horizontal elutriators.

8-2.2 Vertical Elutriators

Lynch [40] developed the vertical elutriator, shown schematically in Fig. 8.8, specifically to obtain a lint-free sample of respirable cotton dust. It is operated as a stationary, general air sampler. At 7.4 liters/min the average linear air

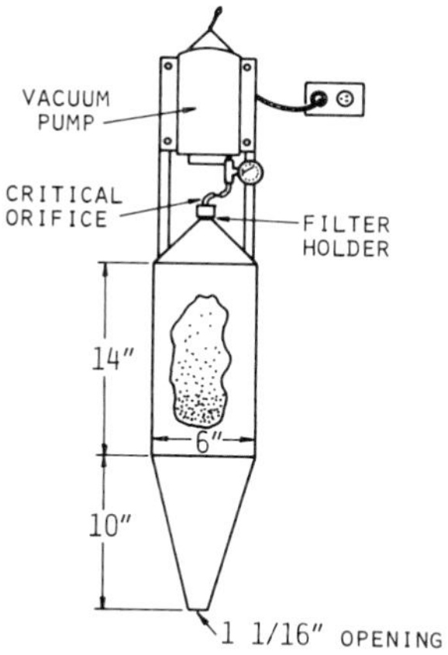

FIG. 8.8. Vertical elutriator cotton dust sampler [40]. Courtesy of J. Lynch.

velocity in the cylindrical section is 0.675 cm/sec, which corresponds to the settling speed of a particle having an aerodynamic diameter equal to 15 μm. Under these conditions [41], the penetration of particles through the elutriator section will be

$$P = 1 - (D_A^2/225).$$

Some values of P are given below.

D_A (μm)	1.0	2.5	5.0	7.5	10.0	12.5
P	0.996	0.972	0.894	0.750	0.556	0.305

8-2.3 Cyclone Separators

Figure 8.9 is a schematic diagram of a conventional cyclone used for air cleaning. The width of the inlet is shown equal to the difference between the radii of the cyclone body and the outlet tube, but this is not necessarily the case. The incoming air spirals down the periphery of the cyclone, changes direction and spirals upward into the exhaust tube. Airborne particles are thrown to the outer edge of the spiral and may be deposited on the cyclone wall from which they subsequently fall into the dust collector. Equations

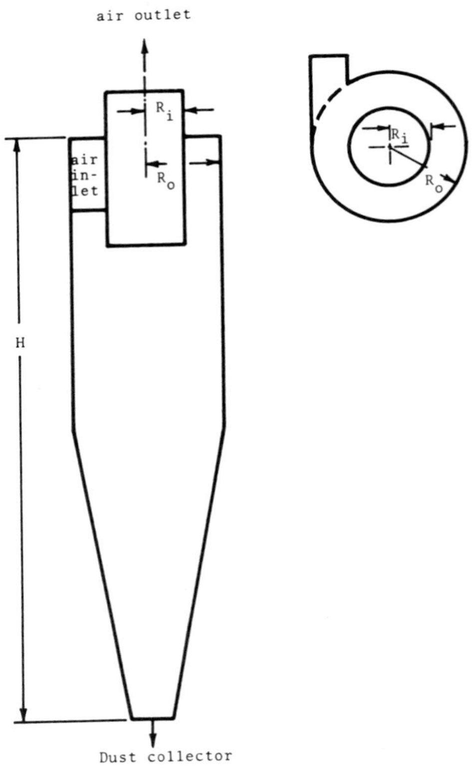

FIG. 8.9. Schematic diagram of conventional cyclone.

describing the collection efficiency of a cyclone have been derived by Rosin *et al.* [42] and by Davies [43], among others. From their equations, the aerodynamic diameter, $D_A(0.5)$, corresponding to a collection efficiency of 0.5 is given by

$$D_A^2(0.5) = (9\eta/2\rho_0 V_0)\cdot f, \tag{8.3}$$

where η is the coefficient of viscosity of air, V_0 is the average linear air velocity at the inlet, and

$$f = (R_0 - R_i)/\pi N \qquad \text{(Rosin } et\ al.\text{)}, \tag{8.3a}$$

or

$$f = R_0^{\,2}(1-\alpha^4)/H \qquad \text{(Davies)}. \tag{8.3b}$$

N is the effective number of turns (usually <5) made by the external spiral and $\alpha = 0.5 + (R_i/2R_0)$. A slip factor of unity has been assumed for all particles. Davies' equation has the advantage that all quantities in it can be determined directly, whereas the value of N in Rosin's equation must be estimated from experimental data.

Diagrams of several of the cyclones that have been used as respirable activity samplers are shown in Fig. 8.10. Pertinent dimensions and operating conditions are given in Table 8.1, with $D_A(0.5)$ as observed and as calculated using Eq. (8.3). The dimensions for the BAT cyclone were estimated from the diagram, which fortunately has a built-in scale, and are subject to greater error than the others. The derived diameters are correct within a factor of about 2, which is not bad considering that the theoretical equations are only approximate, even for the air-cleaning cyclones for which they were derived. Cyclone samplers operate at linear input velocities that are in the same range as those of cyclone air cleaners, but the dimensions of their inlet ports are so much smaller that the Reynolds number of the input air flow is usually in the region of laminar flow.

TABLE 8.1

DIMENSIONS AND OPERATING CONDITIONS OF SEVERAL CYCLONE SAMPLERS

Type	R_o (cm)	R_i (cm)	H (cm)	Flow rate (liters/min)	Flow rate V_0 (cm/sec)	D_A (0.5) according to Eq. (8.3b)	D_A (0.5) according to Expt.	Ref.
HASL 10 mm nylon	0.5	0.28	5	1.4	483	2.3	3.5	[48]
HASL 10 mm nylon	0.5	0.28	5	1.7	586	2.1	3.5	[39]
HASL 10 mm nylon	0.5	0.28	5	2.0	636	1.9	3.5[a]	[52]
HASL ½ in. steel	0.585	0.385[b]	5.2	8.0	175	4.9	3.5	[48]
HASL ½ in. steel	0.585	0.385	5.2	10.0	219	4.4	3.5	[39]
HASL ½ in. steel	0.585	0.385	5.2	13.0	285	3.9	—	[47]
BCIRA	0.475	0.315[b]	3.75	1.9	210	4.4	5.0	[44]
BAT[c]	1.5	1.0	11	200	2900	1.7	1.8	[30]

[a] Equivalent volume diameter for coal particles.
[b] $R_i = R_0 -$ width of inlet port.
[c] Dimensions estimated from diagram. Experimental D_A is for slate dust corrected for density but not for shape.

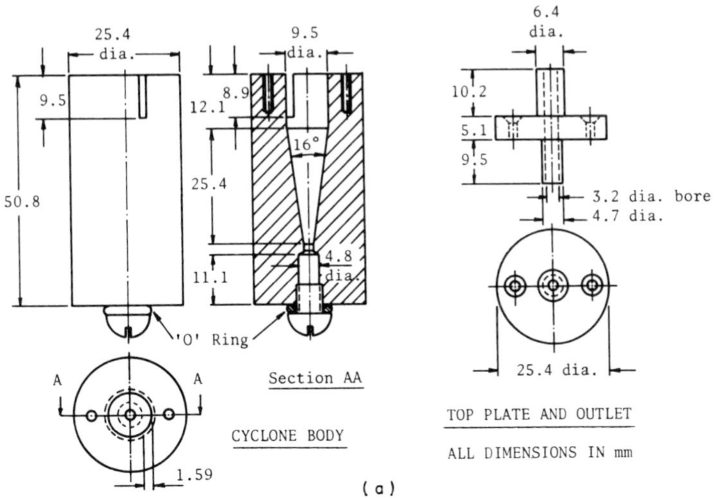

(a)

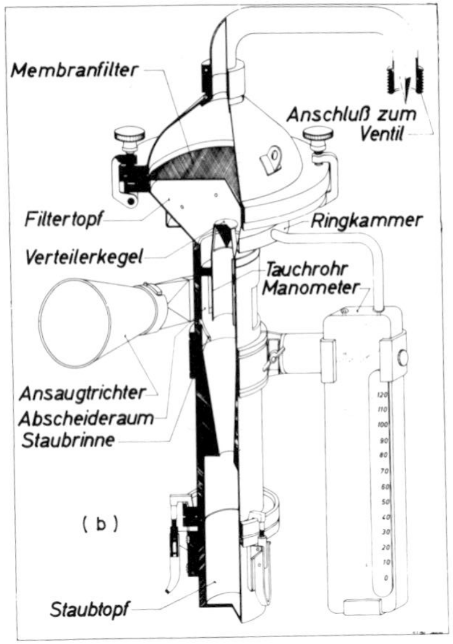

Membranfilter	= Membrane Filter
Filtertopf	= Filter Head
Verteilerkegel	= Baffle Cone
Ansaugtrichter	= Inlet Cone
Abscheideraum	= Separation Region
Staubrinne	= Dust Channel
Staubtopf	= Dust Container
Anschluss zum Ventil	= Valve Coupling
Ringkammer	= Annular Chamber
Tauchrohr Manometer	= Snorkel Manometer

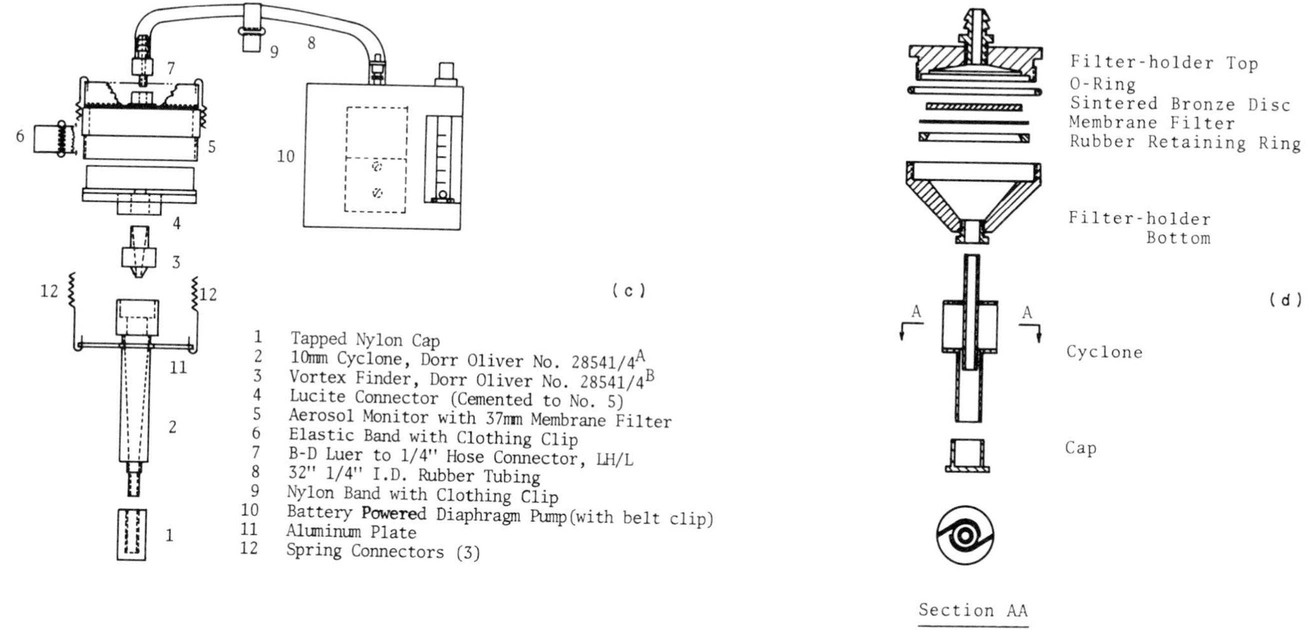

FIG. 8.10. Schematic diagrams of some cyclone samplers. (a) The BCIRA sampler [44]; courtesy of Pergamon Press, Book Division. (b) The BAT sampler [45]; courtesy of Dr. Hans Breuer. (c) The HASL 10-mm nylon sampler [46]; courtesy of the American Industrial Hygiene Association. (d) The HASL 1/2 in. steel sampler [47]; courtesy of the British Occupational Hygiene Society.

Equation (8.3b) can be generalized to calculate the diameter $D_A(E)$ corresponding to collection efficiency E, by setting $\alpha = 1 - E(1 - R_i/R_0)$. Figure 8.11a,b shows the theoretical efficiencies for cyclones having $R_i/R_0 = 0.67$ and 0.56, respectively, in addition to experimental data for some of the cyclones in Table 8.1. There is qualitative agreement between theory and experiment up to $D_A(E)/D_A(0.5) \simeq 1$, but the data points diverge rapidly from theory at higher values. Up to 1.8 liters/min, the data of Lippmann and Kydonieus [49] for charge-equilibrated, monodisperse particles (plotted in Fig. 8.11c) fit the theoretical curve much better and are generally more consistent than those shown in Figs. 8.11a,b. In all cases, however, the effect of flow rate has been

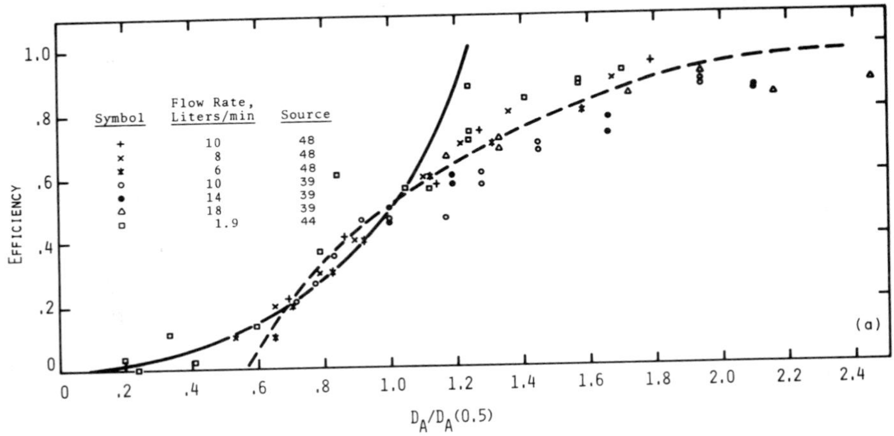

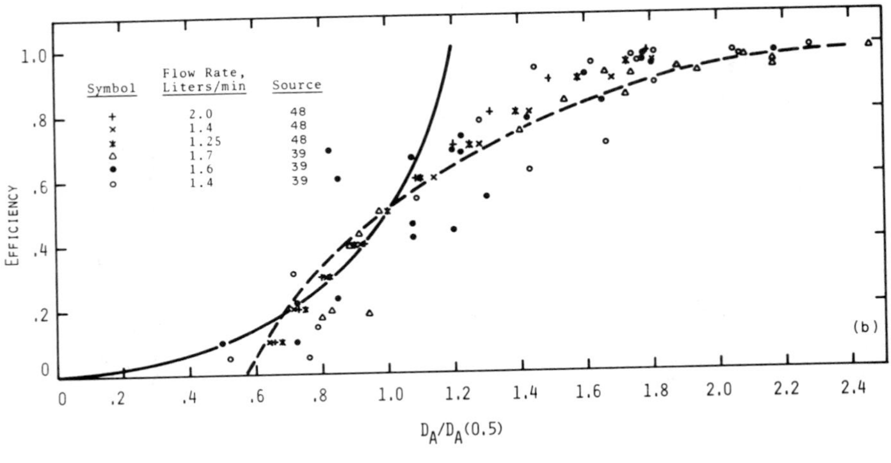

masked by the method of plotting. The theoretical equation predicts that for a given cyclone $D_A(0.5)$ should vary inversely with the square root of the flow rate. That this is not so is apparent from Fig. 8.12, where the available data for three of the cyclones have been plotted on a log–log scale. The data related to coal particles [52] have been adjusted for shape factor as suggested by Ettinger [39]. The results indicate that $D_A(0.5)$ is inversely proportional to flow rate for both the HASL 1/2 in. steel and 10-mm nylon cyclones. The BAT-1 cyclone, however, shows an inverse relationship between $D_A(0.5)$ and $F^{3/2}$.

The calibrations by Knuth [48] and by Ettinger [39] do not agree with respect to the flow rates at which the 1/2 in. steel and 10-mm nylon cyclones should be operated to meet the AEC criteria. Knuth recommends 8 and 1.4 liters/min respectively, whereas Ettinger recommends 10 and 1.7 liters/min. Both studies were made using monodisperse aerosols which were not charge-equilibrated, although the methods of generation could have produced particles with large charges. Knuth measured the aerodynamic diameters of his test aerosols using a sedimentation method; Ettinger calculated his from electron microscope measurements, a technique that Knuth considered unreliable. For the 10-mm cyclone, the fit to the data points of Fig. 8.12 shows that $D_A = 3.5\ \mu m$ when $F = 1.6$ liters/min. This is in better agreement with Ettinger's results than with Knuth's. Additional evidence pointed out below tends to favor Ettinger's recommended flow rate.

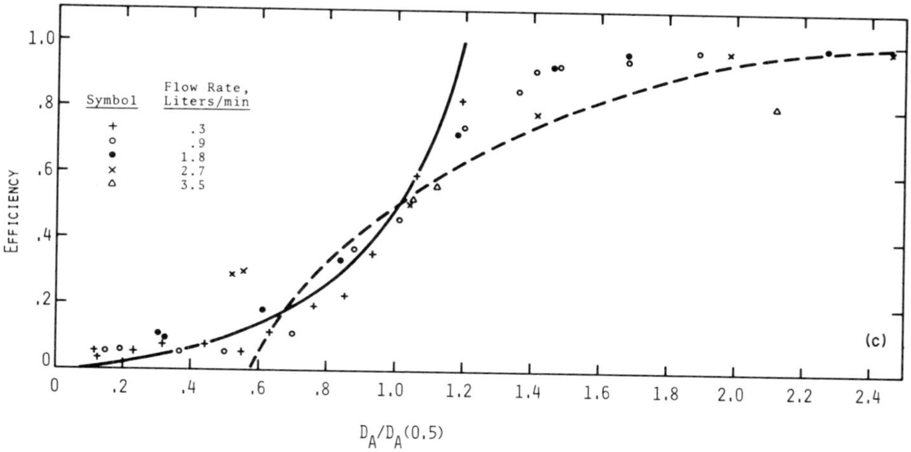

FIG. 8.11. (a) Cyclone collection efficiencies for $R_i/R_o = 0.67$. Solid curve, theory; dashed curve, AEC curve, Fig. 8.3. (b) Cyclone collection efficiencies for $R_i/R_o = 0.56$. Solid curve, theory; dashed curve, AEC curve, Fig. 8.3. (c) Cyclone collection efficiencies for $R_i/R_o = 0.56$. Solid curve, theory; dashed curve, AEC curve, Fig. 8.3. Data from Lippmann and Kydonieus [49].

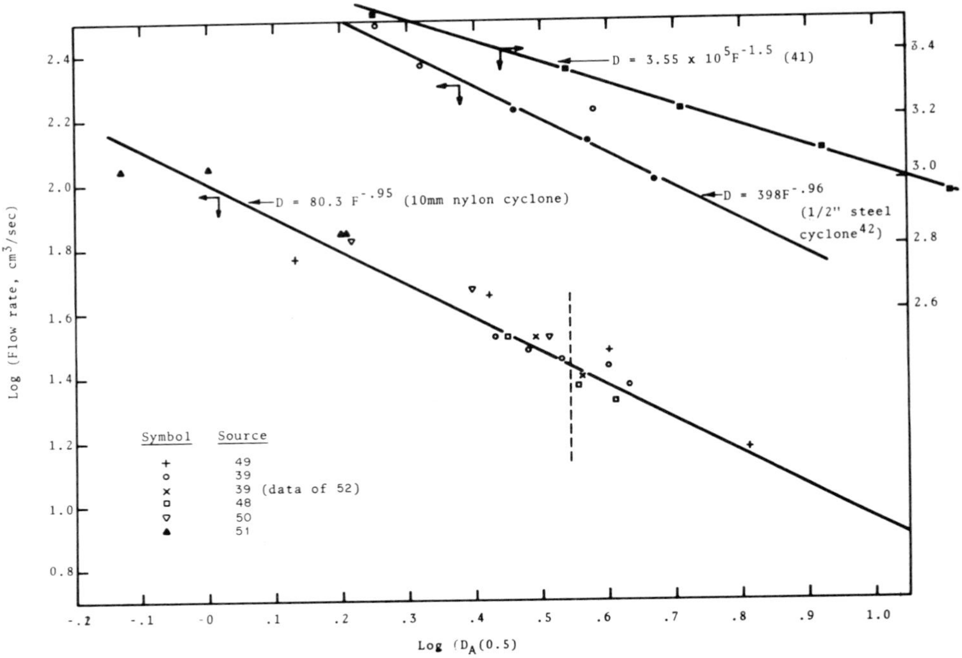

FIG. 8.12. Variation of $D_A(0.5)$ in μm with flow rate.

8-2.4 Comparisons between Cyclones and Elutriators

Since neither type of sampler provides a measure of respirable activity in accordance with Eq. (8.1), the factor converting collected activity to pulmonary deposit is a function of the size distribution of the sampled aerosol. The growing application of both samplers in the fields of industrial hygiene and health physics has made it desirable to establish the proportionality factors for each and to compare results from the two instruments when sampling the same aerosol.

(a) *Determination of Proportionality Factors.* To calculate the desired proportionality factors, it is assumed that an aerosol having a given size distribution is simultaneously sampled by each instrument and inhaled by an individual. The quantities of interest are then calculated as

$$\text{Respirable activity} = \int_0^\infty P(D) \cdot f(A) \cdot dD \qquad \text{(sampler)}, \qquad (8.4a)$$

$$\text{Pulmonary deposit} = \int_0^\infty P'(D) \cdot f(A) \cdot dD \qquad \text{(individual)}, \qquad (8.4b)$$

where P and P' have the same meaning as in Eq. (8.1) and $f(A) \cdot dD$ is the relative amount of activity associated with particles having diameters in the increment defined by dD. The desired proportionality factor is the ratio between respirable activity and pulmonary deposit according to the two equations. Both $P(D)$ and $P'(D)$ are precisely defined so that the calculated respirable activities are as accurate as the definition of the aerosol and are free of experimental errors related to sampling conditions or to sampler shortcomings. In the discussion below, $P(D)$ refers to the BMRC curve of respirable activity for the horizontal elutriator and to the AEC curve for the cyclone (except for Watson's study).

Watson [24] applied this method to eight aerosols of coal dust and four of rock dust. His samplers were the standard horizontal elutriator and a cyclone of his own design that had a collection efficiency curve closely resembling the AEC respirable dust curve. His pulmonary deposition curve was that of Brown *et al.*, properly adjusted for particle density and shape (see Fig. 8.2). His aerosol distributions were described in terms of four broad size intervals. Plotted as cumulative distributions on log-probability coordinates, they are comparable to lognormal distributions having geometric standard deviations between about 2.2 and 2.8 and median diameters varying over more than an order of magnitude. For respirable activity referred to mass, the elutriator passed, on the average, 3.7 times as much material as would have been deposited in the pulmonary region. The ratio between respirable masses as defined by elutriator and by cyclone was 1.15. He concluded that these samplers provide "an index of concentration proportional to the amount of dust likely to be deposited in the lung" and that the index of concentration "is almost independent of the dust, its size distribution, density, and particle shape."

Beeckmans [28] made similar calculations for hypothetical aerosols having either lognormal or power-law distributions. The former had geometric standard deviations between 1.1 and 1.8 and median diameters between 0.2 and 10.0 μm. He compared the elutriator and cyclone respirable activity curves with curve B of Fig. 8.4. The geometric standard deviations he considered are mostly outside the range of values encountered in practice and only his results for $\sigma_g = 1.8$ will be mentioned. For median diameters between 0.5 and 5.0 μm, the elutriator estimated the respirable mass as 2.5 to 3.5 times the pulmonary deposition; the corresponding limiting values for the cyclone sampler were 2.5 and 4.8. Over the full range of median diameters for which calculations were made, the ratio between respirable masses defined by elutriator and by cyclone was 1.23, with a relative standard deviation of about 13%. Beeckmans felt that the ranges of values for the proportionality factors were unacceptably large, but he recognized the practical difficulties in attempting to apply the correct proportionality factor. This factor could only be determined for a

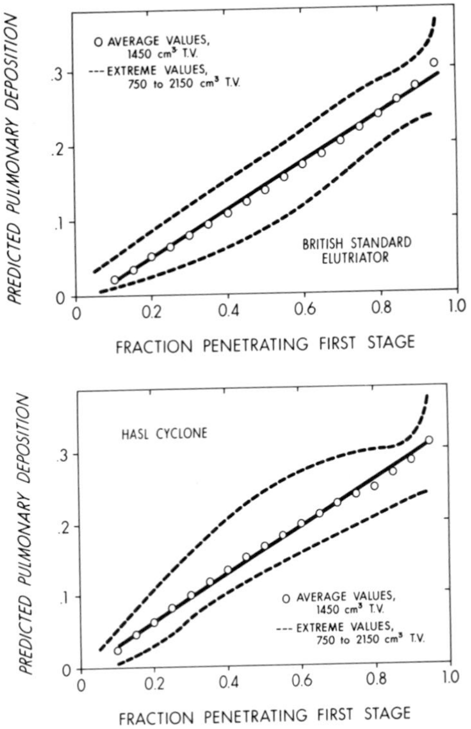

FIG. 8.13. Relation between pulmonary deposition according to Task Group and respirable activity according to horizontal elutriator (top) and cyclone (bottom) [53].

given aerosol by obtaining both a particle size distribution and a sample of respirable activity.

Mercer [53] compared respirable activity with pulmonary deposition as defined by the ICRP Task Group for each of the assumed conditions of respiration. He assumed activity was distributed lognormally with respect to aerodynamic diameter and made comparisons for six values of σ_g between 1.5 and 4.0. The range of activity median aerodynamic diameter (AMAD) for which calculations were made yielded respirable activity fractions from about 0.05 to about 0.95. The results for the elutriator and cyclone are shown in Fig. 8.13. Average values are given for a tidal volume of 1450 cm^3 and extreme values for tidal volumes of 750 cm^3 and 2150 cm^3. For respirable activity fractions between 0.1 and 0.9, extreme values of deposition at a given penetration value do not differ from the average by more than 0.1. In line with the Task Group's other recommendations, Mercer recommended that the average curve for 1450 cm^3 tidal volume be used when estimating pulmonary deposition with the aid of respirable activity samplers. He pointed out, also, that the average curve was not limited to lognormal distributions. For penetration values between about 0.1 and 0.9, the deposition-penetration curves can be represented as

$$P'(\text{AMAD}, \sigma_g) = a + b \cdot P(\text{AMAD}, \sigma_g). \tag{8.5}$$

If an activity distribution is not lognormal itself, but can be expressed as the summation of several lognormal distributions, each of which contributes a fraction F_i to the total amount penetrating the first stage, and if the penetration P_i of each individual distribution lies in the range for which Eq. (8.5) is valid, then the total predicted pulmonary deposition is

$$\begin{aligned} P'(\text{total}) &= \sum P_i' F_i = \sum (a + b \cdot P_i) F_i \\ &= a \sum F_i + b \cdot \sum F_i P_i \\ &= a + b \cdot P(\text{total}), \end{aligned}$$

and Eq. (8.5) is valid for any activity distribution satisfying the conditions described above.

The quantities a and b are -0.012 and 0.312 for the elutriator and 0 and 0.326 for the cyclone. For a given value of pulmonary deposition, the corresponding respirable activities according to elutriator and cyclone are in the ratio

$$P_H/P_C = 1.012 + 0.038/P_C.$$

This ratio varies between 1.39 and 1.05 for $0.1 \leqslant P_C \leqslant 0.9$.

(b) *Elutriator Respirable Activity/Cyclone Respirable Activity—Theoretical.* In this case, calculations are similar to those described above but are limited to a comparison between respirable fractions as defined by the two samplers. Calculations have been reported by Lynch [54] and by Moss [55]. The former found that the elutriator:cyclone ratio of respirable activities was 1.23 ± 0.17 for the usual range of size distributions encountered in industrial dust clouds. This size range included count median diameters from 0.5 to 2.0 μm and geometric standard deviations from 2.0 to 3.0. Moss considered typical industrial aerosols to be those having geometric standard deviations between 2.0 and 4.0, for which he found the elutriator:cyclone ratio to be 1.20 ± 0.15. He used the ACGIH definition of respirable fraction, but the results are undoubtedly applicable to the AEC definition also.

(c) *Elutriator Respirable Activity/Cyclone Respirable Activity—Experimental.* Several workers have reported experimental comparisons of the two types of instuments when used to sample the same aerosol. Their results are summarized in Table 8.2. At 1.9 liters/min, penetration data for the BCIRA cyclone [44] fit the theoretical elutriator curve at least as well as do data for

TABLE 8.2

COMPARISON OF RESPIRABLE ACTIVITIES AS MEASURED BY ELUTRIATOR AND CYCLONE

Type of cyclone	Operating flow rate (liters/min)	Cloud	Ratio of respirable activity concentration, elutriator : cyclone	Ref.
BCIRA	1.9	Quartz dust[a]	1.03 ± 0.05[b]	[44]
BCIRA	1.9	Foundry dust (iron)	0.96	[44]
Simpeds (BCIRA)	1.85	Coal dust	1.03 ± 0.12	[58]
HASL ½ in.	13.0	Pottery dust	1.0	[47]
HASL 10 mm	1.4	Coal, silica, pyrite, mica, glass and asbestos fiber[a]	1.0 ± 0.1	[57]
HASL 10 mm	2.0	Coal dust	1.6	[56]

[a] Laboratory tests.
[b] Error refers to 95% probability level; other errors are standard deviations.

the elutriators themselves, so that the elutriator:cyclone ratio at that flow rate should be unity. On the basis of flow rates, the ratio should be larger for BCIRA than for Simpeds, which employs the same first stage cyclone. The mean values do not show this effect, but the data are, nonetheless, in satisfactory agreement. Bloor *et al.* [47] believed that $D_A(0.5)$ would equal 5 μm for the HASL 1/2 in. cyclone when operated at 13 liters/min. Their belief was supported by their observation that the cyclone gave practically the same respirable activities as the Hexhlet when the latter was operated according to BMRC specifications. The observation is at odds, however, with the calibrations of both Knuth [48] and Ettinger [39] and, when proper allowance is made for particle shape [50], with the original calibrations on which they based their choice of flow rate.

The elutriator:cyclone ratio of 1.6 observed by Jacobson [56] indicates that the cyclone was operated at too high a flow rate (2 liters/min), causing it to pass too small a fraction of respirable activity to the second stage. The observation supports Ettinger's argument that the cyclone calibration with coal particles [52], on which Jacobson's choice of flow rate was based, was in error because equivalent volume diameters were converted to aerodynamic diameters without making allowance for a resistance shape factor.

The observation by Knight and Lichti [57] that the cyclone, when operated at 1.4 liters/min, gave the same results as the elutriator also supports Ettinger's view that Knuth's recommended flow rate is too low, since the elutriator:cyclone ratio could not be unity if the cyclone actually operated according to the AEC specification. Their data yield a ratio of 1.12 ± 0.16 at a flow rate of 1.7 liters/min, which is a more reasonable value when the two samplers are operating according to their respective specifications.

Knight and Lichti also pointed out that cyclones have the advantage that fluctuations in sampling flow rate are partially compensated for by changes in efficiencies. For a given sampling period, the amount of activity collected on the second stage is relatively insensitive to flow rate. The efficiency of the elutriator, on the other hand, decreases with increasing flow rate, compounding the error in the amount of respirable activity collected. The advantage was not observed by Knight and Lichti, however, when sampling asbestos clouds with the cyclone. Instead, the respirable activity collected per unit time increased with flow rate as if the overall collection efficiency remained constant. They imply that the effect occurred because the cloud was made up of some particles for which the cyclone efficiency was very low, even at the highest flow rate, and other particles for which the cyclone efficiency was very high, even at the lowest flow rate, with very few particles in the range of sizes within which efficiency could vary much with flow rate.

8-2.5 Other Samplers

Several other instruments have been designed to measure respirable activity, some of them intended to yield samples that approximate very closely an expected pulmonary deposit. None of them appears likely to supplant either the elutriator or cyclone as a routine sampling device, but they are usually ingenious in design and in some cases have applications of special interest.

(a) *The Conicycle* [20, 59]. The sampling portion of this instrument, which is shown in Fig. 8.14, rotates about the z axis at an angular velocity ω, causing air to be drawn in at A–A′ and expelled at 0–0′. The orifices in the exhaust limit the flow rate to 10 liters/min when the sampler rotates at 8000 rpm ($\omega = 837\ \text{sec}^{-1}$). If air enters the sampling head symmetrically and both air

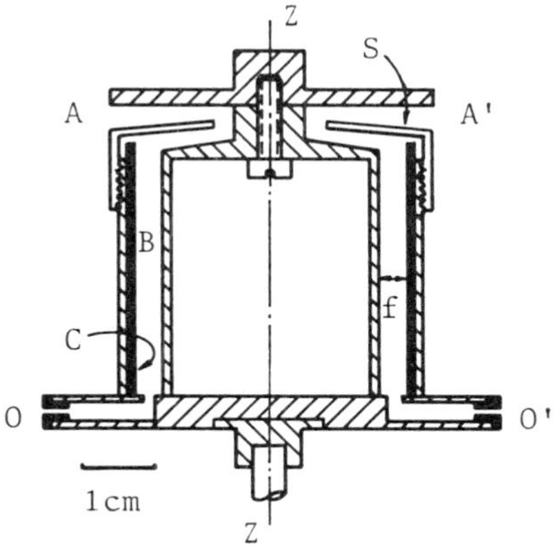

FIG. 8.14. The Conicycle [59]. Courtesy of Pergamon Press, Book Division.

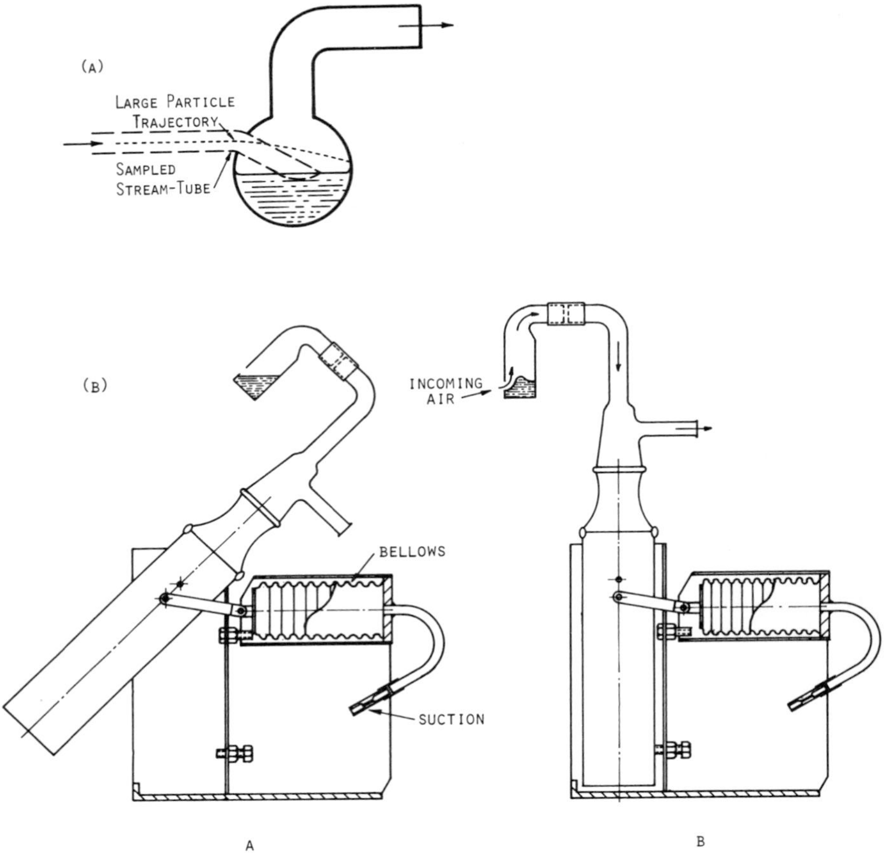

and particles are accelerated instantly to the sampler's rotational speed, a spherical particle of diameter D acquires an outward radial velocity U, due to the centrifugal acceleration $\omega^2 r$, given by

$$U = \rho D^2 \omega^2 r / 18\eta, \tag{8.6}$$

where r is the radius of the sampling head at the entrance. If h is the height of the entrance slit and F is the volumetric flow rate into it, then the average linear velocity into the sampling head is

$$V = F/2\pi rh. \tag{8.7}$$

The number of particles N of a given size that enter the sampling head per unit time is

$$N = 2\pi rhC_0(V-U) = C_0 F[1-(U/V)],$$

where C_0 is their concentration in the air near the entrance slit. The relative

(C)

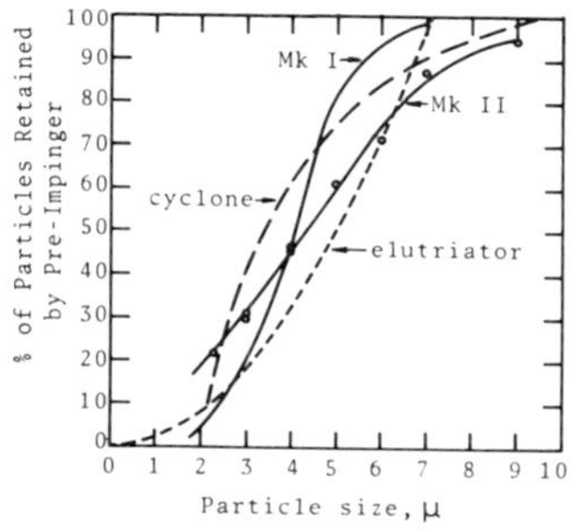

(D)

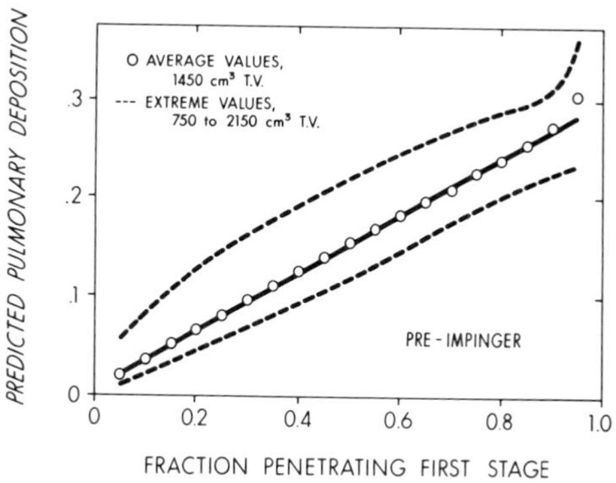

FIG. 8.15. Impingement samplers. (A) The pre-impinger (Mk I) [60]; courtesy of British Medical Journal. (B) The tilting pre-impinger (Mk II) [61]; courtesy of British Occupational Hygiene Society. (C) Collection efficiencies as a function of aerodynamic diameter [61]. (D) Pulmonary deposition in relation to pre-impinger performance [53].

concentration available for deposition in the annular space B is

$$C/C_0 = N/FC_0 = 1 - (U/V),$$

and, provided C_0 is not affected by the action of the sampler, the inlet section of the instrument functions as a horizontal elutriator. The limiting aerodynamic diameter, obtained by setting $U = V$, can be calculated from

$$\rho_0 D_A{}^2 = 9F\eta/\pi\omega^2 r^2 h,$$

and should equal 7.1 μm for the specified operating conditions and $h = 0.06$ cm. Apparently, the instrument has not been calibrated to establish its collection efficiency as a function of size. Comparison of its performance with that of a horizontal elutriator [20] has shown that theory does not accurately predict the cut-off size; however, the two performances can be brought into agreement by adjusting the height of the entrance slit.

At 8000 rpm and 10 liters/min, all incoming particles having aerodynamic diameters >1 μm are deposited on the outer wall of the annular space B. The design of the instrument permits considerable flexibility in the choice of operating conditions, but the size-selecting characteristics of the inlet must be established experimentally.

(b) *The Pre-Impinger* [60] *and Tilting Pre-Impinger* [61]. Diagrams of these two instruments are shown in Fig. 8.15. They were designed to collect viable aerosols in a manner simulating deposition in the upper respiratory tract. They are used ahead of a critical-orifice impinger in which all incoming particles having diameters larger than about 0.5 μm are collected. In the first model, incoming air impinged at low velocity on the horizontal liquid surface. Under certain conditions of ambient air flow, some larger particles had sufficient horizontal momentum to carry over the liquid and deposit on the inner wall of the bulb, so that they were not included in the assay of the pre-impinger liquid. The tilting pre-impinger was designed to ensure that all impinging particles would deposit in the liquid. When not in use, the pre-impinger and the back-up impinger are in the tilted position, leaving the level of the liquid surface below the bottom of the inlet. When suction is applied to the instrument, the bellows contracts pulling the impinger into an upright position. The volume of liquid is just equal to the internal volume of the pre-impinger up to the top of the orifice. The pressure of the incoming air piles up the liquid behind the orifice in the manner shown, preventing leakage and providing that particles impinge only on liquid. When sampling is stopped, the flap valve closes but the presence of the reservoir causes the flow rate to decay gradually, while the rising pressure permits the bellows to return the impinger to its tilted position without loss of liquid.

The collection efficiencies of both impingers are shown in Fig. 8.15(C) with the efficiency curves of elutriator and cyclone. The pre-impinger (MkI) was in-

The collection efficiencies of both impingers are shown in Fig. 8.15c with the efficiency curves of elutriator and cyclone. The pre-impinger (MkI) was included among the samplers for which respirable activity and pulmonary deposit were calculated according to Eqs. (8.4a,b) for $P'(D)$ defined by the ICRP Task Group [26]. The results are shown in Fig. 8.15D.

(c) *The Bubbler Sampler* [51]. In this sampler (Fig. 8.16), the air is bubbled through a water column and particles are lost from the bubbles through sedimentation and diffusion. By proper adjustment of flow rate and column size, a collection efficiency curve for the bubbler can be obtained that is very similar to the curve for total deposition in the respiratory tract (Fig. 8.16B). If the air first passes through a cyclone, the bubbler collects a sample comparable to a pulmonary deposit (Fig. 8.16B). The device has the advantages that it includes the diffusion mechanism in its sampling process and it subjects hygroscopic particles to a relative humidity comparable to that of the lung.

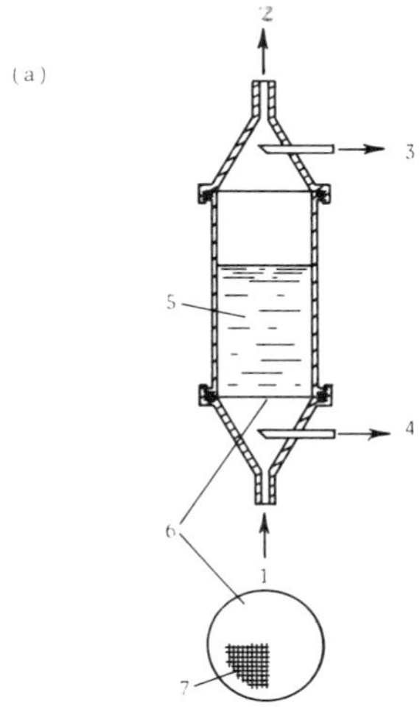

(1) aerosol inlet, (2) aerosol outlet, (3) outlet sampling, (4) inlet sampling, (5) water, (6) sieve plate and (7) holes 0.35mm diameter.

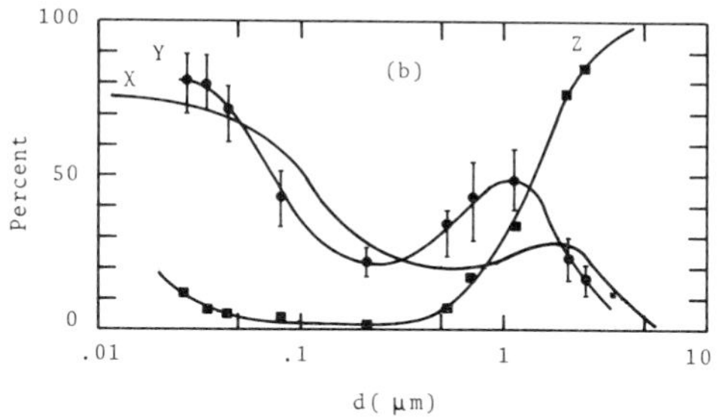

Curve X: pulmonary deposition efficiency for unit density spheres at 2150 cm^3 tidal volume, 15 respirations/min (from Reference 25).
Curve Z: measured deposition efficiency of the stage simulating the nasopharyngeal plus tracheo-bronchial tract, for polystyrene spheres.
Curve Y: deposition efficiency in the bubbling column when a stage of the characteristics of curve Z is added before the column.

FIG. 8.16. The bubbler sampler [51]. (a) Schematic diagram of the sampler. (b) Deposition and efficiency curves. Courtesy of the American Industrial Hygiene Association.

(d) *Multistage Samplers.* Any instrument that yields a value for the AMAD of an aerosol can be used to estimate pulmonary deposition as defined by the ICRP Task Group [26]. For laboratory work, the desired measurement can be made most simply using cascade impactors that have conventional collection surfaces. For field use, however, where large quantities of innocuous material may be accumulated before an adequate amount of activity has been collected, conventional impactor stages may become overloaded, leading to inaccurate results. A few methods to overcome this problem have been described.

Hounam and Sherwood [62] developed a cascade "centripeter" in which the impacted particles are collected on a filter paper at the base of a conical nozzle (Fig. 8.17) set behind a circular orifice. The air flow pattern causes particles to occupy an area smaller than that of the orifice itself when they flow through it; the larger particles occupy smaller areas. This favors the collection of large particles as the air stream diverges to flow around the cone. The pressure difference between the front and back of the cone causes some flow of air through the filter. This flow is less than 3% of the total when a high efficiency, glass-fiber filter paper is used [63]. Four such stages in series provide data with which to estimate the AMAD and σ_g of the active aerosol.

Langmead and O'Connor [63] have described a personal centripeter consisting of a single centripeter stage followed by a high efficiency filter. They obtain an estimate of the AMAD by a two-point log-probability plot. One point is based on their single stage data and the other is based on the arbitrary assumption that 99% of the activity is associated with particles having aerodynamic diameters smaller than 20 μm. The shortcomings of the method are obvious; however, the performance characteristics of their centripeter element are similar to those of the pre-impinger (MkI), so the instrument could be used to obtain an estimate of pulmonary deposit.

Lippmann and Kydonieus [49] use six 10-mm nylon cyclones arranged symmetrically about a single filtration unit, all enclosed in a housing that has a single aerosol inlet tube. Each cyclone is backed with a filter and all of the samplers operate in parallel. The cyclones are run at different flow rates in the range of 0.3 to 4.7 liters/min. The value of $D_A(0.5)$ at the flow rate of a given cyclone is taken as the effective cut-off diameter for that cyclone. The activity concentration measured by its back-up filter relative to the total concentration measured by the single filter is the fraction of available activity on

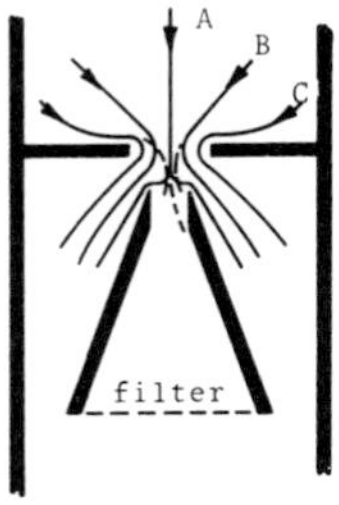

FIG. 8.17. Schematic diagram of a "centripeter" stage [62]. Courtesy of the American Industrial Hygiene Association.

particles smaller than the effective cut-off diameter. The data are then treated in the manner used with cascade impactors.

Schleien *et al.* [64] collect airborne fission products using several different filters arranged in a series of increasing collection efficiency. Their data do not permit them to calculate the size distribution parameters of the aerosol. However, by assuming the aerosol to be made of several discrete sizes, they are able to put a range of values on deposition at each size, using a linear programming technique. The method was applied to deposition curves for the three lung compartments defined by the Task Group on Lung Dynamics, but it is applicable to other deposition curves.

8-3 Some Limitations of Respirable Activity Samplers

Deposition models are defined in terms of aerodynamic diameter because the processes of sedimentation and impaction, which are chiefly responsible for the deposition of particulate material in the respiratory tract, are both functions of aerodynamic diameter. The only other processes accorded any significant role in deposition are interception and diffusion due to Brownian motion. The former is independent of density, while the latter is not only independent of density but varies inversely with particle size. When particles have extreme shapes or very low density, interception may cause their deposition to differ significantly from that predicted according to their aerodynamic diameters. Fibers, which have aerodynamic diameters almost independent of their lengths, will show increased nasal and tracheo-bronchial deposition and sometimes increased pulmonary deposition, although at the larger aerodynamic diameters the increase in the first two will occur at the expense of the latter. Similarly, the effect of diffusion may lead to erroneous predictions concerning the deposition of very small particles of large density. A spherical particle of density ρ and aerodynamic diameter D_A has a geometric diameter D_p given by

$$D_p = (\rho_0 K_{SA}/\rho K_{Sp})^{1/2} \cdot D_A,$$

where the K_S's are slip factors. A particle of $D_A = 0.3$ μm and $\rho = 11$ g/cm^3 has a geometric diameter of about 0.05 μm and its proper predicted pulmonary deposition is at least twice the value assigned to it on the basis of aerodynamic diameter.

References

1. Medical Research Council, *Chronic Pulmonary Disease in S. Wales Coal Miners*, Spec. Rep. Ser., No. 244, HMSO, London, 1943.
2. W. Findeisen, Concerning the Deposition of Small Airborne Particles in the Human Lung during Breathing. *Pflüger's Arch. Physiol., 236:* 367–379 (1935).

3. H. D. Landahl, On the Removal of Airborne Droplets by the Human Respiratory Tract: I. The lung, *Bull. Math. Biophys.*, *12:* 43–56 (1950).
4. A. M. Van Wijk and H. S. Patterson, The Percentages of Particles of Different Sizes Removed from Dust-Laden Air by Breathing, *J. Ind. Hyg. Toxicol.*, *22:* 31–35 (1940).
5. H. D. Landahl and R. G. Hermann, On the Retention of Airborne Particulates in the Human Lung: I, *J. Ind. Hyg. Toxicol.*, *30:* 181–188 (1949).
6. H. D. Landahl and T. N. Tracewell, On the Retention of Airborne Particulates in the Human Lung: II, *A.M.A. Arch. Ind. Health*, *3:* 359–366 (1951).
7. J. H. Brown, K. M. Cook, F. G. Ney and T. Hatch, Influence of Particle Size upon the Retention of Particulate Matter in the Human Lung, *Amer. J. Pub. Health*, *40:* 450–459 (1950).
8. C. N. Davies, Dust Sampling and Lung Disease, *Brit. J. Ind. Med.*, *9:* 120–126 (1952).
9. Medical Research Council Panels, cited by R. J. Hamilton and W. H. Walton, The Selective Sampling of Respirable Dust, in C. N. Davies (Ed.), *Inhaled Particles and Vapors*, Pergamon, Oxford, 1961.
10. A. J. Orenstein (Ed.), *Proc. Pneumoconiosis Conf.*, *Johannesburg*, *1959*, p. 620, Little, Brown, Boston, Massachusetts, 1960.
11. W. B. Harris and M. Eisenbud, Dust Sampler Which Simulates Upper and Lower Lung Deposition, *A.M.A. Arch. Ind. Hyg. Occup. Med.*, *8:* 446–452 (1953).
12. T. J. Burnett, *Sampling Methods and Requirements for Estimating Airborne Radioparticulate Hazards*, Rep. CF–52–11–1, Oak Ridge Nat. Lab. 1952.
13. R. Dennis, R. Coleman, L. Silverman, and M. W. First, *Particle Size Efficiency Studies on a Design 2 Aerotec Tube*, Rep. NYO–1583, Air Cleaning Laboratory, Harvard Univ. School of Public Health, 1952.
14. M. Eisenbud and J. A. Quigley, Industrial Hygiene of Uranium Processing, *A.M.A. Arch. Ind. Health*, *14:* 12–22 (1956).
15. E. C. Hyatt, H. F. Schulte, C. R. Jensen, R. N. Mitchell, and G. H. Ferran, *A Study of Two-Stage Air Samplers Designed to Simulate the Upper and Lower Respiratory Tract*, USAEC Rep. LA–2440, 1960.
16. M. Lippmann and W. B. Harris, Size-Selective Samplers for Estimating "Respirable" Dust Concentrations, *Health Phys.*, *8:* 155–163 (1962).
17. Threshold Limits Committee, Threshold Limit Values of Airborne Contaminants for 1968, *Amer. Conf. Governmental Ind. Hyg.*, *Cincinnati*, *1968*.
18. U. S. Dept. of Labor, Public Contracts and Property Management, *Federal Register*, *34:* 7946 (1969).
19. H. E. Ayer, G. W. Sutton and I. H. Davis, *Report on Size-Selective Gravimetric Sampling in Foundries*, Rep. TR–39, U. S. Dept. of Health, Education, and Welfare, 1967.
20. J. Carver, G. Nagelschmidt, S. A. Roach, C. E. Rossiter, and H. S. Wolff, The Conicycle, a Portable Gravimetric Airborne Dust Sampling Instrument and Its Preliminary Calibration against the Long-Running Thermal Precipitator, *Min. Eng.*, *No. 21:* 601–612 (1962).
21. J. J. Bloomfield and J. M. DallaValle, *The Determination and Control of Industrial Dusts*, Public Health Bull. No. 217, p. 43, Washington, D.C., 1935.
22. K. M. Morse, H. E. Bumsted and W. C. Janes, The Validity of Gravimetric Measurements of Respirable Coal Mine Dust, *Amer. Ind. Hyg. Ass. J.*, *32:* 104–114 (1971).
23. T. Hatch, K. M. Cook, and P. E. Palm, Respiratory Dead Space, *J. Appl. Physiol.*, *5:* 341–347 (1953).
24. H. H. Watson, Dust Sampling to Simulate the Human Lung, *Brit. J. Ind. Med.*, *10:* 93–100 (1953).

25. M. Lippmann, "Respirable" Dust Sampling, *Amer. Ind. Hyg. Ass. J., 31:* 138–159 (1970).
26. Task Group on Lung Dynamics, Deposition and Retention Models for Internal Dosimetry of the Human Respiratory Tract, *Health Phys., 12:* 173–207 (1966).
27. R. E. Pattle, The Retention of Gases and Particles in the Human Nose, in C. N. Davies (Ed.), *Inhaled Particles and Vapors*, Pergamon, Oxford, 1961.
28. J. M. Beeckmans, Correction Factor for Size-Selective Sampling Results, Based on a New Computed Alveolar Deposition Curve, *Ann. Occup. Hyg., 8:* 221–231 (1965).
29. E. E. Weibel, *Morphometry of the Human Lung*, Academic Press, New York, 1963.
30. H. Breuer, Erfahrungen mit dem gravimetrischen Feinstaubfiltergerät BAT, *Staub, 24:* 324–329 (1964).
31. J. Cartwright and G. Nagelschmidt, The Size and Shape of Dust from Human Lungs and its Relation to Selective Sampling, in C. N. Davies (Ed.), *Inhaled Particles and Vapors*, Pergamon, Oxford, 1961.
32. R. J. Hamilton and W. H. Walton, The Selective Sampling of Respirable Dust, in C. N. Davies (Ed.), *Inhaled Particles and Vapors*, Pergamon, Oxford, 1961.
33. B. M. Wright, A Size-Selecting Sampler for Airborne Dust, *Brit. J. Ind. Med., 11:* 284–288 (1954).
34. W. B. Parkes, Measurement of Airborne Dust Concentrations in Foundries, *Amer. Ind. Hyg. Ass. J., 25:* 447–459 (1964).
35. J. H. Dunmore, R. J. Hamilton, and D. S. G. Smith, An Instrument for the Sampling of Respirable Dust for Subsequent Gravimetric Assessment, *J. Sci. Instrum., 41:* 669–672 (1964).
36. J. G. Dawes, G. K. Greenough and J. S. Seager, The Penetration of Irregularly-Shaped Particles through an Airborne-Dust Elutriator, *Brit. J. Appl. Phys., 8:* 236–241 (1957).
37. R. J. Hamilton, A Portable Instrument for Respirable Dust Sampling, *J. Sci. Instrum., 33:* 395–399 (1956).
38. R. J. Hamilton, G. D. Morgan and W. H. Walton, Measurements of Dust by Mass and by Number, in C. N. Davies (Ed.), *Inhaled Particles and Vapors II*, Pergamon, Oxford, 1967.
39. H. J. Ettinger, J. E. Partridge and G. W. Royer, Calibration of Two-Stage Air Samplers, *Amer. Ind. Hyg. Ass. J., 31:* 537–545 (1970).
40. J. R. Lynch, Air Sampling for Cotton Dust, *Trans. Nat. Conf. Cotton Dust Health, Univ. North Carolina*, May 2, 1970.
41. W. H. Walton, Theory of Size Classification of Airborne Dust Clouds by Elutriation, *Brit. J. Appl. Phys., Suppl. 3:* S29–S37 (1954).
42. P. Rosin, E. Rammler and W. Intelmann, Principles and Limits of Cyclone Dust Removal, *Z. Ver. Deut. Ing., 76:* 433–437 (1932).
43. C. N. Davies, The Separation of Airborne Dust and Particles, *Proc. Inst. Mech. Eng. B, 1B:* 185–198 (1952).
44. R. I. Higgins and P. Dewell, A Gravimetric, Size-Selecting Personal Dust Sampler, in C. N. Davies (Ed.), *Inhaled Particles and Vapors II*, p. 575, Pergamon, Oxford, 1967.
45. H. Breuer, Investigations on the Suitability of Cyclones for the Selective Sampling of Respirable Fine Dusts, in C. N. Davies (Ed.), *Inhaled Particles and Vapors II*, p. 523, Pergamon, Oxford, 1967.
46. H. E. Ayer, G. W. Sutton and I. H. Davies, Size-Selective Gravimetric Sampling in Dusty Industries, *Amer. Ind. Hyg. Ass. J., 29:* 336–342 (1968).
47. W. A. Bloor, R. E. Eardley and A. Dinsdale, A Gravimetric Personal Dust-Sampler, *Ann. Occup. Hyg., 11:* 81–86 (1968).

48. R. Knuth, Recalibration of Size-Selective Samplers, *Amer. Ind. Hyg. Ass. J., 30:* 379–385 (1969).
49. M. Lippmann and A. Kydonieus, A Multi-Stage Aerosol Sampler for Extended Sampling Intervals, in T. T. Mercer, P. E. Morrow, and W. Stöber (Eds.), *Assessment of Airborne Particles*, Thomas, Springfield, Illinois, 1972.
50. P. Kotrappa, Revision of Lippmann-Harris Calibration of a Two-Stage Sampler Using Shape Factors, *Health Phys., 20:* 350–352 (1971).
51. C. Melandri and V. Prodi, Simulation of the Regional Deposition of Aerosols in the Respiratory Tract, *Amer. Ind. Hyg. Ass. J., 32:* 52–57 (1971).
52. T. F. Tomb and D. L. Raymond, Evaluation of the Collection Characteristics of Horizontal Elutriator and 10-mm Nylon Cyclone Gravimetric Dust Samplers, presented at *Amer. Ind. Hyg. Ass. Conf., Denver, Colorado, 1969.*
53. T. T. Mercer, Air Sampling Problems Associated with the Proposed New Lung Model, *Proc. Ann. Bioassay Anal. Chem. Meeting, 12th Gatlinburg, Tennessee*, 1966.
54. J. R. Lynch, Evaluation of Size-Selective Presamplers: I. Theoretical Cyclone and Elutriator Relationships, *Amer. Ind. Hyg. Ass. J., 31:* 548–551 (1970).
55. O. R. Moss and H. J. Ettinger, Respirable Dust Characteristics of Polydisperse Aerosols, *Amer. Ind. Hyg. Ass. J., 31:* 546–547 (1970).
56. M. Jacobson, Respirable Dust in Bituminous Coal Mines in the U. S., in W. H. Walton (Ed.), *Inhaled Particles III*, p. 745, Unwin Bros., Woking (1971).
57. G. Knight and K. Lichti, Comparison of Cyclone and Horizontal Elutriator Size Selectors, *Amer. Ind. Hyg. Ass. J., 31:* 437–441 (1970).
58. B. A. Maguire and D. Barker, A Gravimetric Dust Sampling Instrument (Simpeds): Preliminary Underground Trials, *Ann. Occup. Hyg., 12:* 197–201 (1969).
59. H. S. Wolff and S. A. Roach, The Conicycle Selective Sampling System, in C. N. Davies (Ed.), *Inhaled Particles and Vapors*, p. 460, Pergamon, Oxford, 1961.
60. K. R. May and H. A. Druett, The Pre-Impinger—a Selective Aerosol Sampler, *Brit. J. Ind. Med., 10:* 142–151 (1953).
61. K. R. May, A Size-Selective Total Aerosol Sampler, the Tilting Pre-Impinger, *Ann. Occup. Hyg., 2:* 93–106 (1960).
62. R. F. Hounam and R. J. Sherwood, The Cascade Centripeter: a Device for Determining the Concentration and Size Distribution of Aerosols, *Amer. Ind. Hyg. Ass. J., 26:* 122–131 (1965).
63. W. A. Langmead and D. T. O'Connor, The Personal Centripeter—a Particle Size-Selective Personal Air Sampler, *Ann. Occup. Hyg., 12:* 185–195 (1969).
64. B. Schleien, A. G. Friend and H. A. Thomas Jr., A Method for the Estimation of the Respiratory Deposition of Airborne Materials, *Health Phys., 13:* 513–516 (1967).

9

Special Problems

9-1 Production of Test Aerosols

An important part of aerosol technology is the production of test aerosols with which to calibrate instruments, to study the physical and chemical properties of airborne particles, or to carry out experiments related to the inhalation of particles. It is usually advantageous to be able to work with a monodisperse aerosol; that is, one in which all of the particles have nearly the same size. In many cases, however, the nature of the study leaves no choice but to use polydisperse aerosols. Some of the common methods for producing aerosols of both types are discussed below.

9-1.1 Monodisperse Aerosols

Logically, the definition of a monodisperse aerosol should be related to the purpose for which it was produced, because deviations from a mean size are more significant if the process under investigation is related, for instance, to surface area rather than to diameter. In practice, the definition is seldom explicit or, if given at all, is qualitative. An exception is the definition advanced

(A)

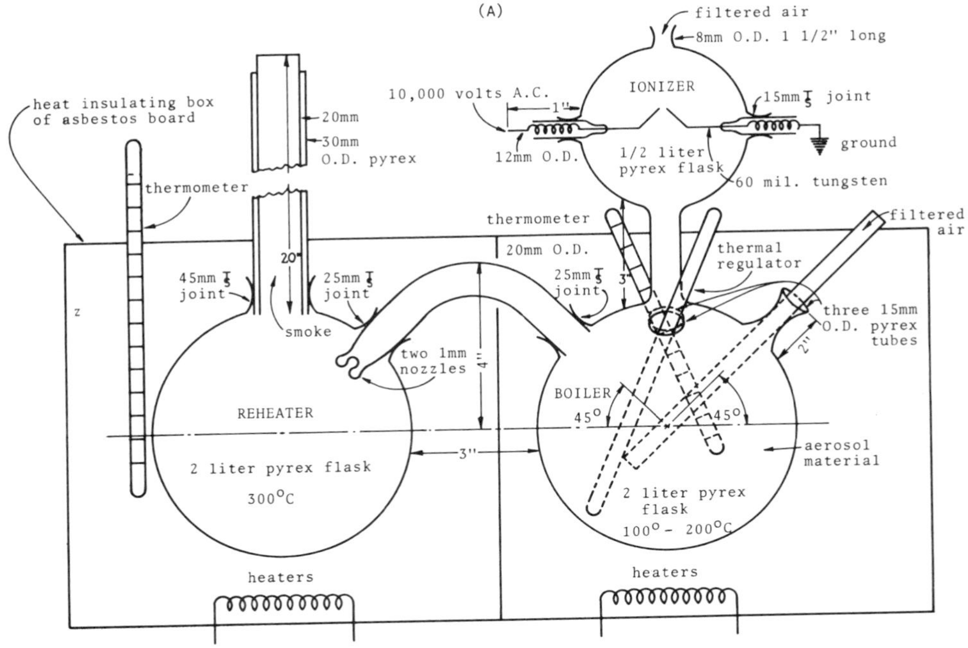

(B)

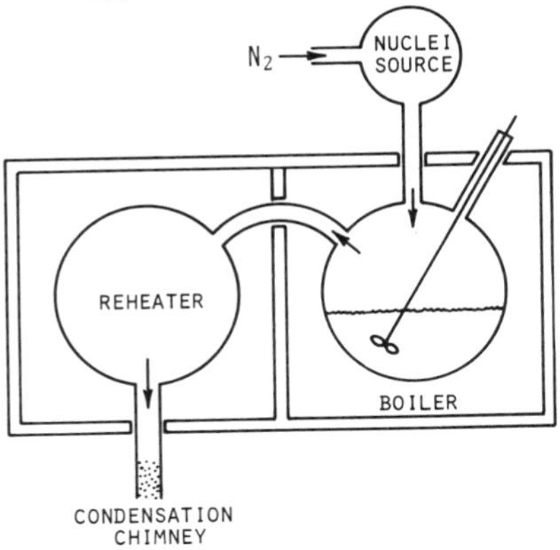

by Fuchs and Sutugin [1], who restricted their review of monodisperse aerosols to those having geometric standard deviations $\leqslant 1.22$ or coefficients of variation $\leqslant 0.2$. This definition may be arbitrary, but it leaves no doubt about the meaning of monodispersity. Several methods of production provide aerosols that satisfy this definition.

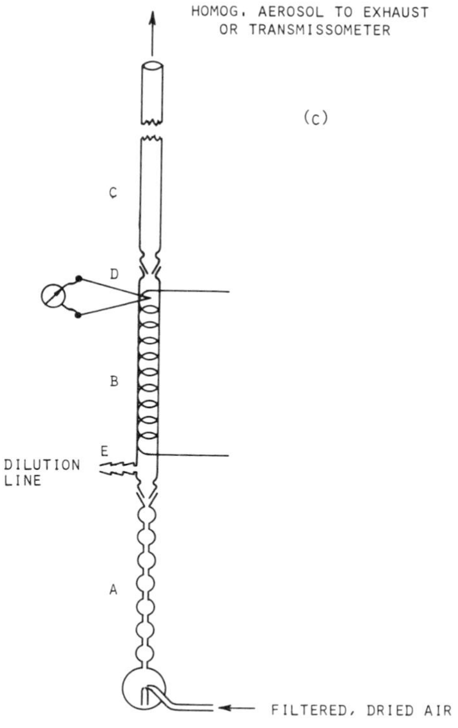

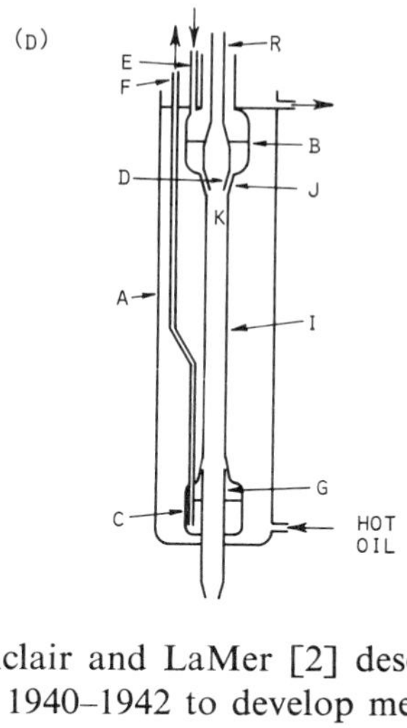

FIG. 9.1. Schematic diagrams of apparatus for the production of monodisperse aerosols by controlled condensation. (A) Sinclair and LaMer [2]; courtesy of the American Chemical Society. (B) Swift [3]; courtesy of the British Occupational Hygiene Society. (C) Rapaport and Weinstock [4]; courtesy of *Experientia*. (D) Nicaloan *et al.* [5]; courtesy of Academic Press, New York.

(a) *Condensation Aerosols.* In 1949, Sinclair and LaMer [2] described work that had been done during the period 1940–1942 to develop methods for the production and measurement of monodisperse aerosols. They had obtained particles in a narrow size range by the controlled condensation of organic vapors in the apparatus shown in Fig. 9.1(A). The basic elements of the apparatus are the nuclei source (ionizer), the boiler, the reheater, and the cooling chimney. The boiler, containing a quantity of the material to be aerosolized, is maintained at a constant temperature in the range 100–200°C. The reheater is maintained at a temperature exceeding that of the boiler by an amount sufficient to ensure the complete vaporization of all aerosol material entering it. The nuclei are produced by an electric spark or by heating a wire coated with an inorganic salt such as sodium chloride. A flow of clean air carries the nuclei into the boiler where they are mixed with droplets and vapor produced by a second flow of clean air that has bubbled through the boiler liquid. The droplets are vaporized in the reheater and nuclei and vapor are thoroughly mixed by the action of the air stream entering through the two nozzles. Then the mixture passes into the condensation chimney where the vapor, cooling at a relatively slow rate, condenses on the nuclei to produce the desired monodisperse aerosol. The condensation process is diffusion controlled, so that the surface area of the growing droplet increases at a constant rate, producing a particle having a diameter D at time t that is related to the initial diameter (of the nucleus) D_0 by

$$D^2 = D_0{}^2 + bt, \tag{9.1}$$

where b is a constant, related to the concentration and diffusivity of the vapor and to the temperature. Since bt is constant for all particles, variations in the size of the nuclei have relatively little effect on the final diameter of the particle. Sinclair and LaMer pointed out that if $bt = 0.64\ \mu\text{m}^2$, $D = 0.8\ \mu\text{m}$ when $D_0 = 0.02\ \mu\text{m}$ and $0.82\ \mu\text{m}$ when $D_0 = 0.20\ \mu\text{m}$. Under ideal operating conditions, there would be no loss of vapor to walls, no vapor or temperature gradients in the cooling chimney and outputs of liquid from the boiler and nuclei from the ionizer would be constant at a given flow rate and boiler temperature. The average diameter of particles issuing from the chimney then would be

$$D = (6M/\pi\rho N)^{1/3}, \tag{9.2}$$

where M is the mass concentration of the vapor and N is the number of nuclei per unit volume of air entering the chimney. Particle size could be varied equally well by varying M or N.

In reality, the coefficient b in Eq. (9.1) varies continuously as the temperature declines in the condensing chimney and the validity of the equation depends on the extent to which b varies in a similar manner along different flow lines. Since the chimney wall functions both as a heat sink and as a vapor sink, both temperature and concentration gradients exist between the chimney axis and the wall, as well as in the direction of flow. When the chimney flow is directed upwards, the temperature gradients may cause convection currents that lead

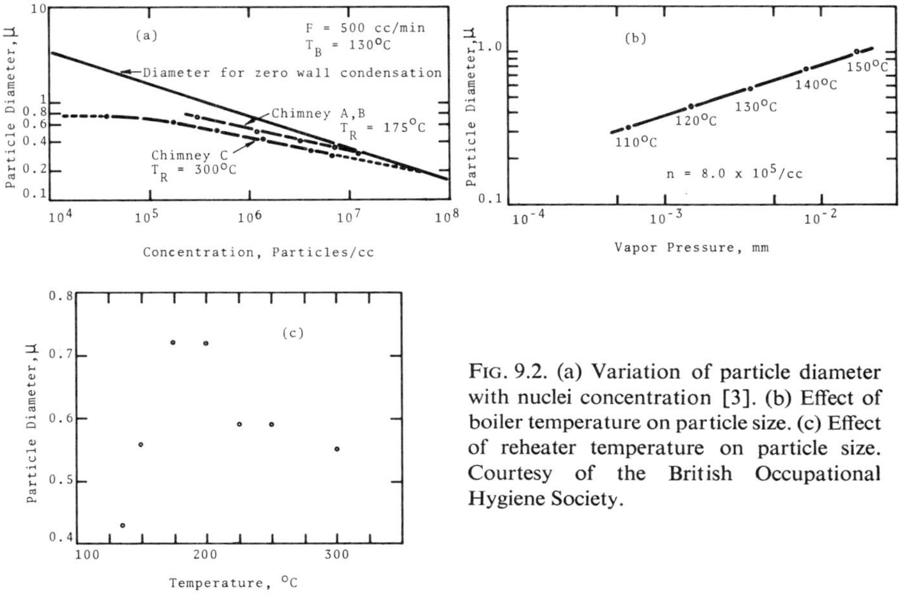

FIG. 9.2. (a) Variation of particle diameter with nuclei concentration [3]. (b) Effect of boiler temperature on particle size. (c) Effect of reheater temperature on particle size. Courtesy of the British Occupational Hygiene Society.

to a broadening of the size distribution of the output aerosol. Muir [6] has shown that these can be avoided if the chimney flow is directed downwards.

Temperature and concentration gradients are primarily responsible for the finite range of diameters in the particles generated, while condensation of vapor on the chimney wall affects the average diameter as calculated using Eq. (9.2). Swift [3] studied the effect of various operating conditions on particle diameter, using the modified generator shown in Fig. 9.1b. He used nitrogen rather than air as the carrier gas, the entire flow entering through the nuclei source. None of it was bubbled through the boiler liquid. Following Muir, the condensation chimney was directed downward, an orientation used originally by Sinclair and LaMer when producing particles of large diameter. Some of Swift's data are shown in Fig. 9.2. Only at large nuclei concentration (Fig. 9.2a) is the expected proportionality between D and $N^{-1/3}$ seen. As the concentration of nuclei decreases, the wall surfaces acquire an increasing fraction of the available vapor until more than 95% of the vapor condenses on walls and particle size becomes independent of nuclei concentration. On the other hand, particle size is proportional to $M^{1/3}$ (Fig. 9.2b) at a given reheater temperature and nuclei concentration. Under the conditions for which the data in Fig. 9.2b apply, about 50% of the vapor condensed on the chimney wall, but this fraction apparently was not a function of the amount of vapor present so that the predicted proportionality between D and $M^{1/3}$ was actually observed. Figure 9.2c shows the effect of reheater temperature at constant boiler temperature and constant nuclei concentration. Swift was not able to explain the presence of a maximum diameter in the range 170°–200°C.

Swift also investigated various methods for producing nuclei. Monodispersity, as judged from Tyndall spectra, was best with nuclei of tungsten oxide (produced by heating a light bulb filament), carbon (produced by an electric arc), and Apiezon M grease (evaporated from a coating on a platinum wire). The last of these was most satisfactory, providing a constant output of nuclei at a concentration up to 10^7 cm^{-3}. By contrast, the output of tungsten nuclei deteriorated rapidly. Adequate Tyndall spectra and a constant output of nuclei at high concentrations were obtained with either sodium chloride or silver chloride evaporated from a combustion boat or from a coating on platinum wire.

Despite the success of the Sinclair–LaMer generator, it had certain disadvantages that stimulated others to develop modified versions of it. The simplest of these was the Rapaport–Weinstock generator [4] shown in Fig. 9.1(C). An organic liquid is aerosolized from a Dautrebande nebulizer (A) and the droplets are vaporized in the heated zone B. The evaporation is not complete, however, leaving a high concentration of residual nuclei on which the vapor recondenses in passing through the chimney C. The source of these nuclei has not been definitely established; they may be due to the presence of less volatile impurities in the organic liquid or they may be droplets of the pure liquid that have reached a state of equilibrium due to their charge. Using dioctyl phthalate (DOP), Rapaport and Weinstock produced aerosols having mean diameters close to 1 μm and relative standard deviations (coefficients of variation) less than 0.1.

Unlike the Sinclair–LaMer generator, the generator developed by Rapaport and Weinstock does not require a long initial period for thermal equilibrium, and the substance being aerosolized is not subjected to possible decomposition due to prolonged heating. For a given substance, however, it cannot readily produce a variety of particle sizes, a disadvantage that was partly responsible for the development of two modified versions of the generator. Lassen [7] replaced the Dautrebande nebulizer with a commercial device and altered the aerosol concentration entering the vaporizing section by adjusting the distance between the atomizing jet and a baffle plate. By preferentially removing larger droplets, this arrangement allowed the mass concentration to be significantly reduced without appreciably changing the number of droplets or residual nuclei. Lassen was able to produce DOP aerosols that had mean diameters between 0.30 and 1.40 μm and were sufficiently monodisperse to show higher order Tyndall spectra. He also provided an annular drain at the base of the heating section to prevent liquid that collected on the heater wall from draining back into the nebulizer.

Liu *et al.* [8] used a Collison nebulizer to produce the primary aerosol and directed the output downward through a vaporizing zone into a condensation chimney. To minimize the effects on monodispersity caused by radial gradients of concentration and temperature in the chimney, they extracted only the central 5% of the aerosol flow. They were able to vary the mean particle size by nebulizing solutions of DOP in ethyl alcohol. The average particle diameter D was related to the DOP concentration in percent C by $D = 0.27C^{0.33}$ μm. For diameters below about 0.6 μm, however, the aerosols did not satisfy the definition of monodispersity given above, the geometric standard deviation increasing from 1.22 at 0.6 μm to 1.50 at 0.036 μm.

Nicolaon *et al.* [5] described a generator (Fig. 9.1D) in which the functions of both boiler and reheater are carried out by a single vertical glass tube (I) immersed in a constant temperature oil bath (A). A thin layer of liquid drains down the inside surface of the tube, some of it vaporizing into the nuclei-laden gas flowing through the tube. The mixture of vapor and nuclei then enters a condensation chimney in horizontal orientation. Using dibutyl phthalate as the aerosol substance, helium as the carrier gas, and nuclei produced by heating sodium chloride in a combustion boat, they generate aerosols that have mean diameters in the range of about 0.4 to 0.6 μm and dispersions equivalent to geometric standard deviations of about 1.18. This design has the advantage that the generator will accurately reproduce a given aerosol even after long periods of disuse.

For many years, the esters of phthalic acid, particularly DOP, have been the liquids of choice for the production of condensation aerosols, although esters of sebacic acid are now becoming popular. A variety of other organic compounds have been used for this purpose, including some (stearic acid, menthol, rosin) that are solids at room temperature. Movilliat [9] has had some success in producing monodisperse aerosols of zinc and cadmium by condensation methods, and Sinclair and LaMer reported the generation of monodisperse aerosols of NH_4Cl.

Table 9.1 summarizes performance data for some of the generators. The data on output and number concentration relate to the useful aerosol at the outlet of the condensation chimney. It is customary to dilute the output to slow the process of coagulation, which degrades the monodispersity of the aerosol. For the size range shown in the table, the half-life of the number concentration is

$$t_{1/2} = 5.5 \times 10^7/N_0 \qquad \text{min},$$

where N_0 is the concentration of the particles emerging from the chimney. In this time, the average number of particles will have been reduced by a factor of 2.

(b) *Spinning Disk Atomizers.* Following the observation that spinning

TABLE 9.1

PERFORMANCE DATA FOR SEVERAL CONDENSATION AEROSOL GENERATORS

Flow rate (liters/min)	Output (mg/liter)	Number concentration (cm^{-3})	Particle size average diameter (μm)	Coefficient of variation or σ_g	Nuclei source	Carrier gas	Ref.
1–4	1–10	10^4–10^7 (est.)	0.1–20	0.10	Electric spark	air	[2]
0.1–1	0.02–0.2	4×10^4–10^7	0.3–1	—	AgCl, NaCl, or Apiezon coating on heated filament	N_2	[3]
0.5–2	<1	1.5×10^6	0.2–1	1.15	Electric spark	N_2	[6]
0.3–2.5	~0.2	3×10^6	0.4–0.6	0.17	NaCl in combustion boat	He	[5]
3.5 (0.175 of useful aerosol)	—	—	0.036–1.3	1.14–1.50	Residual nuclei	air	[8]
3.7–8.7	0.035–7	8.5×10^4–10^7	0.92–1.20	0.095	Residual nuclei	air	[4]
10–20	0.14–1.4 (est.)	10^6–10^7	0.30–1.40	—	Residual nuclei	air	[7]

disk atomizers could produce droplets of uniform size [10], Walton and Prewett [11] made a detailed study of the operating conditions under which monodispersity could be brought about. They used sprayers of the types shown diagrammatically in Fig. 9.3. The motor-driven disk sprayer could be operated at rotational speeds up to 4000 rpm, but the top, spun pneumatically, was

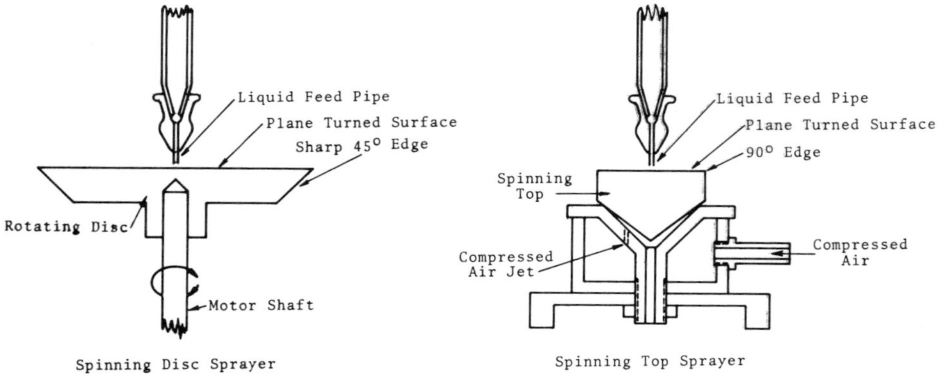

FIG. 9.3. Diagrams of spinning disk and spinning top sprayers [11]. Courtesy of The Institute of Physics and The Physical Society.

operated at speeds up to 98,000 rpm. Uniform droplets are produced when a liquid that wets the disk surface is fed onto the center of the disk at a constant rate. The liquid spreads over the disk surface in a symmetrical thin film accumulating at the rim until the centrifugal force acting to discharge it exceeds the capillary force acting to hold the liquid together and a droplet is thrown off. The centrifugal force is equal to the product of the mass of liquid in the drop and the centrifugal acceleration at the disk rim; the capillary force is proportional to the product of the surface tension of the liquid and the droplet diameter:

$$(\pi/6)\,\rho D^3 \cdot \omega^2 d/2 \propto \sigma D$$

$$D = (B/\omega)(\sigma/d\rho)^{1/2}, \tag{9.3}$$

where D is the droplet diameter, d is the disk diameter, ρ and σ are the liquid density and surface tension, respectively, ω is the angular velocity of the disk, and B is a constant that relates the droplet diameter to the length over which surface tension acts at the instant the mass of liquid detaches from the disk. Walton and Prewett tested the validity of this equation for angular velocities between 50 and 10^4 rad/sec, disk diameters from 2 to 8 cm, densities from 0.9 to 13.6 g/cm^3, and surface tensions from 31 to 465 dyn/cm. Although the liquid viscosity varied between 0.01 and 15 poise, they found, in agreement with Eq. (9.3), that it did not affect the droplet diameter. The constant B ranged from 2.67 to 6.55, with an average value of 3.8. It was generally larger at the higher rotational speeds of the spinning top.

Although most of the atomized liquid went into the formation of droplets of a common diameter, a small fraction of it appeared in the form of much smaller "satellite" droplets. Walton and Prewett designed a system to separate the two sizes on the basis of their stopping distances; however, the problem was handled more simply with an improved version of the spinning top designed by May [12]. He discovered that the rotor was completely stable up to 240,000 rpm if its diameter was smaller than that of the stator (Fig. 9.4). Moreover, the spent air escaping between the stator and rotor discharged downward rather than upward as it had in earlier models. The effect was enhanced by rounding the outer lip of the stator, and the flow of air followed the path indicated by the broken lines of Fig. 9.4. The design has the additional advantage that the spent air, passing at high speed through the annulus between the stator (G) and the housing (J), acts as an ejector, sucking air in between the cap (E) and the housing. This inflow of air entrains the satellite droplets and they are carried out with the exhaust air. The larger droplets have sufficient momentum to carry them out of the apparatus.

May found that the diameter of the satellite droplets is about one-fourth that of the primary droplets and that, on the average, there are about four of the former for each of the latter. Less than 10% by volume of the atomized

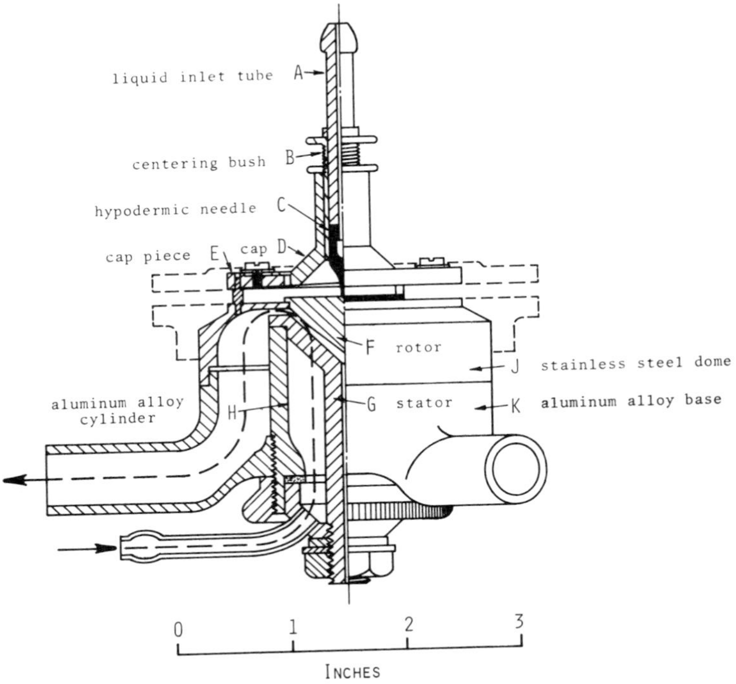

FIG. 9.4. An improved spinning top sprayer [12]. Courtesy of the American Institute of Physics.

liquid occurs as satellites if the liquid flow rate is kept below about 1 cm^3/min. The percentage increases with increasing liquid feed and may become a considerable fraction of the atomized liquid [13].

The tangential velocity of a droplet at the moment that it detaches from the rotor is $\omega d/2$. Assuming that this is also its initial velocity relative to the surrounding air and taking Eq. (9.3) into consideration, the initial Reynolds number for the droplet motion will be

$$\mathrm{Re}_0 = B(\sigma d/\rho)^{1/2}/2\nu, \tag{9.4}$$

where ν is the kinematic viscosity of air. This quantity, which is not a function of either rotational speed or droplet diameter, is well outside the region of Stokes's flow, being generally larger than 100. Under these circumstances, the stopping distance can be calculated using Klyachko's relationship for the drag coefficient [Eq. (2.30)]:

$$S = (\rho_p D/\rho_a)[\mathrm{Re}_0^{1/3} - 2.45 \cdot \arctan(\mathrm{Re}_0^{1/3}/2.45)]. \tag{9.5}$$

The ratio S/D is a constant for a given rotor and liquid. For water ($B = 4.5$)

atomized from a rotor 1 in. in diameter, it is equal to 2495. May determined S/D experimentally for both water and dibutyl phthalate. For the former, he found that if D is expressed in micrometers and S in inches, $D/S = 10$, i.e., $S(\text{in cm})/D(\text{in cm}) = 2540$.

The primary droplets exhibit a good degree of monodispersity. May found that the relative standard deviation was not a function of the mean droplet size and could be as small as 2.3% for oils and organic solvents that readily wet the rotor. With a properly treated rotor surface, he was able to get a relative standard deviation of 5% with water sprays. The data of Walton and Prewett indicate similar values, and Dunskii and Nikitin [13] reported standard deviations between 1.7 and 6.7% of the mean.

Despite the advantages offered by May's spinning top, most investigators have preferred to use the motor-driven spinning disk. Whitby *et al.* [14] employed an air-driven tool grinder motor, capable of rotational speeds up to 70,000 rpm, to spin a 1.83-in. diameter disk. They atomized solutions of methylene blue and uranine in ethyl alcohol or in a mixture of ethyl alcohol and water, obtaining primary droplet sizes in the range 19–100 μm. By using very dilute solutions, they were able to get solid residual particles of 0.6 to 12 μm diameter. An upward air flow of about 15 cfm into a funnel centered over the disk carried away satellite droplets, while the primary droplets, projected beyond the rim of the funnel, were discharged in an air flow of about 100 cfm.

Lippmann and Albert [15] designed the generator shown schematically in Fig. 9.5 for the purpose of producing monodisperse aerosols of insoluble radioactive particles to be used in deposition and retention studies. It is similar in design to the apparatus described by Whitby (see above), but has some significant differences: the flow rate for satellite removal has been reduced to 8.5 cfm and the aerosol flow rate to 5.5 cfm; a frequency-controlled synchronous electric motor, operating at speeds between 21,000 and 60,000 rpm, is used to spin the disk; and a heating chamber has been added to ensure complete evaporation of the droplets. They produce monodisperse aerosols of insoluble particles by atomizing aqueous suspensions of colloidal iron oxide. By varying the concentration of iron oxide in suspension from 0.01 to 10% and spinning the disk at 48,000 rpm, they get residual iron oxide particles having diameters between about 1 and 10 μm.

Performance data on the various generators that have been described are summarized in Table 9.2.

(c) *Atomization of Suspensions of Monodisperse Particles.* Monodisperse aerosols can be produced by atomizing a sufficiently dilute suspension of monodisperse particles using ordinary air-blast atomizers. The method is attractive because of its simplicity and has been used widely since the

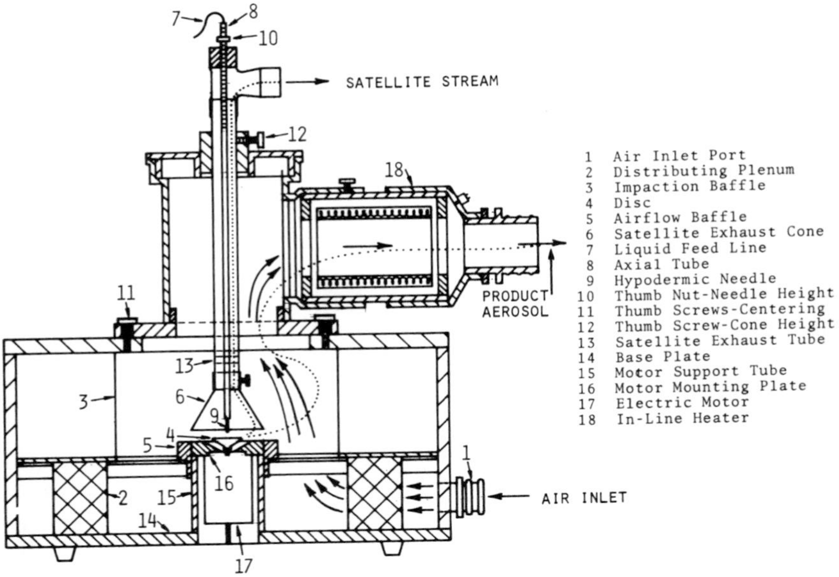

FIG. 9.5. Diagram of spinning system for producing monodisperse aerosols for biological studies [15]. Courtesy of the American Industrial Hygiene Association.

development of polystyrene latexes that have very narrow size distributions. The first of these latexes was apparently produced by accident. Its usefulness in a number of applications created such a demand for it that the Physical Research Laboratory of the Dow Chemical Company [16] made a detailed investigation of the preparation of monodisperse latexes. Carefully controlled emulsion polymerizations led to the production of monodisperse polystyrene latexes in which the particle size distributions had mean diameters between 0.088 and 1.172 μm. The relative standard deviations were generally smaller for the larger sizes, ranging from about 9% at 0.088 μm to about 1.2% at 1.172 μm. Since then, monodisperse polystyrene latexes have been produced that have average diameters up to 1.305 μm. In addition, monodisperse latexes having mean diameters between 1.8 and 3.5 μm have been produced from polyvinyltoluene.

With each sample of their monodisperse latexes, the Dow Chemical Company provides an estimate of the mean diameter of the particles and, in most cases, an estimate of the standard deviation. Several investigators have attempted to verify the Dow statistics for a variety of latexes. A number of the results are given in Table 9.3. Measurements made on the particles while in liquid suspensions consistently yield average diameters smaller than those given by

TABLE 9.2

PERFORMANCE DATA FOR SEVERAL SPINNING DISK AEROSOL GENERATORS

Type	rpm	Disk diameter (cm)	Liquid consumption (ml/min)	$B = D\omega \cdot_{1/2} (d\rho_L/\sigma)$	Primary droplet diameter: mean (μm)	Primary droplet diameter: relative standard deviation	Liquid	Ref.
Top	31,200–61,200	3	—	3.2	16–32	—	Hg	[11]
	19,680–86,400	3	—	5.1	16–80	—	DBP	[11]
	11,000–240,000	2.5	⩽1	4.5	10–200	0.05	H_2O	[12]
	11,000–240,000	2.5	⩽1	4.5	6–150	0.023	DBP	[12]
Disk	400–4000	2–8	up to 168	3.0	280–3000	—	H_2O	[11]
	660–2850	8	up to 168	3.2	210–1020	—	Hg	[11]
	1500–6000	5	up to 168	3.8	180–600	—	DBP	[11]
	7200–19,000	7	1.8–6	4.0	37–90	0.017–0.067	Oil	[13]
	12,000–70,000[a]	4.65	2.5–4	5.3	19–100	0.05–0.1	C_2H_5OH–H_2O solution	[14]
	21,000–60,000[b]	2.8	1.8	3.5 (est)	28–80	0.05–0.12	H_2O suspensions	[15]

[a] Requires 15 cfm satellite removal flow; 100 cfm aerosol carrier flow.
[b] Requires 8.5 cfm satellite removal flow; 5.5 cfm aerosol carrier flow.

TABLE 9–3

PARTICLE SIZE STATISTICS OF DOW LATEXES[a]

		Airborne			Liquid suspension		
Batch	Dow	Rimberg and Thomas [17]	Stöber and Flachsbart [18]	Heard *et al.* [19]	Davidson *et al.* [20]	McCormick [21]	Dezelic and Krahtovil [22]
LS–040A	0.088 ± 0.0080					0.0736	0.0762 ± 0.002
15–N23	0.1380 ± 0.0062					0.1146	0.120 ± 0.003
LS–1044E	0.109 ± 0.0027			0.102			
LS–1045E	0.176 ± 0.0023			0.16			
LS–055A	0.1881 ± 0.0076				0.1770 ± 0.0106	0.1675	0.175 ± 0.005
LS–1047E	0.234 ± 0.0026			0.22			
LS–057A	0.2638 ± 0.006				0.2617 ± 0.0144	0.2482	0.245 ± 0.004
LS–1010E	0.357 ± 0.0056		0.354 ± 0.007	0.33 ± 0.008			
LS–061A	0.3646 ± 0.0079				0.3499 ± 0.0198		0.339 ± 0.008
LS–1029E	0.500 ± 0.0027		0.55 ± 0.0075	0.50 ± 0.009			
LS–063A	0.5567 ± 0.0108	0.582 ± 0.048			0.5101 ± 0.0267		0.534 ± 0.008
LS–1012E	0.714 ± 0.0053			0.74 ± 0.014			
LS–449E	0.7962 ± 0.0083	0.81 ± 0.05			0.7646 ± 0.0624		
LS–1165E	0.810 ± 0.0053		0.84 ± 0.017				
LS–066A	0.8140 ± 0.0105						0.803 ± 0.007
LS–1114B	0.813 ± 0.0063			0.76 ± 0.01			
LS–1028E	1.0992 ± 0.0159	1.12 ± 0.03	1.22 ± 0.018	1.16 ± 0.016	1.1827 ± 0.0636		
LS–067A	1.1710 ± 0.0133						1.167 ± 0.021
LS–464E	1.3046 ± 0.0158				1.0867 ± 0.2348		
866–43	1.857 ± 0.007		1.83 ± 0.025				
866–38	1.947	1.92 ± 0.13					
L–6046–36[b]	2.049 ± 0.0180				2.024 ± 0.1088		
EP–1358–38[b]	2.9583 ± 0.0150				3.0356 ± 0.1904		

[a] Values are for mean diameter and standard deviation according to references cited.

[b] Polyvinyl toluene.

Dow. The standard deviations, as measured by light scattering, are smaller also. Mean diameters estimated from sedimentation in air, however, are usually higher than Dow's values, and the standard deviations are much larger. Mean diameters based on electron microscope data, obtained by Dow [16] and by Heard *et al.* [19], are in reasonably good agreement. Standard deviations according to the latter are two to three times those found by Dow, but they still indicate a high degree of monodispersity. Both investigators supported the particles in replicas of diffraction gratings for electron microscope analysis, eliminating some significant sources of error in the final measurement. The electron microscope measurements of Davidson *et al.* [20] show much higher standard deviations, which they ascribe to the large number of particles counted. They also estimated the average diameter from measurements of number concentrations made on dilute suspensions, using a flow ultramicroscope. These were consistently higher than the diameters obtained by the electron microscope method, a result they attributed to the presence of aggregates in the suspension.

The use of latex particles is complicated by the presence in the suspension of a stabilizing material introduced as an emulsifier during the polymerization process. Dow does not reveal the identity of the stabilizer used in their latexes, but states that it is an anionic detergent having a sulfonate group as the surface active agent. Fuchs and Sutugin [1] give the concentration of the stabilizing agent as 30% of the total solids, but this exceeds by a factor of 6 the maximum amount of emulsifying agent ordinarily used in the polymerization process [23]. Some representative values provided by Dow are given in Table 9.4. Included in that table is an estimate of the increase in particle diameter that would be observed if all of the nonpolymeric solids were evenly distributed over the total particle area. Even at the smallest diameter, where the relative concentration of stabilizer is greatest, the effect on the particle size is negligible. After dilution and atomization of the latex, most of the overall droplet volume will contain no particles, so that the remainder have a proportionately smaller amount of stabilizer available with which to coat the particle. Under the usual conditions of use, then, the stabilizer will not increase the particle diameter by as much as indicated in Table 9.4.

A droplet of diameter D_s that contains no particle will dry to a particle having a diameter D_p that depends on the relative volume concentration of nonpolymeric solids in the original suspension and the dilution factor Y. In general, the total volume of solids is about 10% and the volume of nonpolymeric solids (assuming none is added with the diluent) is about 1–2% of that, so that

$$0.1 \cdot D_s/Y^{1/3} \leqslant D_p \leqslant 0.126 \cdot D_s/Y^{1/3}. \tag{9.6}$$

Residual particles produced in this way create a nuisance background that is

TABLE 9.4

RELATIVE CONCENTRATIONS OF SOLIDS IN LATEX SUSPENSIONS [24]

Batch	Particle diameter (μm)	Percent of total solids in Polymer	Stabilizer	Inorganics	ΔD (μm)[a]
LS–04A	0.088	92.59	5.56	1.85	0.0023
LS–052A	0.126	97.09	2.43	0.48	0.0013
15–N23	0.138	96.44	2.56	—	0.0012
LS–057A	0.264	98.23	1.13	0.64	0.0016
LS–063A	0.557	98.78	0.26	0.95	0.0020
LS–449E	0.796	99.04	0.25	0.71	0.0026
LS–066A	0.814	98.93	0.28	0.79	0.0030
LS–464E	1.305	99.15	0.29	0.56	0.0037
642–6	2.68	99.24	0.01	0.75	0.0070

[a] ΔD is the increase in particle diameter if uniformly coated with nonpolymeric solids.

especially troublesome when light-scattering methods of analysis are being used [25, 26].

The size distributions of droplets produced by air-blast or ultrasonic atomizers are polydisperse, frequently including a very large range of sizes, so that a certain fraction of the suspended particles will appear in the aerosol as doublets, triplets, etc. When the experimental method includes the direct observation of the individual particles, as it does if measurements are made with an aerosol spectrometer or if samples are subjected to microscope analysis, the occurrence of clusters is not a serious problem. On the other hand, when the gross aerosol is analyzed by gravimetric, radiological, or light scattering methods, the presence of multiple particles may introduce serious experimental errors. To preserve monodispersity, it is necessary to dilute the suspension sufficiently to keep the number of multiple particles below some acceptable level. The probability that droplets of a given size will contain x particles is given by the Poisson equation:

$$P(x) = m^x \cdot e^{-m}/x!, \tag{9.7}$$

where $m = C_P \cdot V_s$ is the average number of individual particles expected in a droplet of volume V_s from a suspension containing C_p individual particles per unit volume. To apply this equation to the output of an atomizer, it is necessary to take into account the size distribution of the droplets. For most atomizers, the useful aerosol has a lognormal size distribution, so that

$$P(x) = \frac{1}{\sigma' x! \sqrt{2\pi}} \cdot \int_{-\infty}^{\infty} m^x \cdot e^{-m} \cdot \exp - [(\ln m - \ln m_g)^2/2\sigma'^2] \, d(\ln m), \tag{9.8}$$

where m_g is the value of m corresponding to the count median droplet diameter, $\sigma' = 3 \cdot \ln \sigma_g$, and σ_g is the geometric standard deviation of the distribution. Unfortunately, integration of this equation by expanding e^{-m} leads to an infinite series that cannot be evaluated, so the equation must be integrated by numerical methods. Raabe [27] did this for a number of values of m_g, σ_g, and x, and obtained the following empirical equation for the dilution factor necessary to give a desired singlet ratio R, which is the number of droplets containing single particles relative to the total number of droplets containing particles:

$$Y = f \cdot D_{3g}^3 \cdot \exp(4.5 \ln^2 \sigma_g) \cdot [1 - 0.5 \cdot \exp(\ln^2 \sigma_g)]/(1-R)\, D_i^{\,3}. \qquad (9.9)$$

Here f is the volume fraction of individual particles of diameter D_i in the original suspension, and D_{3g} is the volume median diameter of the droplet distribution. The equation is not accurate for values of $\sigma_g > 2.1$ or for values of $R < 0.9$. The limitations are not serious, however, since most practical situations are included in the region for which Eq. (9.9) is valid.

It is important to note that at least half of the suspended particles will be aerosolized in droplets having diameters greater than D_{3g}, so that serious errors can result if the average droplet diameter is used in Eq. (9.7) to calculate the singlet ratio R. That equation gives results comparable to those of Eq. (9.9) if m is given a value corresponding to a diameter equal to $D_{3g} \cdot \exp(5 \cdot \ln^2 \sigma_g/6)$, which is significantly larger than the average droplet diameter, even at smaller values of σ_g. It is also important to note that if 10% of the airborne particles contain more than one individual particle, then about 20% of the mass or light scattering power will be associated with multiple particles.

Stöber *et al.* [28] have described the production of suspensions of monodisperse silica spheres by reacting tetraesters of silicic acid in water-alcohol mixtures, using ammonia as a catalyst. The tetrapentyl ester, mixed with varying amounts of ammonium hydroxide and either ethanol or a methanol-propanol mixture, produced spherical particles having mean diameters in the range from about 0.08 to 1.5 μm, with geometric standard deviations generally less than 1.10. To facilitate analytical procedures, Flachsbart and Stöber [29] modified the method to incorporate certain radioactive isotopes in the particles. Details for the production of monodisperse particles of different mean sizes are given in the papers cited.

(d) *Other Methods.* Uniform droplets are produced if a filament of liquid is oscillated at a fixed frequency. This principle has been used in a number of methods in which a capillary jet is vibrated mechanically, but the droplet sizes are usually too large for use in aerosol work. A device developed recently by Fulwyler [30], however, uses this principle to produce monodisperse droplets down to a diameter of about 15 μm [31]. A pure liquid, or aqueous solution, is forced through a very small (25 μm or less) orifice by the combined effects of a hydrostatic pressure and ultrasonic vibrations impressed on the liquid by a piezoelectric crystal. The droplet diameter and rate of production depend on the orifice diameter and frequency of the ultrasonic vibrations. The geometric standard deviation of the droplet size distribution is less than 1.01 [31].

If a liquid of low electrical conductivity is placed in a glass tube having a fine capillary at the bottom, it can be dispersed through the capillary as a cloud of small droplets of uniform size by subjecting it to a high positive potential [32]. According to Drozin [33], the specific conductivity must be between 10^{-13} and 10^{-5} ohm^{-1} cm^{-1} and its dipole moment must be greater than about 10^{-18} $dyn^{1/2}-cm^2$. The droplets are highly charged. Yurkstas and Meisenzahl [34] used the method to disperse solutions of polystyrene (0.041%) in acetone, obtaining residual particles having median diameters between 0.11 and 0.27 μm and geometric standard deviations between 1.07 and 1.18.

9-1.2 Polydisperse Aerosols

Atomization of solutions or suspensions has been widely used to provide aerosols for studies of inhalation toxicity. While the method has been applied successfully to the production of insoluble radioactive compounds, it does not yield air concentrations that are sufficiently high for most studies of chemical toxicity, and it becomes necessary to aerosolize particles from the dry powder state. Although both methods of aerosol production have been the subject of numerous papers, the quantitative aspects of maintaining an aerosol at a concentration and size distribution that does not change significantly during the course of an experiment have not received the attention they deserve.

(a) *Air-Blast Atomizers.* Figure 9.6(A) is a diagrammatic representation of the DeVilbiss No. 40 glass nebulizer* (The DeVilbiss Company, Somerset, Pennsylvania), which is one of the simplest and most widely used devices of this sort. Air enters the nebulizer at a mass flow rate that depends chiefly on the diameter of the orifice and the pressure drop across it. The air expands, more or less adiabatically, in a roughly conical airstream that passes the mouth of the liquid inlet tube at high speed, causing a drop in pressure at that point that brings about a flow of liquid into the airstream. The jet of air entrains a filament of liquid, drawing it out to some unstable length at which it breaks up to form droplets of various sizes. If the air vent is plugged, the work done by the expanding jet goes primarily into accelerating the droplets; only a very small fraction is used to create new droplet surface. If the air vent is open, some

* When an atomizer operates within a small container from which only a fraction of the droplets escape, the device is called a "nebulizer."

of the work is used to pump room air through the nebulizer at a flow rate considerably greater than that of the jet air itself.

The droplets are rapidly accelerated to the speed of the airstream, which imparts sufficient momentum to them that more than 99% of the total droplet mass impinges on the wall of the nebulizer and returns to the bulk solution. The energy that went into the production of new surface for these droplets and into their acceleration is retained in the nebulizer as thermal energy. The energy leaving the nebulizer as surface energy of useful aerosol and as kinetic energy of translation of both air and useful aerosol is negligible. On the other hand, evaporation of solvent to saturate the outgoing air causes a substantial loss of thermal energy from the nebulizer, cooling it until an equilibrium temperature is reached at which energy is extracted from the incoming air and conducted to the nebulizer from its surroundings at a rate that just equals that with which energy is removed by the evaporating solvent molecules. The latter comes partly from the droplets traversing the nebulizer and partly from the surface of the bulk liquid, which includes previously impacted droplets that spread in a thin layer over the inner surface of the nebulizer as they drain back into the base.

The evaporation of solvent causes a continuous increase in the concentration of solute in the liquid remaining in the nebulizer. This, in turn, increases the concentration of airborne solute in the useful aerosol and shifts the size distribution of the residual solute particles to larger values. Some improvement can be brought about by saturating the incoming jet air with solvent vapor. The result is not as encouraging as might be expected, however, because the nebulizer does not cool to the same extent as before. If the solvent is water, the rate of evaporation can be reduced by an amount equal, at most, to 37% of the rate at which water vapor enters with the jet air [35]. The problem can be circumvented by continuous replacement of the solvent or by circulating the nebulizer contents through a large reservoir of solution. In the latter case, the contents of the nebulizer reach an equilibrium concentration that exceeds the concentration of the initial solution by an amount that depends on the rate at which solvent vapor and aerosol are carried away.

The empirical equation of Nukiyama and Tanasawa [38] is generally used to estimate the mean size of the primary droplets produced by atomization. There is reason to doubt the validity of their equation, however, when both the liquid atomization rate and the relative velocity between air and liquid are large [35]. A more accurate estimate of the size parameter can be made using the equation given by Glukhov [39] for calculating the diameter $\bar{D}_3$ of the droplet of average mass:

$$\bar{D}_3 = D_j \cdot A_0[1 + B(G_L/G_a)^m](\rho_a V^2 D_j/2\sigma)^{-0.45}, \tag{9.10}$$

where both A_0 and B are functions of the dimensionless parameter, $4\eta_L{}^2/\rho_L \sigma D_j$,

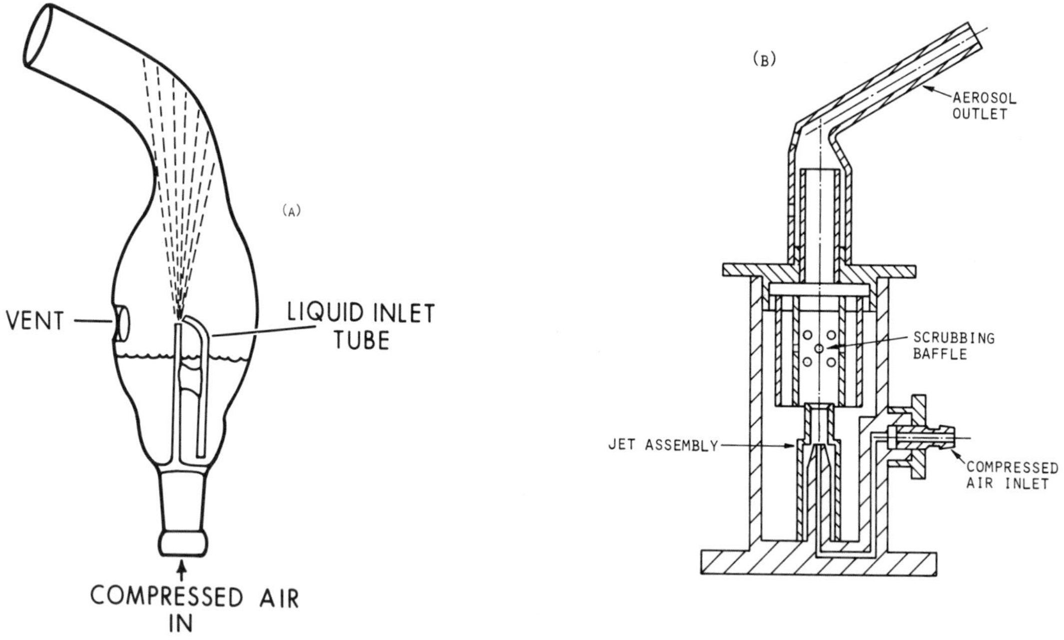

(A)
VENT
LIQUID INLET
TUBE
COMPRESSED AIR
IN
(B)
AEROSOL
OUTLET
SCRUBBING
BAFFLE
JET ASSEMBLY
COMPRESSED
AIR INLET

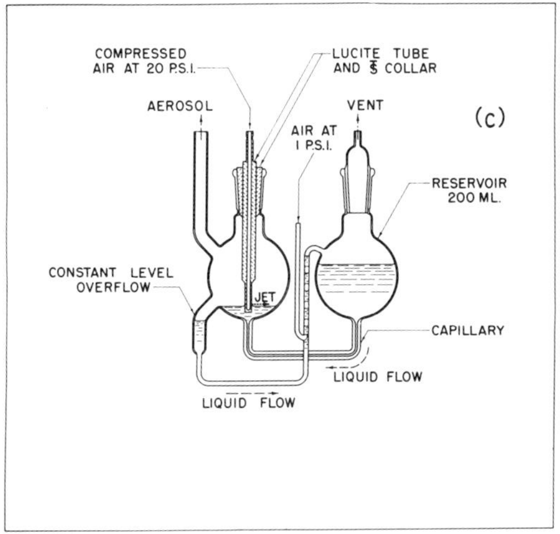

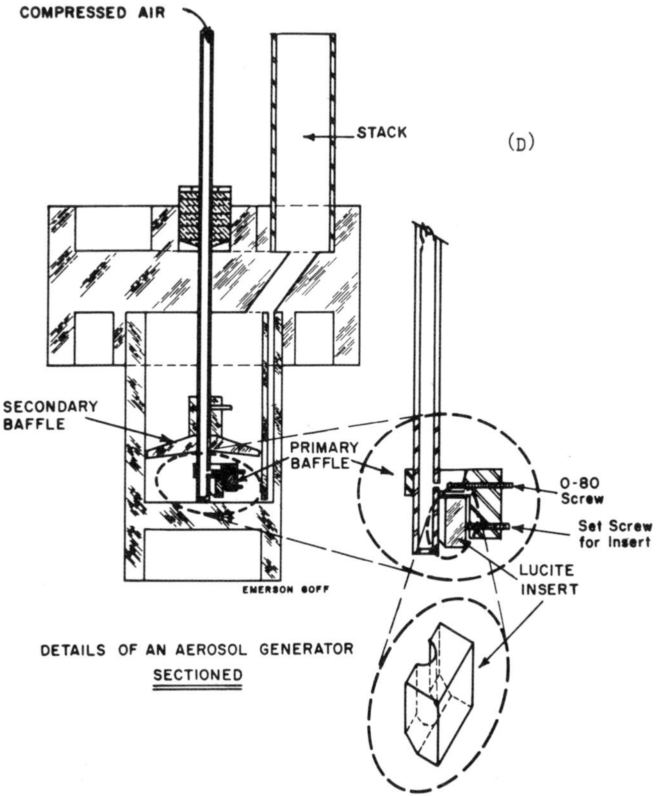

FIG. 9.6. Diagrams of compressed-air nebulizers. (A) DeVilbiss No. 40 glass nebulizer [35]; courtesy of the American Industrial Hygiene Association. (B) Dautrebande generator [36]. (C) Lauterbach generator [37]; courtesy of the American Medical Association. (D) Lovelace generator [35]; courtesy of the American Industrial Hygiene Association.

D_j is the diameter of the liquid flow tube, V is the relative velocity between air and liquid, G_L, ρ_L, and G_a, ρ_a, are, respectively, the mass flow rates and densities of liquid and air, η_L is the viscosity of the liquid, and σ is its surface tension. According to Glukhov, $A_0 = 0.64$ when the dimensionless parameter is less than 0.01, and experimental data give values of 0.011 for B and 2 for m. Under these circumstances, surface tension is the most important liquid property in the atomization process, with viscosity having no significant effect for liquids that are ordinarily used in the production of aerosols. As far as aerosol production is concerned, Eq. (9.10) is useful mostly in that it brings out the significance of the properties of the liquid being atomized.

The various factors described above were investigated when aqueous solutions were atomized using the DeVilbiss, Dautrebande [36], Lauterbach [37] and Lovelace nebulizers [35]. The principle of operation is similar for all of these devices, but they differ widely in structural details. The Dautrebande nebulizer is arranged so that droplets are subjected to a scrubbing action as they pass through small holes over which solution is constantly draining. The Lovelace nebulizer employs a hemispherical baffle, set close to the jet orifice, to enhance the output of useful aerosol. The results of the investigation are summarized below. Diagrams of the nebulizers are included in Fig. 9.6.

(1) Pressure Drop–Mass Flow Rate Relationships. When air initially at absolute temperature T_0 and pressure $P = P_a + \Delta P$ issues from a circular orifice of cross-sectional area a cm^2 into a region at pressure $P_a > \Delta P$, the mass flow rate G in g/min of air can be calculated from [40]

$$\frac{G}{caP_a} = \left(\frac{9.47 \cdot 10^4}{T_0^{1/2}}\right) \times \left[\left(1 + \frac{\Delta P}{P_a}\right)^{0.577} - \left(1 + \frac{\Delta P}{P_a}\right)^{0.289}\right]^{1/2} \quad \text{gm/cm}^2 \cdot \text{min} \cdot \text{atm}, \tag{9.11}$$

where c is a coefficient of contraction relating the area of the orifice to the area of the jet at its vena contracta, and P_a is measured in atmospheres. When $\Delta P \approx P_a$, the jet speed in the vena contracta becomes sonic, and for all larger values of ΔP the relationship is

$$\frac{G}{caP_a} = \left(\frac{2.46 \cdot 10^4}{T_0^{1/2}}\right)\left(1 + \frac{\Delta P}{P_a}\right) \quad \text{gm/cm}^2 \cdot \text{min} \cdot \text{atm}. \tag{9.12}$$

For each of the four nebulizers, a value of c was found that brought experimental data into close agreement with the predictions of Eqs. (9.11) and (9.12). The value of c varied between 0.84 and 0.92, but an average value of 0.88 could be used to predict flow rate through a given orifice at a given pressure drop with an accuracy adequate for most design purposes.

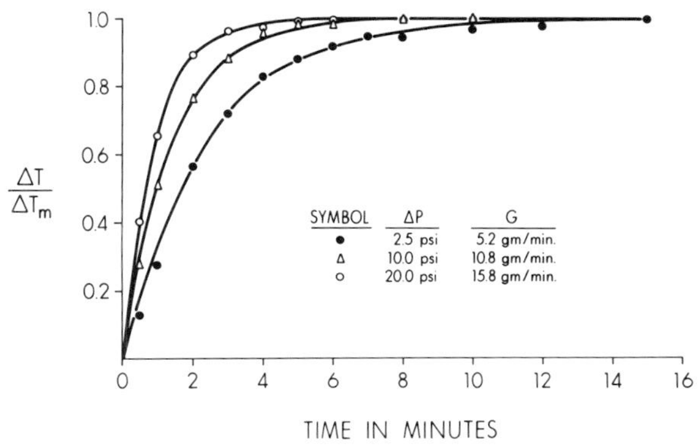

FIG. 9.7. Rate of cooling of the DeVilbiss nebulizer [35]. Courtesy of the American Industrial Hygiene Association.

(2) Cooling of the Nebulizer. The difference in temperature between the nebulizer and its surroundings approaches an equilibrium value ΔT_m at a rate that depends primarily on the mass flow rate of air through the nebulizer and the overall heat capacity of the nebulizer and its contents. Examples for the DeVilbiss nebulizer are shown in Fig. 9.7. Equilibrium temperatures for three of the nebulizers are given in Table 9.5.

TABLE 9.5

EQUILIBRIUM TEMPERATURE AND FLOW RATE AT VARIOUS PRESSURE DROPS ACROSS JET ORIFICE

Pressure drop (psi)	DeVilbiss[a]		Dautrebande[a]		Lovelace[a]	
	G	*T*	*G*	*T*	*G*	*T*
2.5	5.2	13.6	—	—	—	—
5	7.5	12.0	13.4	8.5	—	—
10	10.8	11.0	17.9	7.9	1.2	14.6
15	13.5	10.4	—	—	—	—
20	15.8	10.0	25.4	7.4	1.4	12.1
30	20.5	9.3	32.7	7.3	1.9	10.6

[a] *G* is the jet flow rate in grams per minute; *T* is the equilibrium solution temperature in degrees centigrade.

(3) Concentration of the Nebulizer Solution. If it is assumed that each liter of jet air produces an amount of useful aerosol equivalent to *A* volume

units of solution and carries away, as water vapor, an additional W volume units of solvent from the solution, then at time t the concentration C relative to the initial concentration C_0 is

$$\frac{C}{C_0} = \left[\frac{V_0}{V_0 - (A+W)Ft}\right]^{W/(A+W)}. \tag{9.13}$$

V_0 is the initial volume of solution and F is the jet flow rate of air at ambient

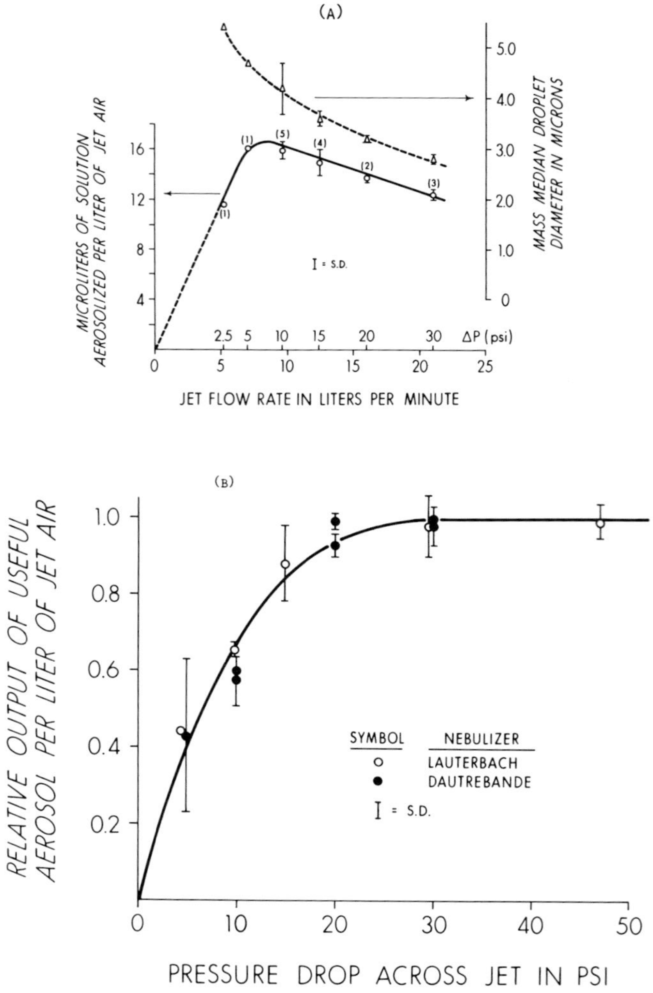

temperature and pressure. This equation permits the experimental determination of A and W from measured values of C, C_0, V_0, and the residual volume of solution V, which is equal to $V_0 - (A+W)Ft$. The value of A is a measure of the efficiency with which the nebulizer produces useful aerosol.

(4) Output of Useful Aerosol and Water Vapor. Output of useful aerosol for the four nebulizers is shown in Fig. 9.8. For the Lauterbach nebulizer (Fig. 9.8B), a relative value of 1.0 represents an output of about 6 μl/liter of jet air, although this quantity varies somewhat with the design of the unit. A similar representative value for the Dautrebande is about 2.4 μl/liter of jet air. The output of the Lovelace nebulizer is a function of both baffle setting and pressure drop. Figure 9.8(C) is an example of the manner in which output varies with the distance of separation between the jet orifice and the baffle. A relative output of 1.0 corresponds to 30.4, 34.7, and 36.2 μl/liter of jet air, respectively, at pressure drops of 20, 30, and 50 psi.

The output of water vapor and aerosol and the jet flow rates (dry air) at several pressure drops are given in Table 9.6. The ratio $W/(A+W)$, which is the significant parameter for the concentration of the nebulizer solution, is included in the table. It is apparent from Eq. (9.13) that it is advantageous to keep this ratio as small as possible for a given initial volume of nebulizer solution.

(5) Parameters of the Size Distributions. The mass median diameters of the droplet distributions of useful aerosol, referred to the time of formation,

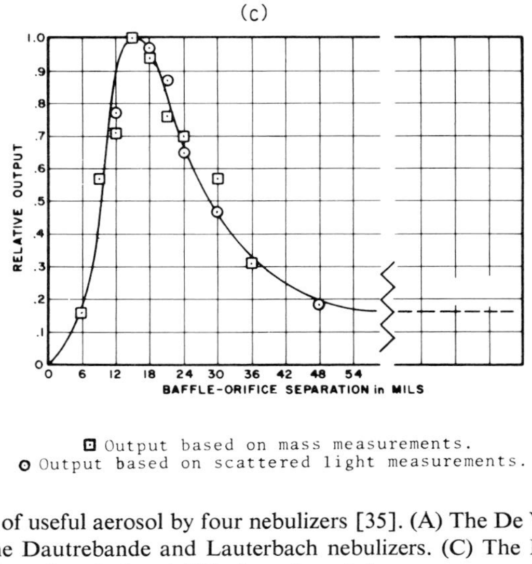

FIG. 9.8. Output of useful aerosol by four nebulizers [35]. (A) The De Vilbiss No. 40 glass nebulizer. (B) The Dautrebande and Lauterbach nebulizers. (C) The Lovelace nebulizer. Courtesy of the American Industrial Hygiene Association.

TABLE 9.6

SUMMARY OF OUTPUT DATA FOR THREE NEBULIZERS

Nebulizers	ΔP (psi)	W (μl/liter)	A (μl/liter)	$W/(A+W)$	F (liters/min)
DeVilbiss	15	8.6	15.5	0.36	12.4
	20	7.0	14.0	0.33	16.0
	30	7.2	12.1	0.37	20.9
Dautrebande	10	9.6	1.6	0.86	14.9
	20	8.6	2.3	0.79	21.2
	30	8.2	2.4	0.77	27.3
Lovelace	20	10.1	30.4	0.25	1.3

are included in Fig. 9.8(A) for the DeVilbiss nebulizer. Values for the Dautrebande and Lauterbach nebulizers are given in Table 9.7. For the Lovelace, the mass median droplet diameter at 20 psi is 5.4 μm. The geometric standard deviations are 1.8–1.9 for the DeVilbiss nebulizer, 1.6–1.7 for the Dautrebande, and 2.0–2.1 for the Lauterbach. The distributions produced by the Lovelace nebulizer are only approximately lognormal with a geometric standard deviation of about 1.9.

If there are C grams of solute per milliliter of solution in the droplets, the size distribution of the residual dry particles of solute will have a mass median diameter $\bar{D}_p$ given by

$$\bar{D}_p = \bar{D}_s(C/\rho)^{1/3}, \tag{9.14}$$

where $\bar{D}_s$ is the mass median diameter of the droplet distribution and ρ is the density of the dry particles. If measurements are made at several values of C, then a plot of $\bar{D}_p$ vs. $(C/\rho)^{1/3}$ should be a straight line of slope $\bar{D}_s$ passing through the origin. Data for all four nebulizers are shown in Fig. 9.9. The results verify the predicted relationship of Eq. (9.14), except for the Dautrebande nebulizer at high concentrations. Otherwise, there seems to be little effect of solute concentration on the size distribution of the droplets.

TABLE 9.7

MASS MEDIAN DROPLET DIAMETER (MICROMETERS) FOR DAUTREBANDE AND LAUTERBACH NEBULIZERS

ΔP (psi)	Dautrebande	Lauterbach
10	1.7	3.8
20	1.4	2.4
30	1.3	2.4

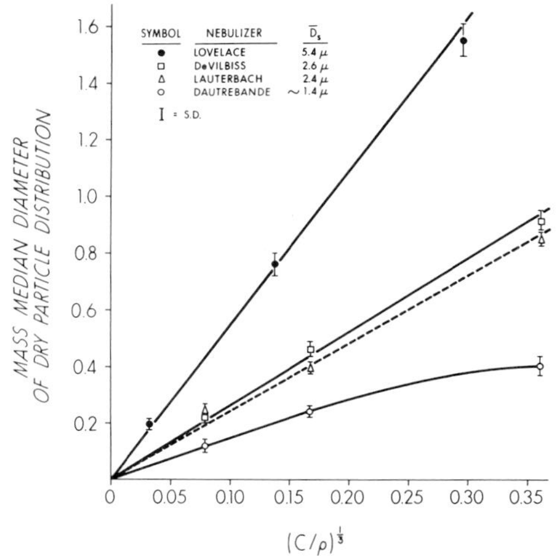

FIG. 9.9. Effect of solute concentration on mass median diameter [35]. Courtesy of the American Industrial Hygiene Association.

(6) Effect of Vent Air. The gross output of useful aerosol per minute for the DeVilbiss nebulizer is shown in Fig. 9.10 for operation both with the vent open and with it closed. The maximum total flow rate was obtained when the vent was left open to room air; other data classified as "vent open" were obtained by introducing dry air through the vent at a controlled rate. The mass median diameter of the droplet distribution increases by 25 to 30% when

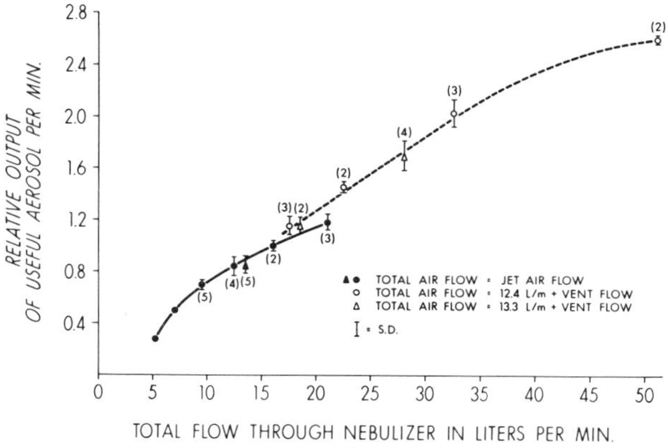

FIG. 9.10. Effect of vent air flow on output of useful aerosol [35]. Courtesy of the American Industrial Hygiene Association.

vent air is added. These results show that there is a good deal more aerosol produced in the useful size range than is carried away by the jet air itself. On the other hand, the added vent air does not carry away relatively as much as the jet air, so that there is a steady decrease in concentration of useful aerosol as the vent airflow increases.

(7) Effect of Baffling. The baffle arrangement of the Lovelace nebulizer is similar to that described by Wright [41]. Provided the orifice-baffle arrangement is symmetrical, different baffle shapes and sizes yield results that are qualitatively similar to those of Fig. 9.8(C), but that show peaks having different absolute outputs and occurring at different orifice-baffle separations.

(b) *Ultrasonic Nebulizers.* In ultrasonic nebulizers, the mechanical energy necessary to atomize a liquid comes from a piezoelectric crystal vibrating under the influence of an alternating electric field produced by an electronic high-frequency oscillator. The vibrations are transmitted through a coupling liquid to a nebulizer cup containing the solution to be aerosolized (Fig. 9.11). At low intensities, capillary waves are formed at the air–liquid interface. As the intensity increases, a fountain of liquid appears at the center of the cup. The fountain is conical in shape and from its apex a jet of large droplets is emitted periodically with the formation of a fog of very small droplets around the lower part of the jet [42]. At still higher intensities, the jet becomes essentially a cylinder of liquid from which the fog is continuously emitted.

Söllner's [43] early work on ultrasonic atomization led him to conclude that the process was a cavitation phenomenon. However, the mean size of the

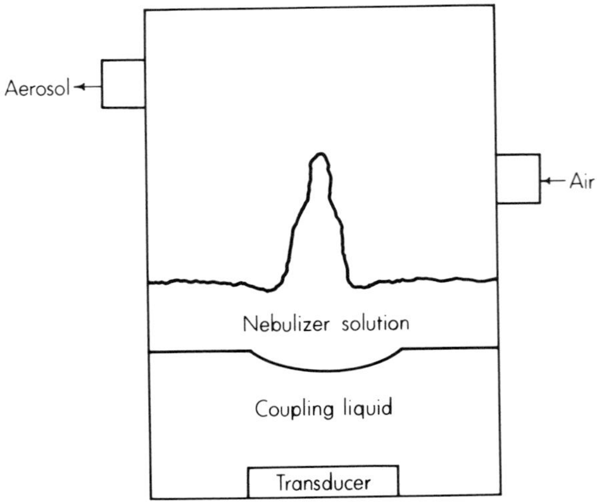

FIG. 9.11. Schematic diagram of an ultrasonic generator.

droplets in a fog produced at a given frequency is closely related to the wavelength of the capillary waves formed on the surface of a liquid at the same frequency [44]. Because of the excellent agreement between theory and experiment in the matter of droplet formation, the hypothesis that ultrasonic atomization is a capillary wave phenomenon has been generally accepted [45], even though the relatively placid formation of droplets from a vibrating surface seems hardly to fit the violent eruptions occurring in ultrasonic atomization of liquids. The anomaly was recently resolved by Boguslavskii and Eknadiosyants [45], who showed that shock waves formed by cavitation can excite standing capillary waves of finite amplitude on the surface of the jet emitted from the ultrasonic fountain. Subsequently, aerosols having an average droplet diameter related to the wavelength of the capillary waves are formed.

The capillary wavelength λ on the surface of a liquid of surface tension σ and density ρ that is vibrated ultrasonically, is

$$\lambda = (8\pi\sigma/\rho\omega^2)^{1/3}, \tag{9.15}$$

where ω is the frequency of the ultrasound. Lang [46] found that the count median diameter of the droplets produced in this way is given by:

$$D_{\text{og}} = 0.34\lambda. \tag{9.16}$$

The proportionality between D and λ has been found to hold for frequencies between 12 kHz and 3 MHz, with some tendency for droplets to be smaller than predicted at frequencies above about 1 MHz [45]. Size distribution data for droplets produced by three ultrasonic nebulizers used for inhalation therapy show good agreement with Eq. (9.16) [47]. The data are shown in Table 9.8.

The rate at which aerosol is produced by an ultrasonic nebulizer is not related

TABLE 9.8

SIZE DISTRIBUTION PARAMETERS OF SOME ULTRASONIC NEBULIZERS

		Droplet size distribution parameters		Count median diameter (μm)	
Nebulizer	Operating frequency (kHz)	Mass median diameter (μm)	Approximate standard deviation[c]	Experimental	Eq. (9.16)
DeVilbiss[a]	1350	6.9	1.6	3.8	3.5
Mist-0_2-Gen[b]	1400	6.5	1.4	3.7	3.9
Mead Johnson	800	9.0	1.7	4.9	4.5

[a] At maximum intensity.
[b] Operated with reservoir.
[c] Geometric.

to the flow of air through it, so that the concentration of useful aerosol leaving the nebulizer varies inversely with flow rate. The concentration of water vapor in the nebulizer output is equal to the saturation concentration at the temperature of the nebulizer solution, but the total output is relatively constant, being fixed by the rate of energy input to the transducer. Some typical output data are shown in Table 9.9.

TABLE 9.9

OUTPUT OF WATER VAPOR AND USEFUL AEROSOL FROM ULTRASONIC NEBULIZERS

Nebulizer	Air throughput (liters/min)	Aerosol output		Evaporative water loss	
		ml solution per minute	μl solution per liter of air	ml/min	μl/liter
DeVilbiss[a]	41.0	6.16	150.0	1.36	33.1
Mist-0_2–Gen[b]	24.7	1.63	66.0	0.55	22.2
Mead Johnson [c]	8.0	1.19	149.0	—	—

[a] At maximum intensity.
[b] Operated with reservoir.
[c] Peak output on sinusoidal flow at 41 $\min^{-1}$.

(c) *Some Design Considerations.* When aqueous solutions are atomized, it must be remembered that most of the liquid appearing as aerosol is water and the air flow must be sufficient to keep the relative humidity well below 100% to allow for proper drying of the particles and to prevent condensation in ducts. At room temperature, 10 μl of water vapor per liter of air will yield a relative humidity of about 50%. Air blast nebulizers evaporate 7–10 μl per liter of jet air on the average (see Table 9.6), leaving essentially all of the water associated with the aerosol to be diluted to a level of 10 μl per liter. Ultrasonic nebulizers can require very large flows of diluting air to handle the volumes of water aerosolized.

The aerodynamic diameters of the residual particles can be estimated using Eq. (9.14):

$$(\rho_0 K_{SA})^{1/2} \cdot D_A = (\rho K_{SP})^{1/2} \cdot D_p = \rho^{1/6} K_{SP}{}^{1/2} \cdot C^{1/3} \cdot D_s . \qquad (9.17)$$

For a given mass concentration of solute in the nebulizer, increasing values of ρ produce a relatively slow increase in the aerodynamic diameters of the residual particles. The relationship holds for suspensions of insoluble colloids, if ρ is the effective density of the dry aggregates of colloidal particles.

In some cases, aerosols of insoluble spherical particles can be produced by passing the nebulizer output through a heating column. The method was developed for use with clay particles that had been impregnated with one or another radioisotope by ion exchange [48]. A suspension of the particles was nebulized and the output aerosol subjected to temperatures up to 1100°C. The clay particles fused into spheres with the radioactivity effectively sealed within. The method has been extended to the formation of spherical oxide particles by the thermal degradation of oxalate compounds [49, 50].

(d) *Dispersion of Dry Powders.* Studies involving the inhalation of insoluble, toxic particles usually require a method for dispersing the particles from the dry powder form. This can be done most simply by blowing air through a loose mass of the powder, but the concentration and particle size distribution of the output aerosol cannot be maintained at constant levels. Devices that operate on this principle include the "dustshaker" [51] and the apparatus shown schematically in Fig. 9.12(A) [52]. In the latter, the powder is supported by a sintered disk (H) in the chamber (I), which is constantly shaken by the motor (B). A stream of dry air passing through I entrains some of the particles and then mixes with a stream of clean, dry air at D, where loose aggregates are broken up during acceleration. The concentration of the aerosol tends to drop off with time, but the designers obtained reasonably good results for periods up to about 2 hours. To maintain constant output conditions using this principle of dispersion, however, requires quite elaborate experimental arrangements [56].

Figure 9.12b shows a modification of the apparatus described above, in which the powder available for dispersion is constantly being renewed. Devices that achieve this same effect by other means have been described by Burdekin and others [57] and by Stead *et al.* [58].

The most successful devices for dispersing dry powders have been those in which the surface of a plug of compacted powder is abraded at a controlled rate, either by jets of compressed air [59] or by mechanical scrapers [54, 60]. A diagram of the Wright dust feed, which operates in this way, is shown in Fig. 9.12(C). The dust tube, which contains the compressed powder to be aerosolized, revolves about the scraper head, losing a small amount of dust to the scraper blade with each revolution. Compressed dry air enters the dust tube through the annular space in the spindle, passes through the slot in the scraper head, picking up the loose dust as it does so, and flows out through the axial hole leading from the center of the scraper. The air is directed at high velocity against the impactor plate, breaking up aggregated particles. The concentration of the output aerosol can be controlled over a rather wide range by varying the air flow rates and gear ratios. The device has been used for many years to produce an aerosol of UO_2 for studies of the effects of chronic exposure on experimental animals [55]. The data in Table 9.10 show how the instrument performs over long periods.

The device described by Timbrell and others [60] was designed specifically for the production of aerosols of asbestos fibers in the respirable size range.

In this case, fibers are scraped from the surface of a compacted plug by blades rotating at 1500 rpm within a cylindrical chamber. A flow of air through the chamber, aided by the centrifuging action of the rotating blades carrying the detached fibers around the inner surface of the chamber, selectively entrains the respirable fraction and carries it away as useful aerosol. When the standard reference samples of asbestos, for which this dispenser was designed, is used

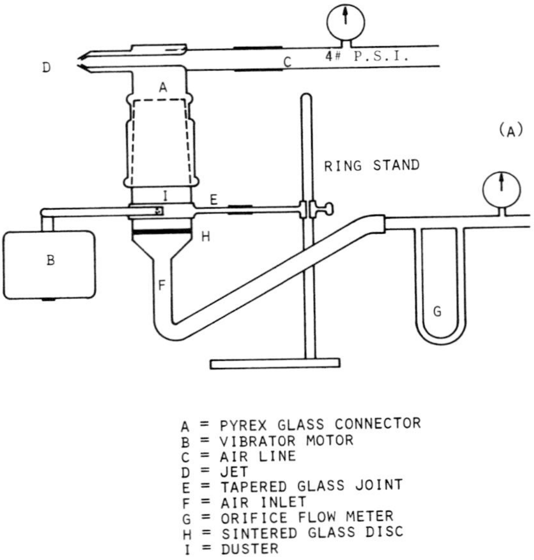

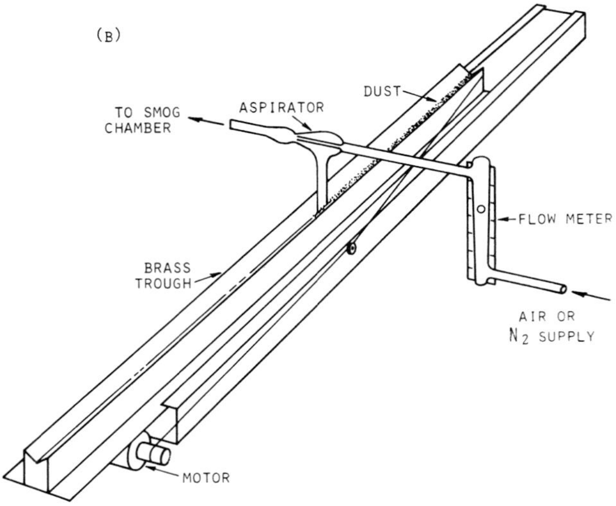

TABLE 9.10

TWO-YEAR AVERAGE CONCENTRATIONS AND PARTICLE SIZE DATA FOR UO_2 DUST

Chamber	Number of concentration measurements	Mean dust concentration (mg/m^3)	Average MMD (μm)	Average σ_g
1	481	5.59	0.98	2.61
2	482	5.77	0.95	2.47
3	480	5.59	0.96	2.52
4	481	5.69	0.98	2.56

with the Wright dust feed, good dispersion is also obtained. This is brought out by the data in Table 9.11 obtained by Leach [61]. "Percent respirable" was measured using a Hexhlet horizontal elutriator (Sect. 8-2.1). Under the same conditions, Timbrell's generator produced 48 and 70%, respectively, for the respirable masses of chrysotile and amosite.

TABLE 9.11

CONCENTRATION GENERATED BY WRIGHT DUST FEED USING STANDARD REFERENCE SAMPLES

Standard reference sample	Number of measurements	Concentration, mg/m^3		Percent respirable
		chamber average[a]	Hexhlet[b]	
Chrysotile "B"	6	17.2 ± 1.3	16.7 ± 1.3	24 ± 2
Amosite	8	15.3 ± 0.7	15.5 ± 1.0	71 ± 4

[a] Routine membrane filter samples. Errors are standard deviations.
[b] Fiberglass filter samples at Hexhlet entrance.

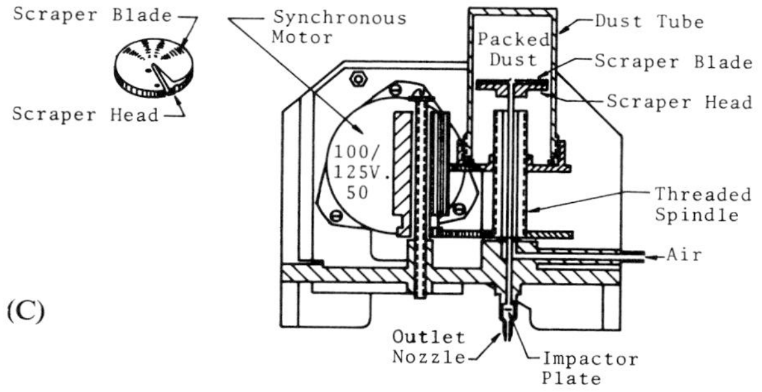

FIG. 9.12. Apparatus for dispersing dry particles. (A) Sonkin [52]; courtesy of the American Industrial Hygiene Association. (B) Cadle and Magill [53]; courtesy of the American Chemical Society. (C) The Wright dust feed [54] (diagram from Ref. [55]). Courtesy of the American Industrial Hygiene Association.

9-2 Flow Measurement

The usual methods of measuring flow rate are based on Bernoulli's theorem, which expresses the conservation of energy for a fluid in motion. If the fluid is a gas of negligible viscosity undergoing adiabatic changes, the theorem can be represented adequately by

$$\Phi + \tfrac{1}{2}V^2 = \Phi_0, \tag{9.18a}$$

where

$$\Phi = \frac{\gamma}{\gamma-1} \cdot \frac{P_0}{\rho_0} \cdot \left(\frac{P}{P_0}\right)^{(\gamma-1)/\gamma}. \tag{9.18b}$$

γ is the ratio of the specific heats of the gas, P and ρ are the gas pressure and density, V is the gas velocity, and Φ_0 refers to the initial state of the gas at rest. The magnitudes of Φ and V change when the cross-sectional area A of the moving fluid changes. Conservation of mass imposes on the fluid flow a second condition, which is expressed in the continuity equation

$$\rho_1 V_1 A_1 = \rho_2 V_2 A_2. \tag{9.19}$$

For two positions defined by A_1 and A_2,

$$\Phi_1 + \tfrac{1}{2}V_1{}^2 = \Phi_2 + \tfrac{1}{2}V_2{}^2. \tag{9.20}$$

Combining this equation with (9.18b) and (9.19),

$$\begin{aligned} V_2{}^2 &= \frac{2\gamma}{\gamma-1} \cdot \frac{P_1}{\rho_1}\left[1 - \left(\frac{P_2}{P_1}\right)^{(\gamma-1)/\gamma}\right] \Bigg/ \left[1 - \left(\frac{\rho_2 A_2}{\rho_1 A_1}\right)^2\right] \\ &= \frac{2\delta P}{\rho_1}\left[1 + \frac{1}{2\gamma} \cdot \frac{\delta P}{P_1} + \frac{\gamma+1}{6\gamma^2} \cdot \frac{\delta P^2}{P_1{}^2}\right. \\ &\quad \left. + \frac{(\gamma+1)(2\gamma+1)}{24\gamma^3} \cdot \frac{\delta P^3}{P_1{}^3} + \cdots\right] \Bigg/ \left[1 - \left(\frac{\rho_2 A_2}{\rho_1 A_1}\right)^2\right], \end{aligned} \tag{9.21}$$

where $\delta P^i = (P_1 - P_2)^i$. For small values of $\delta P/P_1$, Eq. (9.21) reduces to the expression for an incompressible fluid,

$$V_2{}^2 = (2\delta P/\rho)/[1-(A_2{}^2/A_1{}^2)], \tag{9.21a}$$

which can be applied to gases with negligible error as long as $V_2 \gtrsim C/3$, where C is the speed of sound at P_1, ρ_1.

The mass flow rate G of the gas is given by Eq. (9.19):

$$G = \rho_2 V_2 A_2,$$

and the volumetric flow rate, referred to ambient air, is

$$F_a = (\rho_2/\rho_a) \cdot V_2 A_2. \tag{9.22}$$

Adiabatic expansion requires that $\rho_2/\rho_1 = (P_2/P_1)^{1/\gamma}$ and $T_2/T_1 = (P_2/P_1)^{(\gamma-1)/\gamma}$. When this is the case, the velocity V_2 has an upper limit equal to the speed of sound C_2 given by

$$C_2 = (\gamma P_2/\rho_2)^{1/2}.$$

For air ($\gamma = 1.41$), this occurs when $P_2 = 0.53P_1$, $\rho_2 = 0.64\rho_1$, $T_2 = 0.83T_1$, and $C_2 = 0.91C_1$. If P_1 is equivalent to 1 atm and $T_1 = 293°\text{K}$,

$$C_2 = 3.1 \times 10^4 \quad \text{cm/sec,}$$

In general, if $V_2 = C_2$,

$$F_a = \frac{0.58A_2}{\rho_a} \cdot (\gamma\rho_1 P_1)^{1/2}. \tag{9.22a}$$

ρ_1 and the ambient air density ρ_a are not necessarily identical.

Instruments that operate according to Bernoulli's theorem are termed inferential flow meters. They usually comprise two distinct parts: a primary element that is in contact with the fluid, and a secondary element that registers the degree of interaction between fluid and primary element. Devices of this sort include venturi and orifice meters and rotameters.

Venturi Meters. The primary element in this device (Fig. 9.13a) is a constriction in the form of an accurately machined throat section. At $V_2 < C/3$, the volumetric flow rate, referred to ambient air, is

$$F_a = (kA_2/\rho_a)\{2\delta P\rho_1/[1-(A_2{}^2/A_1{}^2)]\}^{1/2}. \tag{9.23}$$

Under real conditions, the discharge coefficient, k, is slightly less than the ideal value of unity.

Orifice Meters. The primary element in this device is a thin plate having a sharp-edged, usually circular, orifice at the center (Fig. 9.13b). In this case A_2, which is generally believed to be the area of the fluid stream at the vena contracta, cannot be measured directly, so it is customary to replace it in Eq. (9.23) with the orifice area A_0, adjusting the discharge coefficient as experience dictates so that the equation gives the correct value of F_a. For the arrangement pictured in Fig. 9.13b, $k = 0.61$ when $A_0/A_1 \lesssim 0.6$ [64].

Rotameters. Unlike the meters described above, in which the secondary element is an external sensor of pressure differences, the rotameter combines both elements in a single float, which rides on a column of air in a vertical, tapered tube (Fig. 9.13c). As the flow rate increases, the float rises, always coming to rest in a position at which

$$\delta P = P_1 - P_2 = v_f(\rho_f - \rho)g/A_f.$$

v_f is the volume of the float, ρ_f is its density, A_f is its cross-sectional area normal to the direction of flow, ρ is the density of fluid within the rotameter, and g is the acceleration due to gravity. Rotameters are designed to operate at fluid velocities for which Eq. (9.23) holds. Taking $A_1^2 \gg A_2^2$ and $\rho_f \gg \rho$, this equation gives

$$F_a = (kA_0/\rho_a)\cdot(2\delta P\rho)^{1/2} = (kA_0/\rho_a)\cdot(2v_f\,\rho_f\,g\rho/A_f)^{1/2}, \qquad (9.24)$$

where A_0 is the cross-sectional area of the annulus defined by the tube wall and A_f. As indicated by Fig. 9.13c, the discharge coefficient k, which includes a factor to relate A_2 to A_0, is a function of the Reynolds number for flow about the float. The effect of Reynolds number depends on the shape of the float. For the type shown in Fig. 9.13c, k is 0.77 when Re > 300, diminishing to 0.50 at Re = 10 [65].

If a rotameter is calibrated at a density ρ_c, which may or may not equal ρ_a,

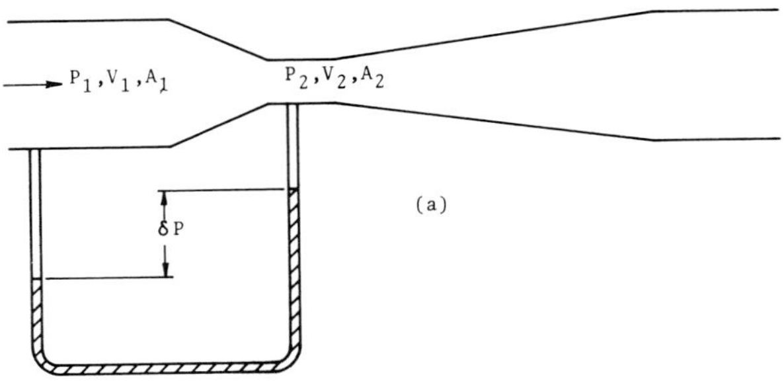

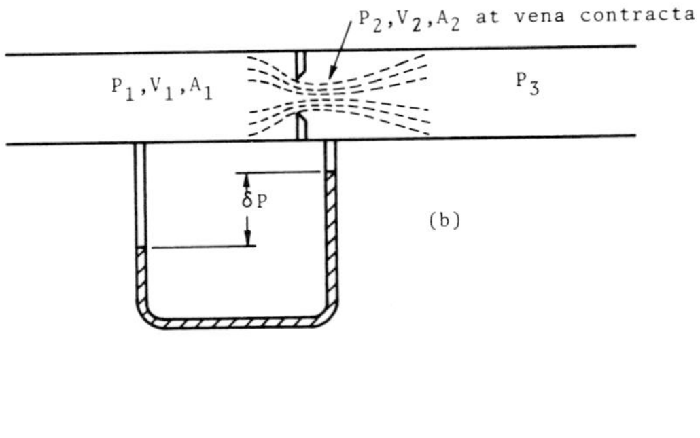

a given reading R represents a flow rate

$$F_a(R, \rho_c) \propto \sqrt{\rho_c}.$$

At some other density ρ, the same reading represents a flow rate

$$F_a(R, \rho) \propto \sqrt{\rho}.$$

If the change in Re due to the change in density does not affect the value of k,

$$F_a(R, \rho) = F_a(R, \rho_c) \cdot (\rho/\rho_c)^{1/2} = F_a(R, \rho_c)(P/P_c)^{1/2}.$$

The determination of P permits an accurate measurement of flow rate when rotameters are used in a system involving samplers across which significant pressure drops may occur, provided the pressure changes do not significantly alter k.

Spherical floats in tubes designed to guide them axially permit accurate measurement of flow rates down to 0.25 cm^3/min [65]. At low flow rates and small float sizes, however, the Reynolds numbers fall below the values for which k is a constant. The effect becomes significant when the float diameter is less than 0.25 in. Craig [66] recently showed that such small float diameters

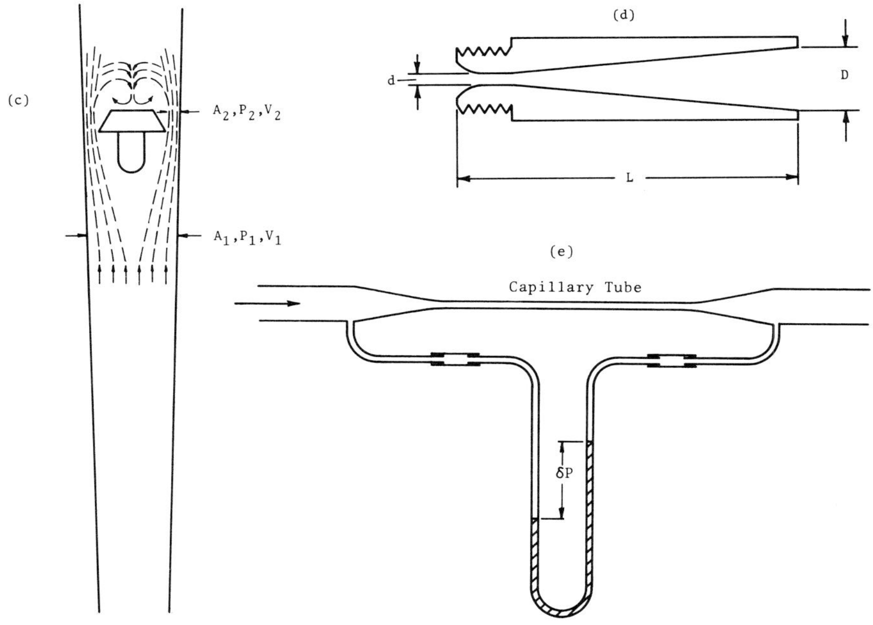

FIG. 9.13. Inferential flow meters. (a) Venturi meter. (b) Orifice meter. (c) Rotameter (d) Critical orifice [62]; courtesy of the British Occupational Hygiene Society. (e) Resistance meter [63].

make it necessary to have calibration curves for a wide range of values of P, if accurate measurements of flow are desired.

Valves for adjusting the flow through rotameters should be installed between the meter and the source of air movement, i.e., downstream when used with vacuum and upstream when used with compressed air.

Critical Orifices. When $P_2 = 0.53P_1$ [Eq. (9.21)], V_2 equals the speed of sound at the air temperature and density in effect at A_2. Since V_2 cannot exceed that value (except for certain special nozzle designs), an orifice such as that of Fig. 9.13b can be used to establish a fixed, well-controlled flow rate merely by making P_3 sufficiently small that P_2 can be maintained at $0.53P_1$. The flow rate through an orifice of cross-sectional area A_0 is then given by Eq. (9.22a), with the inclusion of a discharge coefficient:

$$F_a = (0.58kA_0/\rho_a)(\gamma\rho_1 P_1)^{1/2}. \tag{9.25}$$

By setting k $(=A_2/A_0)$ equal to 0.61, Eq. (9.25) can be used to estimate the orifice diameter necessary to yield a given limiting volumetric flow rate, F_a.

It is not easy to construct critical orifices for small values of F_a. Corn and Bell [67] have described a technique for producing critical orifices from steel hypodermic needles. They recommended the use of a standard length of 1 in. and measured the flow rates through needles of that length and various bores at $P_3 = 0.45P_1$. At small diameters, however, such long orifices lead to viscous losses that make it impossible to achieve sonic velocities. The data of Corn and Bell indicate that the effects of viscous losses become significant when the needle diameter is less than about 0.01 in.

With proper design of the orifice approach and outlet, some of the pressure loss can be regained downstream of the orifice. The orifice design of Fig. 9.13d allows sonic velocities to be achieved when $P_3 = 0.9P_1$, if $D \simeq 1.9d$ and $L \simeq 15d$ [62].

Resistance Flow Meters. The pressure drop developed across capillary tubes (Fig. 9.13e) or porous plugs as a result of viscous resistance to flow can be used to meter flow rates. The pressure taps are situated to minimize pressure changes due to the Bernoulli effect. Unlike the meters described above, resistance flow meters show a linear relationship between flow rate and pressure drop:

$$F = \pi d^4(P_1 - P_2)/128\eta L. \tag{9.26}$$

This is the Hagen–Poiseuille equation for laminar flow through a tube of circular cross section. L is the tube length, d is its diameter, and η is the viscosity of the fluid. Since density does not appear in Eq. (9.26) and gas viscosities are independent of pressure under conditions of interest here, F is the volumetric flow rate at the average pressure in the meter. The flow rate referred to ambient conditions is

$$F_a = (P_1 + P_2) \cdot F/2P_a.$$

If $d^4(P_1 - P_2)/L^2 \lesssim 20.8\eta^2/\rho$, it is necessary to take into consideration the pressure losses occurring at the ends of the tube if relative errors $\gtrsim 2.5\%$ are to be avoided. Benton [68] gives the following equation for the correct flow rate F_c when the value calculated from Eq. (9.26) is F_P:

$$F_c = (32/\pi)k_1 vL[(1 + k_2 \pi F_P/16vL)^{1/2} - 1]. \tag{9.27}$$

The values of the empirical quantities are $k_1 = 1.015$ and $k_2 = 0.973$ and v is the kinematic viscosity of the fluid. For porous plugs, Eq. (9.26) takes the form of Eq. (7.22).

The derivation of Eq. (9.26) requires laminar flow, so that resistance meters should be operated at Reynolds numbers below about 2000.

9-3 Calibration of Flow Meters

Figure 9.14 provides schematic diagrams of several devices with which flow meters can be calibrated. The instruments of Figs. 9.14(A)–(C) collect all of the metered gas in an expandable chamber and flow rate is calculated from the volume collected during a measured time interval. The instruments of Figs. 9.14(D), (E) are called positive displacement meters. They operate by counting the number of times that containers of known volume are filled and emptied in a given time interval. The pressure drop across any of these instruments is of little significance when they are used to calibrate inferential flow meters; it is of significance, however, when they are used to meter flows in systems in which flow is produced by small pressure differences.

Spirometers. The expanding chamber in a spirometer is sealed against gas leakage by a layer of water between the surfaces of the two cylinders forming the chamber. The movable cylinder is counterbalanced to minimize the pressure difference between the gas within the chamber and ambient air. The measured volume by which the chamber increases within a given time interval must be corrected for the saturation pressure of water vapor at the temperature of the measurement. Spirometers are available that have capacities as small as a few hundred milliliters.

The Vol-u-meter [69]. In this instrument, the gas flow moves a plastic piston within a calibrated glass cylinder of uniform bore. A ring of mercury, carried in a groove about the piston, provides the necessary gas seal between the piston and the cylinder. Accurate measurements of flow rate can be made very rapidly with this device. In a meter having a capacity of 1200 cm^3, the pressure due to the weight of the piston is equivalent to ~ 3.2 cm H_2O.

Soap-Film Meters [70]. The "piston" in this instrument is a film of soap formed when soap solution, introduced into the burette from the bulb **B** is allowed to drain back into the bulb. If the inner surface of the burette is clean and moist, the films are stable and can be moved at speeds at least as high as 200 cm/sec [72]. Barr [70], who developed the method, used burettes of 3-cm diameter and ordinary soap solutions; larger diameter films require special solutions.

The pressure necessary to move a soap film of surface tension σ in a tube of diameter d is

(A)

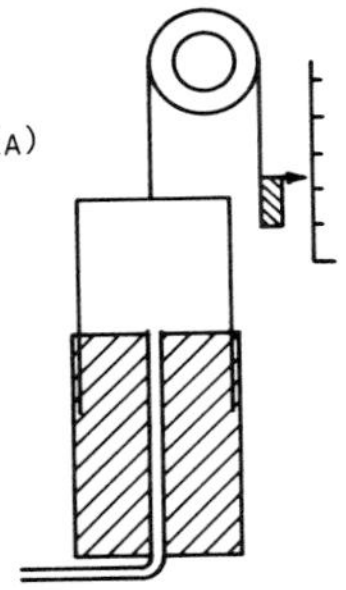

(B)

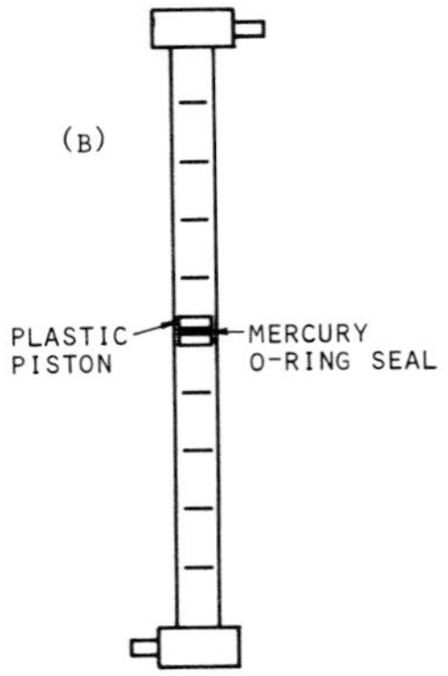

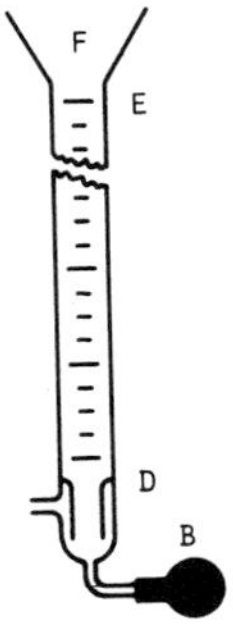

(C)

$$P = 0.04\sigma/d \qquad \text{mm } H_2O,$$

for σ/d in dynes per square centimeter. Because of their low resistance, soap-film meters are especially valuable for measuring flows at low pressure differences, such as the flow of aerosol into a conifuge [73].

Gas Meters. The dry gas meter contains a pair of bellows that are alternately filled and emptied by the metered gas. The process is illustrated in Fig. 9.14(D), in which the stippled areas define the volumes that have just been filled. Movement of the bellows controls the valve action and operates a series of dial indicators that register the total volume of gas that has passed through the meter. The housing of the dry test meter cannot support a very large

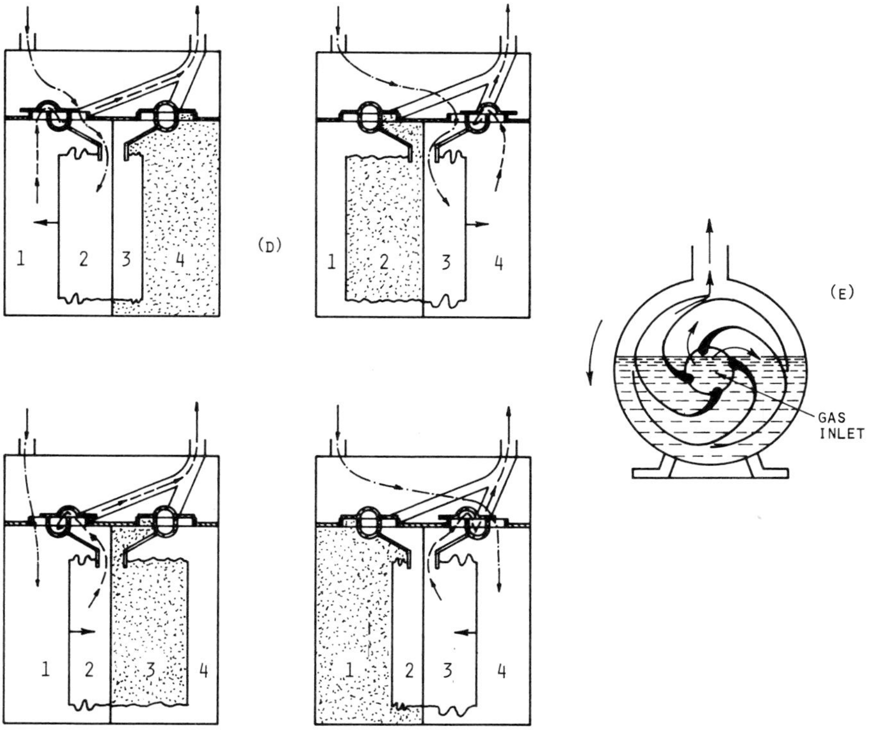

FIG. 9.14. Volume meters. (A) Spirometer. (B) Vol-u-meter [69]. (C) Soap-film meter [70]; courtesy of the Institute of Physics. (d) Dry gas meter [63]; courtesy of the American Society of Mechanical Engineers. (e) Water-sealed gas meter [71]; courtesy of McGraw-Hill Book Company.

pressure difference, so it is wise to operate them at all times with one inlet open to atmosphere. The dial indicators do not move at a constant rate, so that one complete revolution (of the most sensitive indicator) does not always represent the same volume. To avoid any significant error due to this effect, a measurement should include at least ten revolutions.

In the wet gas meter, the flowing gas rotates several cups about a common axis. The water level seals the cups so that no air can escape during the filling process and acts as a valve that stops flow into a given cup just as it becomes filled (Fig. 9.14E). This meter requires more attention than the dry gas meter and measured volumes must be corrected for water vapor content. On the other hand, it is more sturdily built than the dry gas meter and can be built into a closed system.

9-4 Isokinetic Sampling

To ensure collection of a representative sample of particles suspended in a moving air stream, it is necessary that the flow pattern upstream of the sampler be unaffected by the sampler's presence. This can be achieved by setting the plane of the sampling orifice normal to the direction of air flow and adjusting the flow rate of the sampler so that the linear air speed into its orifice is equal to that of the approaching air stream. When this is done, sampling is said to be isokinetic, and a true sample enters the orifice. The validity of the sample may be modified subsequently, however, by the structure of the sampler behind the orifice. Errors associated with anisokinetic sampling are due to particle inertia. If the sampling velocity is less than the air stream velocity (see Fig. 9.15a), some particles, originally contained in air that passes around the sampler, are projected into the flow of air entering the orifice, causing concentration to be overestimated. The opposite effect occurs (Fig. 9.15b) when the sampling velocity exceeds the air stream velocity. If the plane of the orifice is not properly aligned, isokinetic sampling cannot be achieved.

Efforts to establish factors with which to correct data obtained by anisokinetic sampling have led to expressions of the following form [74]:

$$C_s/C_a = 1 + [(V_a/V_s) - 1] \cdot f(\mathrm{Stk}), \tag{9.28}$$

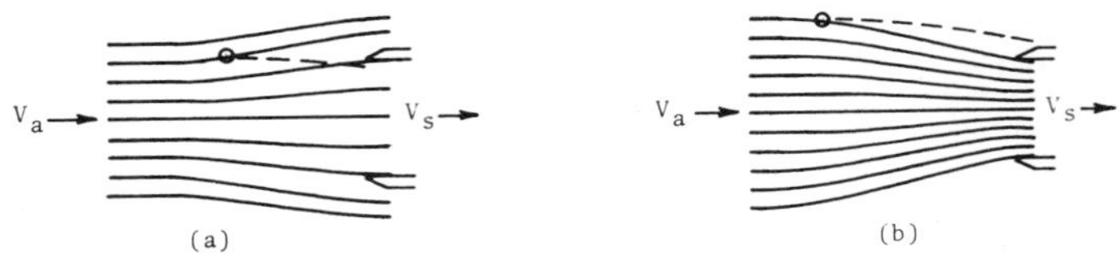

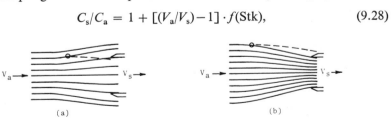

FIG. 9.15. Examples of anisokinetic sampling. (a) $V_s < V_a$; (b) $V_s > V_a$. Solid curves, air flow lines; dashed curves, particle trajectories.

where C is concentration, V is air speed, the subscripts a and s refer respectively to air and sampler, and $f(\text{Stk})$ is a function of the particle's Stokes number, $\text{Stk} = 2mZV_a/d_0$, where m is the particle's mass, Z is its mechanical mobility, and d_0 is the diameter of the orifice. Davies proposed

$$f(\text{Stk}) = 2\text{Stk}/(1+2\text{Stk}). \tag{9.28a}$$

He also showed that an equation derived earlier by Watson [75] could be put in the same form if

$$f(\text{Stk}) = (1-f_1)\left[1+f_1\left(\frac{1-(V_a/V_s)^{1/2}}{1+(V_a/V_s)^{1/2}}\right)\right], \tag{9.28b}$$

where f_1 is an empirical function of Stk that Watson obtained from his data. Watson did not define f_1 for $\text{Stk} > 6.4$, but it is reasonable to assume that $f_1 \to 0$ at large values of Stk. At $\text{Stk} = 0$, $f_1 = 1$. Both forms of Eq. (9.28)

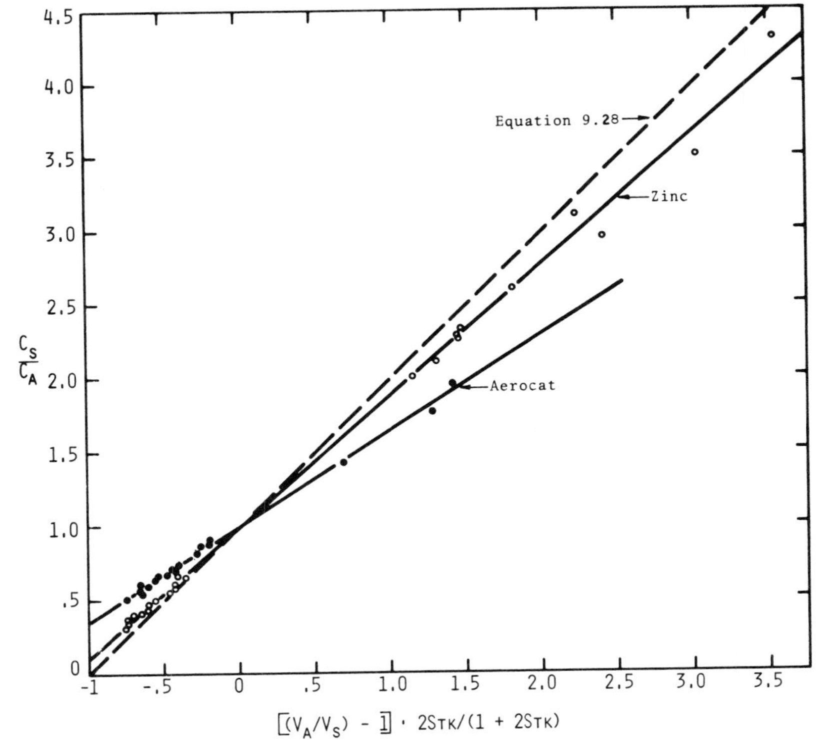

FIG. 9.16. Comparison of Badzioch's data [76] with Davies' form of Eq. (9.28). Open circles, zinc; closed circles, Aerocat.

then predict

$$C_s/C_a = 1, \qquad \text{when} \quad V_a = 0;$$

$$C_s/C_a \simeq 1, \qquad \text{when} \quad V_a > 0 \quad \text{and} \quad \text{Stk} \simeq 0;$$

and

$$C_s/C_a = V_a/V_s, \qquad \text{when} \quad \text{Stk} = \infty.$$

The second of these conditions merely says that very small particles follow the deflecting air stream, while the third says that very large particles do not deviate from the original direction of flow.

Badzioch [76] made an extensive study of anisokinetic sampling. He used particles of Aerocat, a substance made up mostly of amorphous silica (having a density of 1.4 g/cm^3), and of zinc (density, 6.9). He used several orifice diameters in the range 0.65–1.9 cm. The particles were separated into narrow size ranges by centrifugation, and their mass distributions were determined as a function of settling velocity. A given test dust was characterized by its mass median settling velocity. The geometric diameters of both types of dust were very similar. All of his data are plotted in Fig. 9.16 according to Davies' form of Eq. (9.28). The data for each dust show a satisfactory linear relationship, but the lines differ significantly from each other and from Davies' relationship. Some of the differences in slope may be due to losses in the sampling lines, which included a 90° bend between the orifice and the particle collector, but such

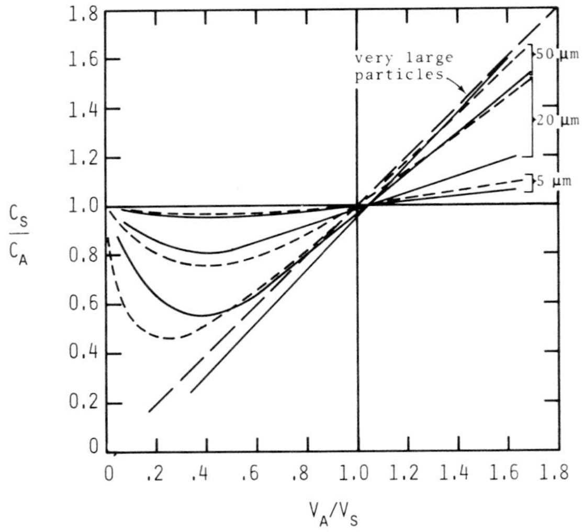

FIG. 9.17. Effect of anisokinetic sampling on measurement of concentration. Solid lines from May [77], broken lines, Eq. (9.28) using Davies' [74] form of f(Stk).

48. S. Posner and J. Bennick, *Preparation of Insoluble Aerosols Containing Mixed Fission Products*, USAEC Report LF–31, 1966.
49. P. Kotrappa, Production of Relatively Monodisperse Aerosols for Inhalation Experiments by Aerosol Centrifugation, in *Fission Product Inhalation Program Annual Report*, LF–43, 1970.
50. G. J. Newton, Some Characteristics of Ruthenium Aerosols Generated from Ru-Organic Complexes and Subjected to Thermal Degradation, in *Fission Product Inhalation Program Annual Report*, LF–43, 1970.
51. W. B. Diechmann, A "Dustshaker," *J. Ind. Hyg. Toxicol., 26:* 334–335 (1944).
52. L. S. Sonkin, M. A. Lipton and D. Van Hoesen, An Apparatus for Dispersing Finely Divided Dusts, *J. Ind. Hyg. Toxicol., 28:* 273–275 (1946).
53. R. O. Cadle and P. L. Magill, Preparation of Solid- and Liquid-in-Air Suspensions, *Ind. Eng. Chem., 43:* 1331–1335 (1951).
54. B. M. Wright, A New Dust-Feed Mechanism, *J. Sci. Instrum., 27:* 12–15 (1950).
55. L. J. Leach, C. J. Spiegl, R. H. Wilson, G. E. Sylvester and K. E. Lauterbach, A Multiple Chamber Exposure Unit Designed for Chronic Inhalation Studies, *Amer. Ind. Hyg. Ass. J., 20:* 13–20 (1959).
56. N. A. Fuchs and F. I. Murashkevich. Laboratoriums–Pulverzerstäuber (Staub–generator), *Staub, 30:* 447–449 (1970).
57. J. T. Burdekin, J. G. Dawes and A. Slack, *A Standard Dust Cloud for Testing Purposes*, S.M.R.E. Res. Rep. No. 141, 1957.
58. F. M. Stead, C. U. Dernehl, and C. A. Nau, A Dust Feed Apparatus Useful for Exposure of Small Animals to Small and Fixed Concentrations of Dust, *J. Ind. Hyg. Toxicol., 26:* 90–93 (1944).
59. R. L. Dimmick, Jet Disperser for Compacted Powders in the One- to Ten-Micron Range, *A.M.A. Arch. Ind. Health, 20:* 8–14 (1959).
60. V. Timbrell, A. W. Hyett and J. W. Skidmore, A Simple Dispenser for Generating Dust Clouds from Standard Reference Samples of Asbestos, *Ann. Occup. Hyg., 11:* 273–281 (1968).
61. L. J. Leach, Univ. of Rochester, Rochester, N.Y., personal communication, 1971.
62. V. Balashov and J. G. Brading, Study of Operating Performance of a Critical Orifice of the Hexhlet Gravimetric Dust Sampler, *Ann. Occup. Hyg., 7:* 307–310 (1964).
63. A.S.M.E. Research Committee on Fluid Meters, *Fluid Meters: Their Theory and Application*, 5th ed., Amer. Soc. of Mech. Eng., New York, 1959.
64. D. F. Boucher, Fluid Metering, in C. E. Lapple (Ed.), *Fluid and Particle Mechanics*, Univ. of Delaware, Newark, Delaware, 1956.
65. M. C. Coleman, Variable Area Flow Meters, *Trans. Inst. Chem. Eng., 34:* 339–350 (1956).
66. D. K. Craig, The Interpretation of Rotameter Air Flow Readings, *Health Phys., 21:* 328–332 (1971).
67. M. Corn and W. Bell, A Technique for Construction of Predictable Low-Capacity Critical Orifices, *Amer. Ind. Hyg. Ass. J., 24:* 502–504 (1963).
68. A. F. Benton, The End Correction in the Determination of Gas Viscosity by the Capillary Tube Method, *Phys. Rev., 14:* 403–408 (1919).
69. Operating Instructions for Vol-u-Meters, Model No. 1056, Brooks Instrum. Div., Emerson Electric Co., North Wales, Pennsylvania.
70. G. Barr, Two Designs of Flow Meter, and a Method of Calibration, *J. Sci. Instrum., 11:* 321–324 (1934).
71. J. H. Perry (Ed.), *Chemical Engineers' Handbook*, 3rd ed., p. 1283, McGraw-Hill, New York, 1950.

72. B. D. Bloomfield and L. Silverman, Characteristics of the Suction Soap Film Meter, *A.M.A. Arch. Ind. Hyg. Occup. Med.*, *4:* 446–457 (1951).
73. O. R. Moss, *Shape Factors for Airborne Particles*, Master of Science Thesis, Univ. of Rochester, 1969.
74. C. N. Davies, The Entry of Aerosols into Sampling Tubes and Heads, *Brit. J. Appl. Phys. Ser. 2*, 1: 921–932 (1968).
75. H. H. Watson, Errors Due to Anisokinetic Sampling of Aerosols, *Amer. Ind. Hyg. Ass. Quart.*, *15:* 21–25 (1954).
76. S. Badzioch, Collection of Gas-Borne Dust Particles by Means of an Aspirated Sampling Nozzle, *Brit. J. Appl. Phys.*, *10:* 26–32 (1959).
77. K. R. May, Physical Aspects of Sampling Airborne Microbes, in P. H. Gregory and J. L. Monteith (Eds.), *Airborne Microbes*, Cambridge Univ. Press, London and New York, 1967.
78. V. Vitols, Theoretical Limits of Errors Due to Anisokinetic Sampling of Particulate Matter, *J. Air Pollution Control Ass.*, *16:* 79–84 (1966).
79. K. R. May and H. A. Druett, The Pre-Impinger: a Selective Aerosol Sampler, *Brit. J Ind. Med.* *10:* 142–151 (1953).

Glossary

Activity. A measurable property of a particulate population; e.g., mass, number, radioactivity, surface area.

Adsorption isotherm. A function relating the volume of vapor adsorped on a surface at a given temperature to the pressure of the vapor in the gas phase.

Aerodynamic diameter. For any particle, the diameter of a sphere of unit density that has the same terminal settling velocity as the particle.

Aerosol. A suspension of solid or liquid particles in air.

Aerosol photometer. A device for estimating the concentration of airborne particles by measuring the amount of light scattered by a known volume of aerosol.

Air centrifuge. A device in which air in laminar flow passes through channels that are rotating at high speed about a fixed axis, creating a centrifugal force that causes particles to deposit on the outer wall of the channel.

Activity Median Aerodynamic Diameter (AMAD). Fifty percent of airborne activity is associated with particles having aerodynamic diameters smaller than the AMAD.

Antithetic Variates Technique. A method for reducing the effort required to obtain a given precision by simultaneously measuring two quantities that estimate the same parameter but are negatively correlated.

Atomizer. A device with which droplets are produced by mechanical disruption of a bulk liquid.

BET (Brunauer–Emmett–Teller) method. A procedure for analyzing an adsorption isotherm to determine the surface area of a particulate sample.

Bipolar ion field. A region in which there are ions of both polarities.

Breathing zone sample. A sample taken at a point as close to the subject's face as working conditions permit.

Brownian motion. Random movement of particles due to their thermal energy.

Cascade impactor. An instrument consisting of a series of impaction stages of increasing efficiency with which particles can be segregated into relatively narrow intervals of aerodynamic diameters.

Centrifugal spectrometer. An air centrifuge in which airborne particles, confined to a thin layer of incoming air, are deposited in narrow bands that represent discrete intervals of aerodynamic diameter.

Coagulation. The collision and adherence of two particles to form a single larger particle.

Condensation coefficient. The fraction of molecules striking a surface that condense.

Conicycle. An air centrifuge respirable activity sampler.

Conifuge. A centrifugal spectrometer in which sampled air flows between two coaxial rotating cones.

Confidence limits. A range of values about a sample statistic; if calculated for a great many samples, the parameter being estimated will fall within the range with a specified relative frequency.

Corona. A region of intense ionization formed in air about an electrode of small diameter when a sufficiently high voltage is applied to it.

Coulter counter. An instrument for determining the volumes of individual particles by measuring the change in resistivity of fluid flowing through a small orifice when a particle passes through the orifice.

Critical orifice. A device that provides a constant air flow rate when the pressure drop across it is sufficient to maintain sonic velocity in the orifice.

Cyclone sampler. A device that collects large particles by centrifugation from a spiraling air stream, permitting only respirable activity to pass to a back-up filter.

Diffusion battery. A device in which a number of ducts, of either circular or rectangular cross section, are arranged in parallel so that significant diffusion losses are possible at relatively large sampling flow rates.

Diffusion charging. A process by which airborne particles acquire a charge when struck by ions that are in random motion due to their thermal energy.

Diffusion. The net movement of particles due to Brownian motion when a concentration gradient exists.

DOP. Dioctyl phthalate. An organic fluid frequently used in the generation of monodisperse aerosols.

Drag coefficient. A function of Reynolds number that makes it possible to express fluid resistance in a general form.

Dust generator. A device for the controlled dispersion of dry powders.

Effective Cut-Off Aerodynamic Diameter (ECAD). The aerodynamic diameter for which the collection efficiency of an impactor stage is 50%.

Electric particle analyzer. A device with which particles are segregated on the basis of their electrical mobility.

Electrostatic precipitator. A device in which airborne particles are charged in a unipolar ion field and deposited on a suitable collecting surface by a strong electric field.

Elutriation. The separation of particles according to aerodynamic diameter by allowing them to settle through a moving air stream.

Elutriation spectrometer. A horizontal elutriator in which airborne particles, confined to a thin layer of incoming air, are deposited on the lower plate in narrow bands that represent discrete intervals of aerodynamic diameter.

Extinction coefficient. The amount of light scattered and absorbed by a particle relative to the amount incident on it.

Field charging. A process by which airborne particles acquire a charge as a result of their "bombardment" by ions in a strong electric field.

Figure of merit. One criterion with which to judge the adequacy of a method for collecting samples for particle size measurement.

Filter. A mat of fibers or a porous membrane used to collect airborne particles.

General area sample. A sample taken in a fixed position where conditions are assumed to be representative, on the average, of the region under surveillance.

Geometric standard deviation. A measure of dispersion in a lognormal distribution.

Hatch–Choate equations. Expressions relating the parameters of one lognormal activity distribution to those of another form of activity for the same population of particles.

Horizontal elutriator. A device in which air passes in laminar flow between two parallel horizontal plates and particles are collected by sedimentation onto the lower plate.

Hydraulic diameter. A hypothetical diameter of a particle equal to four times the ratio of its cross-sectional area to the perimeter of that area.

Hygroscopicity. The tendency of airborne soluble particles to retain condensing water molecules.

Ideal fluid. A hypothetical fluid having no viscosity.

Impaction. The process in which the inertia of particles in an air stream that is deflected about an obstacle causes them to strike the obstacle.

Impinger. A device in which a jet of air, moving at high speed, is directed normally against a surface under water, causing airborne particles of sufficient inertia to deposit on the surface.

Index of filtration. The effective efficiency of a filter mat per unit of thickness.

Interception. The collision of a particle with an obstacle when the air with which the particle moves passes the obstacle at a distance equal to or less than the particle radius.

Isokinetic sampling. The condition in which ambient air flow has the same speed and direction as air flowing into a sampling inlet.

Lognormal distribution. A particle size distribution in which activity is normally distributed with respect to the logarithms of particle diameters.

Lung model. A representation of the respiratory system in which air flow patterns and lung geometry have been made sufficiently simple and orderly that deposition can be described quantitatively.

Median diameter. A distribution parameter such that half of the particulate activity is associated with particles of smaller size.

Mobility. The ratio between a particle's velocity and the force producing that velocity.

Monodisperse. Having a very small range of diameters.

Nasopharyngeal compartment. That portion of the respiratory tract between the anterior nares and epiglottis.

Nebulizer. An atomizer that operates within a small container from which only a fraction of the droplets escape.

Owl. An instrument with which the diameter of particles in a monodisperse aerosol can be related to the position and number of higher order Tyndall spectra produced when the aerosol scatters a beam of white light.

Packing density. The ratio between the fiber (or membrane) volume of a filter and its total volume.

Particle size distribution. A mathematical relationship expressing the relative amount of activity associated with particles in a given element of size.

Permeability method. A procedure for determining specific surface by measuring the pressure drop per unit flow rate when a fluid of known viscosity passes through a particulate sample of fixed porosity.

Personal sampler. A device attached to an individual to sample air in his immediate vicinity during periods of possible inhalation exposure.

Photospectrometer. An instrument for measuring the diameter of a particle in terms of the amount of light it scatters into a fixed solid angle.

Point-to-plane precipitator. An electrostatic precipitator in which the corona is formed about a needle point and particles are collected on a grounded surface normal to the axis of the needle.

Poiseuille's law. The relationship between volumetric flow rate and pressure drop for a viscous fluid in laminar flow through a tube having a length much greater than its diameter. [Also called the Hagen–Poiseuille law.]

Poisson distribution. A function relating the probability that a given number of particles will be observed in a small element of volume to the average concentration of particles in random distribution throughout the volume.

Polydisperse (heterodisperse). Including a range of diameters.

Pre-impinger. A device with which large particles are collected by impingement on a liquid surface, permitting only respirable activity to pass to a back-up impinger.

Pulmonary compartment. That portion of the respiratory tract, including the respiratory bronchioles, in which gas exchange occurs.

Rebound. Return of particles to an air stream when they fail to adhere after striking a collecting surface.

Recombination coefficient. A factor of proportionality between the rate at which ions of opposite sign recombine and the product of their concentrations.

Re-entrainment. Return of particles to an air stream some time after their deposition on a collecting surface.

Relaxation time. The time required for a particle accelerating under a constant force to reach $1/e$ of its terminal velocity.

Respirable activity. The fraction of airborne activity that will be deposited, according to one or another lung model, in the pulmonary compartment.

Reynolds number. A dimensionless parameter related to the ratio of fluid inertial forces to viscous forces.

Scattering cross section. The ratio of the total light scattered (and absorbed) by a particle to the light incident on it per square centimeter.

Sedimentation. Movement of particles in the direction of the earth's center brought about by the force of gravity.

Shape factors. Constants of proportionality (in a statistical sense) that relate some property of a particle to one of the statistical diameters raised to an appropriate power.

Sinclair–LaMer generator. A device for producing monodisperse aerosols by the controlled condensation of a vapor onto nuclei.

Slip factor. A correction to Stokes law made necessary by the existence of a finite net gas velocity at a particle surface.

Specific surface. Particulate surface area per unit mass (or volume) of particles.

Spinning disk (top) atomizer. A device with which monodisperse droplets are produced when a thin film of liquid breaks up on being thrown from the rim of a spinning disk.

Statistic. An estimate of a size distribution parameter calculated from measurements on a random sample of particles.

Statistical geometric diameter. An arbitrary linear dimension defined as particle size and valid only in a statistical sense because its magnitude is a function of particle orientation.

Stokes diameter. For any particle, the diameter of a sphere having the same bulk density and same terminal settling velocity as the particle.

Stokes law. A relationship describing the force exerted by a viscous fluid of negligible inertia on a sphere moving at a constant velocity relative to the fluid.

Stokes number. A dimensionless inertial parameter relating a particle's stopping distance to some characteristic length of a system under examination.

Stopping distance. The distance that a particle, projected at a finite velocity into still air, would travel before being brought to rest by the resistance of the air.

Terminal settling velocity. The equilibrium velocity approached by a particle falling under the influence of gravity and fluid resistance.

Thermal precipitator. A device in which particles suspended in air flowing between two surfaces at different temperatures are subjected to a thermal force that causes them to deposit on the colder surface.

Tubes of flow. Volume elements bounded by trajectories of inertia-less particles of a given terminal settling velocity.

Unipolar ion field. A region in which only ions of one polarity can be found.

Wall loss. The deposition of particles on sampler surfaces other than those designed to collect particles.

Author Index

Numbers in parentheses are reference numbers and indicate that an author's work is referred to although his name is not cited in the text. Numbers in italics show the page on which the complete reference is listed.

L

Mc

M

N

O

P

Q

R

S

T

Subject Index

B

C

D

E

F